Python Web Development with Sanic

An in-depth guide for Python web developers to improve the speed and scalability of web applications

Adam Hopkins

SANIC IS AN IMPRINT OF PACKT PUBLISHING

Python Web Development with Sanic

Group Product Managers: Richa Tripathi
Publishing Product Manager: Sathyanarayanan Ellapulli
Senior Editor: Nisha Cleetus
Content Development Editor: Nithya Sadanandan
Technical Editor: Pradeep Sahu
Copy Editor: Safis Editing
Project Coordinator: Ajesh Devavaram
Proofreader: Safis Editing
Indexer: Hemangini Bari
Production Designer: Shankar Kalbhor
Marketing Coordinator: Teny Thomas

First published: February 2022

Production reference: 1240222

Published by Packt Publishing Ltd.
Livery Place
35 Livery Street
Birmingham
B3 2PB, UK.

ISBN 978-1-80181-441-6

www.packt.com

This book is for everyone who has inspired me on my journey, but in particular, to my sister, who has always believed in me, to my parents for giving me the confidence to become who I am, to my children for reminding me what it is that I am working toward, and to my wife, Rachel, without whom I wouldn't achieve anything. She is my partner who provides me with the daily drive and joy of life, and the one who ultimately sanctioned the writing of this book.

– Adam Hopkins

Contributors

About the author

Adam Hopkins is a self-taught programmer. He started programming in high school and has been using Python and building websites for more than 25 years. He is a licensed attorney and practiced law before transitioning into software engineering as a second career.

Adam is the core developer and project maintainer of Sanic, a popular asynchronous Python framework and web server. Currently, he is a software engineering manager at PacketFabric, where he leads a team building backend web services. He is passionate about open source contributions and helping other developers to grow and learn. Adam lives in his desert home in the south of Israel with his wife and five children.

About the reviewers

Stephen Sadowski is a core developer and steering council member for the Sanic project. He has been involved with Sanic since 2017, first as an early adopter, and subsequently helping Adam Hopkins revitalize the project in 2018.

Outside of the Sanic community, Stephen is the vice president of professional services and DevOps and modernization practice lead for Uturn Data Solutions. He enjoys cooking, reading, and spending time with his wife, Jamie, and their two Portuguese Water Dogs, King and Prince.

Josha Inglis is a data architect at PacketFabric. He has worked extensively with Python and PostgreSQL for more than a decade, starting his career in data science and bioinformatics, analyzing large protein interaction networks involved in metastatic breast cancer, and has published papers in the fields of bioinformatics and archaeology.

He has since worked for data center and NaaS start-ups and commercial mathematics consulting, building web apps for business process optimization. Josha now focuses on architecting safe and low-friction data storage and access patterns.

He lives and works by the beach in Byron Bay, Australia.

Table of Contents

Part 2: Hands-On Sanic

3

Routing and Intaking HTTP Requests

4

Ingesting HTTP Data

5

Building Response Handlers

6

Operating Outside the Response Handler

7

Dealing with Security Concerns

8

Running a Sanic Server

Part 3: Putting It All together

9

Best Practices to Improve Your Web Applications

10

Implementing Common Use Cases with Sanic

11

A Complete Real-World Example

Preface

"What do you want to do when you graduate college?" asked a family friend. "I'm not sure. I really like building web applications," I said. "Maybe I can do that." The response: "No, there's no future in that. Pick something else." Wow, was he wrong!

That was a conversation I had about 20 years ago. It was truly a demoralizing comment. I had begun experimenting with the web and programming in high school in the late 90s. However, burdened with this defeat, I naively accepted it as truth and kept web development as a hobby. Ultimately, I went on to law school and launched a career as a lawyer. Don't get me wrong, I loved being a lawyer and I loved the work that I did. But my years as an attorney drove me back to software development, ultimately turning my hobby into a career. This quite unusual career path was made possible by the open source community. Through the help and guidance of the community at large, I taught myself the skills I would need to become a professional. It is now my turn to help others.

I share this story because it highlights something that I think is applicable not only to my life but also to web application development in general. This bad advice is a reminder that *not all guidance is good*, and that *the best course of action is the one that fits my needs, not those of someone else*. This book is devoted to that concept.

To become better at what we do, we must constantly be moving forward on our journey, learning new things and polishing existing skills. We must take the advice, the design patterns, and the code snippets that others provide us and internalize them. Some of it will be good, and some of it will be bad. By knowing this, we can carefully select *the good* to build something that meets our challenges and is truly extraordinary.

Shortly, we will start a journey together learning about web development. By the end of this book, I hope that you feel empowered to build what you want and need, and not be constrained by bad advice. And maybe—just maybe—you might walk away with just a little more passion and respect for Sanic, for Python, and for open source software. I truly wish you the best of luck on your own personal journey.

Who this book is for

This book is for Python web developers who have basic- to intermediate-level knowledge of how web technologies work and are looking to enhance their skills by taking their applications to the next level using the power of the Sanic framework. Working knowledge of Python web development with frameworks such as Django and/or Flask may be helpful, but is not required.

A basic- to intermediate-level understanding of Python 3, HTTP, RESTful API patterns, and modern development practices and tools, such as type annotations, pytest, and virtual environments, is also helpful.

What this book covers

Chapter 1, Introduction to Sanic and Async Frameworks, covers the background information on why Sanic was built, how it is developed, where it is headed, and who should use it. Important takeaways include the difference between WSGI, Async, and ASGI servers; what is a framework versus a server; and what developers should do to set up their project for success.

Chapter 2, Organizing a Project, takes you through the common approaches to organizing a project, and ideas to help decide an appropriate solution that fits the needs of what you are building.

Chapter 3, Routing and Intaking HTTP Requests, focuses upon the first interactions that the server has with an incoming web request. You'll learn about: how requests are structured; what choices Sanic makes for us and what choices it leaves to us; and other issues involved in turning an HTTP request into actionable code.

Chapter 4, Ingesting HTTP Data, focuses on the types of data that can be received and some effective tools for handling them. You will learn how to extract (and use) data from headers, cookies, paths, body, and query arguments; and also some useful tools to make it easier to build good applications with the data.

Chapter 5, Building Response Handlers, explores different techniques for sending content back to the clients. Modern applications often need several techniques to provide high-quality experiences and dynamic content. You will learn about when to apply each technique and how to optimize Sanic for each use case.

Chapter 6, Operating Outside the Response Handler, goes through all of the other stuff that Sanic can do besides just simple response handlers. You will learn about how to use listeners, middleware, signals, and background tasks to customize web applications. The goal is to learn to recognize patterns that can be applied in a variety of use cases.

Chapter 7, Dealing with Security Concerns, focuses upon common security concerns and how to deal with them in Sanic. Since this topic alone is quite extensive, the approach will be to introduce the concern and explain the issue as it relates to Sanic. Then we will look at common mitigation strategies and what they would look like in Sanic.

Chapter 8, Running a Sanic Server, focuses upon setting up a server for both local development and production-ready environments. Again, this topic can be extremely lengthy by itself. Therefore, the goal is not to be a tutorial on how to use tools such as Docker and Kubernetes. After the introduction, this chapter assumes some working knowledge, provides other materials for reference, and focuses on what these tools mean for Sanic. You will learn about some common deployment patterns and when/how you should decide to use them.

Chapter 9, Best Practices to Improve Your Web Applications, covers some practical tips on how to make your web applications better. The areas covered are the sorts of finishing touches that make an application feel "professional," and most importantly easier to maintain for future iterations.

Chapter 10, Implementing Common Use Cases with Sanic, analyzes common to slightly advanced use cases in depth and how to implement them in Sanic. We will start each section with a description of the problems encountered. After providing solutions, we will do a review of the implementations and try to draw some conclusions and generalizations that can be taken by you to other applications.

Chapter 11, A Complete Real-World Example, provides you with a complete web application that you can learn from. The source code is available for you to review, and the application itself is hosted online so that you can interact with it while trying to learn from it. You are encouraged to check it out at `https://sanicbook.com`.

To get the most out of this book

The code examples in this book assume that you are running the latest **long-term support** (**LTS**) release. As of the time of writing, that is **Sanic v21.12**. We assume that you have a modern installation of Python available. While most of the examples will work fine in Python 3.7, to be able to run all of the code, you will need at least Python 3.8.

You are highly encouraged to follow along and run the code examples in this book. In addition, there are additional patterns and snippets that will only be available in the GitHub repository (see the next section for the link). You are highly encouraged to review these as well.

Software/hardware covered in the book	Operating system requirements
Sanic v21.12 LTS	Windows, macOS, or Linux
Python 3.8+	

If you are using the digital version of this book, we advise you to type the code yourself or access the code from the book's GitHub repository (a link is available in the next section). Doing so will help you avoid any potential errors related to the copying and pasting of code.

The project is fast-moving and adopts new patterns all the time. Where possible, the GitHub repo will be kept up to date. If you encounter any problems while following the examples, you are encouraged to seek help at `https://sanic.dev/help`.

Download the example code files

You can download the example code files for this book from GitHub at `https://github.com/PacktPublishing/Python-Web-Development-with-Sanic`. If there's an update to the code, it will be updated in the GitHub repository.

We also have other code bundles from our rich catalog of books and videos available at `https://github.com/PacktPublishing/`. Check them out!

Download the color images

We also provide a PDF file that has color images of the screenshots and diagrams used in this book. You can download it here: `https://static.packt-cdn.com/downloads/9781801814416_ColorImages.pdf`.

Conventions used

There are a number of text conventions used throughout this book.

`Code in text`: Indicates code words in text, database table names, folder names, filenames, file extensions, pathnames, dummy URLs, user input, and Twitter handles. Here is an example: "When an exception is raised, Sanic stops the regular route handling process and moves it over to an `ErrorHandler` instance."

A block of code is set as follows:

```
@app.exception(PinkElephantError)
async def handle_pink_elephants(request: Request, exception:
Exception):
    ...
```

When we wish to draw your attention to a particular part of a code block, the relevant lines or items are set in bold:

```
@app.exception(PinkElephantError)
async def handle_pink_elephants(request: Request, exception:
Exception):
    ...
```

Any command-line input or output is written as follows:

```
$ pip install sanic-testing pytest
```

Bold: Indicates a new term, an important word, or words that you see onscreen. For instance, words in menus or dialog boxes appear in **bold**. Here is an example: "Log in to the web portal, click on **Kubernetes** on the main dashboard, and set up a cluster."

> **Tips or important notes**
> Appear like this.

Get in touch

Feedback from our readers is always welcome.

Sanic help: If you run into any problems with your web applications or have questions about Sanic, you should turn to https://sanic.dev/help.

Author: To get in touch with the author, please contact him via Twitter @admhpkns or visit his web page at https://amhopkins.com.

General feedback: If you have questions about any aspect of this book, email us at customercare@packtpub.com and mention the book title in the subject of your message.

Errata: Although we have taken every care to ensure the accuracy of our content, mistakes do happen. If you have found a mistake in this book, we would be grateful if you would report this to us. Please visit www.packtpub.com/support/errata and fill in the form.

Piracy: If you come across any illegal copies of our works in any form on the internet, we would be grateful if you would provide us with the location address or website name. Please contact us at copyright@packt.com with a link to the material.

If you are interested in becoming an author: If there is a topic that you have expertise in and you are interested in either writing or contributing to a book, please visit authors.packtpub.com.

Share Your Thoughts

Once you've read *Python Web Development with Sanic*, we'd love to hear your thoughts! Scan the QR code below to go straight to the Amazon review page for this book and share your feedback.

https://packt.link/r/1801814414

Your review is important to us and the tech community and will help us make sure we're delivering excellent quality content.

Part 1: Getting Started with Sanic

Our journey begins. This first part will provide an understanding of why Sanic is different, why you would use it, and how to set yourself and your project up for success.

This part contains the following chapters:

- *Chapter 1, Introduction to Sanic and Async Frameworks*
- *Chapter 2, Organizing a Project*

1
Introduction to Sanic and Async Frameworks

There should be one—and preferably only one—obvious way to do it.

– Tim Peters, The Zen of Python

Too often, this maxim of **Python** is taken to mean that there *must* be only one way to do something. Any Python web developer can simply look at the number of web frameworks that exist and tell you that the choice is not so simple. There are dozens of web frameworks on **PyPI**, and within the ecosystem of any single framework, you will find even more options to solve a single problem. Go ahead and type `authentication` into the search bar at `https://pypi.org`. Looking at the number of results, stating that there is only "[one] obvious way to do it" does not seem so obvious. Maybe this sentence needs to be changed. Perhaps it could read, "There should be one … obvious way *for you* to do it." Why? Because adding to the context that we are talking about your specific application brings us to the next level.

This is **Sanic**, and this is the goal of this book.

What may be obvious for someone building a stock portfolio tracker will not be obvious to someone building a streaming media player. So, to figure out what the *obvious* solution is, we must first understand the problem. And to understand the problem, we must be hyper-aware of our specific use case.

When we're trying to find a solution to a problem, many other tools and frameworks respond by saying: here is how you should do it. Do you want to read data from your web request? Here's how to validate it. Do you need **cross-site request forgery** (**CSRF**) protection? Here's the snippet you need to add. This approach fails to make you a better developer and fails to find the optimal solution for your use case.

Why should I validate my data this way? Why do I need this snippet to protect myself? Because someone else decided for you. You cannot answer these questions. All you know is that the framework documentation—or some blog on the internet—told you to do this, so you did it.

And this is why Sanic—and indeed, this book—takes a different approach. By the end of this book, we want you to know how to spot peculiar use cases, as well as how to bend the tooling to meet your needs. You should be able to think through different types of implementations and select one that is most meaningful for your needs. This will be the *obvious* solution.

Sanic prides itself on being unopinionated. That is not to say that the maintainers of the project do not have strong opinions. I welcome you to engage me or anyone in the community in a discussion about proxy forwarding, deployment strategies, authentication schemes, and more. You will certainly find passionate opinions. By "unopinionated," we mean to say that Sanic's job is to take care of the plumbing so that all you need to do is build the logic. The decision of how to tackle problems is not the domain of the framework.

You will find that Sanic developers are mostly keen to find solutions that are hyper-focused on solving the particular challenges that they face. Developers use Sanic because it is fast and simple. However, you will also find that using Sanic means the *obvious* solution to a problem is not based upon Sanic but upon your unique application requirements.

The other side of the story is that sometimes, your use case doesn't need a hyper-focused solution. This is also fine. For this reason, you will find several plugins (many of which are supported by active members of the Sanic core developer team) or off-the-shelf solutions. We wholly support your adoption of them and their patterns. Throughout this book, our examples will steer away from implementations that require plugins. However, where there are popular solutions that include plugins, we will also point them out to you for reference.

Our goal in this book is to help you learn how to identify your unique application requirements and match them with the tools at your disposal. This will help make your applications better and make you a better developer.

In this chapter, we will begin by building the foundational understanding that's needed to read this book. To do this, we will cover the following topics:

- What is Sanic?
- Leveling up
- Framework versus server
- Why we use Sanic—build fast, run fast

Technical requirements

This chapter will include some basic Python and Terminal usage. To follow along with the examples, make sure that your computer is setup with at least Python version 3.7, but Python version 3.8 or newer is recommended. You will also need to have curl, or a similar program, installed so that you can easily make and inspect HTTP requests. If you are unfamiliar with curl, it is a program that's executed from a Terminal session that allows you to make HTTP requests. It should be available on most macOS and Linux installations by default and can be installed on Windows machines.

What is Sanic?

Since you're reading this book, you're probably familiar with Python and may even know about some of the popular tools that are used to build web applications using Python. As for Sanic, you've either heard of it or have used it and want to improve your skills and understanding of it. You may know that Sanic is unlike traditional tools. Built 100% from the ground up, it's been developed with the idea of asynchronous Python in mind. From the very beginning, Sanic set out to be fast. It is one of the critical decisions that drives much of its development.

To truly understand the Sanic project, we'd benefit from a history lesson. Sanic was the first legitimate attempt to bring asynchronous Python to a web framework. It started as a proof-of-concept, as a hobby project. So, let's set the stage.

The most fundamental building block of Sanic is the **asyncio module** from Python's standard library. The Sanic project was born during the early stages of the module's release and has matured alongside the module.

Python 3.4—released in early 2014—was the first step to introduce the concept of **coroutines** into the standard library in the newly added **asyncio** module. Using standard Python generators, a function's execution can be halted while something else happens, and then data can be injected back into that function to allow it to resume execution. If there was an object that "looped" through a list of tasks that needed work, we could duck in and out of the execution of multiple functions at the same time. This could achieve "concurrency" in a single thread and is the basis of the idea of asyncio.

In the early days, this was mainly a toy and there was no widespread adoption of coroutines. Of course, legitimate application needs were being solved, but the concept was still very early in its development and not fully developed. The concept was refined over the next several Python releases to give us the asyncio module we know today.

Here's a quick look at what asynchronous programming looked like in Python 3.4:

```python
import asyncio
@asyncio.coroutine
def get_value():
    yield from asyncio.sleep(1)
    return 123


@asyncio.coroutine
def slow_operation():
    value = yield from get_value()
    print(">>", value)


loop = asyncio.get_event_loop()
loop.run_until_complete(slow_operation())
loop.close()
```

While we will not delve into how this works, it is worth mentioning that asynchronous Python is built on the premise of generators. The idea is that a generator function can yield back to some "loop" to allow it to duck in and out of execution.

> **Important Note**
>
> Sanic can be run with the standard library asyncio loop. However, out of the box, it uses an alternative loop called **uvloop** that operates significantly faster.

The language and syntax from the new asyncio module were both extremely powerful and a bit clunky. Generators are usually a bit mysterious and difficult for less seasoned Python developers. What exactly is `yield from`? This stuff looked alien to many people; Python needed a better syntax.

Before continuing, it is worth providing a quick side note if you are unfamiliar with generators. Python has a special type of function that returns a generator. That generator can be used to partially execute a function and suspend its operation by *yielding* a value until it is needed again. Generators can also be bi-directional, in that data can be sent back into them during execution. Again, the specifics of this are beyond the scope of this book since it is not entirely pertinent, but it is helpful to know that this is what `yield from` helps us achieve. Using the bidirectional functionality of generators, Python was able to build the capability for asynchronous coroutines.

Because this implementation was complex and slightly more difficult for newcomers when Python 3.5 was released, it included a much simpler and cleaner version:

```
async def get_value():
    await asyncio.sleep(1)
    return 123


async def slow_operation():
    value = await get_value()
    print(">>", value)
```

One of the main benefits of this style of programming is that it helps you block code due to input and output. This is known as **I/O-bound**. A classical example of an I/O-bound application is a web server. So, it was a natural fit for this new asyncio module to be built with the idea of creating protocols for interacting with networking traffic.

At the time, however, there were no frameworks or web servers that adopted this approach. The Python web ecosystem was built upon a premise that is fundamentally at odds with asynchronous Python.

> **Important Note**
>
> The classical Python pattern for integrating an application with a Python web server is known as the **Web Server Gateway Interface (WSGI)**. This is not within the scope of this book. While it is possible to shoehorn Sanic into WSGI, it is generally frowned upon. The problem with WSGI is that the entire premise of it is blocking. A web server receives a request, processes it, and sends a response all within a single execution. This means that the server can only process one request at a time. Using Sanic with a WSGI server destroys the asynchronous ability to efficiently handle multiple requests concurrently.

Classical **Django** and **Flask** could not adopt this pattern. Looking back after 6 years, those projects eventually did find ways to introduce *async/await*. However, it is not the natural use pattern for these frameworks and came at the expense of an extraordinary effort over many years.

As the asyncio module was being released, there was a lack of web frameworks to fill this new use case. Even the concept that eventually came to be known as **Asynchronous Server Gateway Interface (ASGI)** did not exist.

> **Important Note**
>
> The ASGI is the corollary to WSGI, except for asynchronous Python. It is possible—although not required—to use it for Sanic, and we will discuss it further in *Chapter 8, Running a Sanic Server*.

In the summer of 2016, Sanic was built to explore the gap. The idea was simple: could we take an application with a simplistic-looking API from Flask and make it *async/await*?

Somehow, the idea took off and gained traction. It was not a project that was initially set out with the goal of redoing how Python applications handled web requests. It very much was a case of accidental exposure. The project exploded and created excitement very quickly. There was a lot of appeal in making Flask adopt this new pattern; but, since Flask was incapable of doing so itself, many people thought that Sanic could be the async version of Flask.

Developers were excited for this new opportunity to use the latest in Python to bring a whole new level of performance to their applications. Early benchmarks showed Sanic running circles around Flask and Django.

Like so many projects, however, the initial burst of energy died off. The original project was meant to answer the question, "could a Flask-like framework exist?" The answer was a resounding yes. However, being a one-person project that had no intention of handling the level of support and attention that it received, the project started to gather dust. Pull requests started piling up. Issues went unanswered.

Through 2017 and 2018, the ecosystem for asynchronous Python was still very immature. It lacked production-worthy platforms that would be supported, maintained, and viable for both personal and professional web applications. Furthermore, there was still a bit of an identity question about Sanic. Some people felt that it should be a very niche piece of software, while others felt it could have broad applicability.

The result of the months of frustration and lack of responses from the maintainers of the project resulted in the Sanic Task Force. This precursor to the **Sanic Community Organization (SCO)** was a loose collective of developers that were interested in finding a future for the project, which was on the verge of failing. There was a desire to stabilize the API and answer all of the outstanding identity questions. For several months in mid-2018, a debate brewed about how to move the project forward, and how to make sure the project would not suffer the same fate again.

One of the most fundamental questions was whether the project should be forked. Since no one on the Sanic Task Force had admin access to the repository or other assets—and the only person who did was non-responsive—the only option was to fork and rebrand the project. However, Sanic had already existed for 2 years at that time and was known within the Python community as a viable (and fast) option for building asynchronous web applications. Dropping the existing project name would have been a huge blow toward ever getting the new project back up off the ground. Nonetheless, this was the only remaining solution. On the eve of forking the project to a new GitHub repository, the original maintainer offered up access to the repository and the SCO was born. The team worked to reorganize the operation of the community with the following goals:

- Regular and predictable releases, as well as deprecations
- Accountability and responsibility for responding to issues and support
- Structure for reviewing code and making decisions

In December 2018, the SCO released its first community version of Sanic: 18.12 LTS.

With this new structure in place, the SCO turned to its next question: what is Sanic? Ultimately, the decision was to break with any attempt at Flask parity. While it may be said that the original project was a "Flask clone," this is no longer true. You will still hear it being called "Flask-like," but that is a fair comparison, only because they look similar on the surface. Their features and behaviors are fundamentally different, and the likeness stops there. I try to steer away from this comparison because it diminishes the effort and improvements that hundreds of contributors have made to let Sanic stand on its own. Now that you know some of the history behind Sanic, we turn to one of the goals of this book: making you a better developer.

Leveling up

Sanic encourages experimentation, customization, and, most of all, curiosity.

It is probably not the best tool for Python beginners since it assumes that they have some knowledge of both Python and web development. That is not to say that the project discourages newcomers, or people just getting into web development. Indeed, Sanic is a wonderful place to start learning about web development for those that truly want to understand the practice. Much of web development is learning to balance competing decisions. Sanic development often includes learning about these decision points.

This touches upon the goal of this book: answering the question, "what is the *obvious* path to building my Sanic application?" This book intends to explore different patterns that *could* be used in Sanic. While we will learn about the concepts that relate to Sanic, the principles can be abstracted and applied to building an application with any other framework or language. It is critical to remember that there is no "right" or "wrong" way. Online forums are filled with "what is the right way?" questions that assume that there is standard practice. The answers to these questions point to the implication that if they are not following a particular pattern, then their application is wrong.

For example, someone may ask the following questions:

- What is the right way to serve internationalized content?
- What is the right way to deploy my web app?
- What is the right way to handle long-running operations?

These "right way" questions suffer from a critical flaw: the belief that there is only a *single* obvious solution. Structuring a question like this is the domain of opinionated frameworks that do not teach developers to think on their own. Ask these same questions on the Sanic forums, and you'll probably receive an "it depends" as your answer. Opinionated frameworks hinder creativity and design. They ultimately drive the developer into making choices based upon the constraints provided by the framework and not the constraints of the application's needs.

Instead, Sanic provides a set of tools to help developers craft the solutions that work for their use cases without mandating certain practices. This is why the built-in features of Sanic focus on functionality and the request/response cycle, and not on implementation details such as validation, **cross-origin resource sharing** (**CORS**), CSRF, and database management. All of *that other stuff* is, of course, important, and we will explore this in later chapters. But this book intends to look at the issues and see how you *might* solve the problem. Web applications and web APIs have differing needs, so you, as a developer, should be allowed to make the choices that are best suited (and *obvious*) to solving your problems.

Revisiting the previous questions, a better way to phrase them would be as follows:

- How *can* I serve internationalized content?

- Which deployment strategy will work for me, considering my constraints?

- What are the trade-offs to consider for handling long-running operations?

By the end of this book, you will be able to spot these questions and use your creativity to come up with appropriate solutions. You will learn about some of the powerful strategies that Sanic has to offer. But by no means think that any given solution is the only way to do something. Just because it is outlined here does not mean that it is the *right* way, or that the tools being used only apply to one particular situation.

The right way is to use the tools that Sanic and Python provide to solve your problems. If you were tasked with making soup, you would find that there is no single way to do it. You could look up some recipes and try to learn some basic patterns: boil water, add ingredients, and so on. Ultimately, to master the art of soup-making, you need to cook within the constraints of your kitchen, equipment, ingredients, and the people you are serving. This is what I want you to learn about web development: master your tools, environment, and requirements to build what you need for your specific users.

While reading this book, you should both be learning and analyzing the patterns and code. Try to generalize the concepts and think about how to use similar ideas down the road in your code. We encourage you to read this book in its entirety or duck in and out of chapters if they apply to you. Both are valid approaches.

Let's take a closer look at the question about internationalization to see how framing the question differently impacts our application and our knowledge:

- **BAD**: What is the right way to serve internationalized content?

- **GOOD**: How *can* I serve internationalized content?

The answer to the first question would likely include a snippet of code from someone who has had this problem in the past. The developer would copy/paste and move on, not learning anything in the process (and therefore will be unable to generalize and apply knowledge to similar problems in the future). At best, the solution is passably acceptable. At worst, the solution is harmful to the overall design of the application and the developer doesn't even know it.

Framing the question as "How can I…" leaves open the idea that there may be multiple avenues to the same destination. The bad question style is narrow and drives our attention to a single approach. The good style opens up the possibilities to explore different solutions and weigh them against their merits. After asking the question, our job now is to figure out those possible solutions, determine the potential trade-offs, and then come to a conclusion. In this process, we can draw upon our own past experiences, examples from other developers, and resource materials such as this book.

We may think about possible solutions involving the following:

- **Middleware** (catch the headers or path and reroute the request to a different handler)

- **Routing** (embed the language code in the URL path and extract the language from the route)

- **Functional programming** (use different functional handlers to generate separate responses)

- **Decorators** (execute some logic before [or after] the actual handler is run)

But *which* solution should be used? We need to know about the specifics of our application. Here are some important questions to keep in mind:

- Who is developing it? What is their experience level? How many developers are on the team? What tools will they be working with? Who will be maintaining it?

- Who will be using the application? Will it be consumed from a frontend **JavaScript (JS)** framework? A mobile app? Third-party integrations?

- How big will it scale? What sort of content needs to be delivered?

These questions are the domain of this book. We intend to ask these questions and determine the reasons behind making particular design pattern decisions. Of course, we cannot be exhaustive. Instead, we hope to inspire your journey to use as many tools and creative solutions as you can.

Few projects can match the amount of literature that has been written about Django. But precisely because Sanic does not require specific patterns, there is no need for such expansive amounts of documentation. The only prerequisite is knowing Python. The depth of API-specific knowledge that's needed to be successful with Sanic is not large. Do you know how to instantiate objects, pass values, and access properties? Of course, you do—it's just Python!

Two obvious valuable Sanic-specific resources are the User Guide (`https://sanic.dev`) and the API documentation (`https://sanic.readthedocs.io`). We will refer to both of these heavily in this book. But equally important is any other source on the web or in print that you have used up to this point to learn Python.

Coming back to the question of the *obvious* way to handle some tasks within Sanic: use the resources and tools that exist. There is a wealth of information on StackOverflow and the Sanic Community Forums. The Discord server is an active live discussion channel. Make yourself known, make your voice heard.

Don't ask the *right way* questions. Instead, ask the *how can I* questions. Next, we will learn a little bit about where Sanic falls in the web ecosystem, and its unusual place as both a framework and server.

Framework versus server

Sanic calls itself both a web framework and a web server. What does this mean? And more importantly, why is this important? Before we can explore this, we must understand what these terms mean, and why they exist.

Web server

A web server is a piece of software that is designed to deliver documents and data via the **HTTP** protocol. Its function is to accept an incoming HTTP request, decode the message to understand what the request is trying to accomplish, and deliver an appropriate response. The language of web servers is the HTTP protocol.

We will get into the specifics later, but for now, we will set up a simple Sanic server, issue a request from `curl`, and look at the message:

1. Create a file called `server.py` and run it in your Terminal:

    ```python
    from sanic import Sanic, text, Request


    app = Sanic(__name__)


    @app.post("/")
    async def handler(request: Request):
        message = (
            request.head + b"\n\n" + request.body
        ).decode("utf-8")
        print(message)
    ```

```
        return text("Done")


    app.run(port=9999, debug=True)
```

2. Now, we will send a request to our API:

```
$ curl localhost:9999 -d '{"foo": "bar"}'
```

In our console, we should see the following HTTP request message:

```
POST / HTTP/1.1
Host: localhost:9999
User-Agent: curl/7.76.1
Accept: */*
Content-Length: 14
Content-Type: application/x-www-form-urlencoded

{"foo": "bar"}
```

What we can see here is three components:

*   The first line contains the HTTP method, the path, and the HTTP protocol that's being used.
*   Next is a list of HTTP headers, one per line in key: value format.
*   Last is the HTTP body, preceded by a blank line.

HTTP responses are very similar:

```
HTTP/1.1 200 OK
content-length: 4
connection: keep-alive
content-type: text/plain; charset=utf-8

Done
```

The three components are now as follows:

*   The first line contains the HTTP protocol, followed by the HTTP status, and then a status description.

- Next is a list of HTTP headers, one per line in `key: value` format.
- Last is the HTTP body (if there is one), preceded by a blank line.

Though this is the language of web servers, it is very cumbersome to write all of that. This is why tools such as web browsers and HTTP client libraries were created—to build and parse these messages for us. Next we will explore how servers deal with this same problem, and how web frameworks solve it.

Web framework

We could, of course, write a program in Python that receives these raw HTTP messages, decodes them, and returns an appropriate HTTP response message. However, this would require a lot of boilerplate, be difficult to scale, and be prone to mistakes.

Certain tools do this for us: web frameworks. The job of a web framework is to build the HTTP message and handle the request appropriately. Many frameworks go further by providing conveniences and utilities to make the process simpler.

There are many web frameworks in the Python ecosystem that do this work to varying degrees. Some provide a huge number of features, while some are very sparse in terms of what they offer. Some are very strict, while some are more open. Sanic tries to fall on the continuum of being feature-rich, but only so far as what's required to not get in the way of the developer.

One of the features that Sanic provides is that it is both a web framework and a web server.

If you survey the web frameworks on PyPI, you will find that most of them require a separate web server to be installed. When it comes to deploying *most* Python applications, there is a hard line between the persistent operation that runs on the machine and the tooling that's used to develop response handlers. We will not delve too deeply into WSGI since it doesn't apply to Sanic. However, the paradigm that there is a server that calls a single input function, passes information about the request, and then expects a response is important to understand. Everything that happens in-between is the framework.

Narrowing our focus to projects that support *async/await* style coroutine handlers, the vast majority require you to run an ASGI server. It follows a similar pattern: an ASGI-ready server calls into an ASGI ready framework. These two components operate with one another using a specific protocol. There are currently three popular ASGI servers: uvicorn, hypercorn, and daphne.

Precisely because Sanic was born during the era that predated ASGI, it needed a server. Over time, this has become one of its greatest assets and is in large part why it outperforms most other Python frameworks. Development of the Sanic server is hyper-focused on performance and minimizing the request/response cycle. However, in recent years, Sanic has also adopted an ASGI interface to allow it to be run by an ASGI web server.

However, for the majority of this book, you can assume that when we are talking about running Sanic, we mean using the internal web server. It is production-ready and remains one of the best methods for deploying Sanic. Later, in *Chapter 8*, *Running a Sanic Server*, we will discuss all of the potential choices and help you come up with the questions to ask when you're deciding which solution is *obvious* for your needs. Now that you know the what of Sanic, we turn to the why.

Why we use Sanic—build fast, run fast

Let's begin by looking at Sanic's goal:

> *To provide a simple way to get a highly performant HTTP server up and running that is easy to build, expand, and, ultimately, scale.*

Source: `https://sanic.dev/en/guide/#goal`.

Most people who are familiar with the Sanic project will tell you that its defining feature is its performance. While this is important, it is only a single part of the core philosophies of the Sanic project.

The Sanic tagline is: "Build fast. Run fast." This highlights the performance orientation of the project. It also speaks to the goal that building an application in Sanic is meant to be intuitive. Getting an application up and running should not mean learning a complex set of APIs and having a near-constant second browser window open to the documentation. While other tools make heavy usage of "black box" type features such as global variables, "magic" imports, and monkey patching, Sanic generally prefers to head in the direction of well-written, clean, and idiomatic Python (that is, **Pythonic code**). If you know Python, you can build a web API with Sanic.

If performance alone is not the defining feature, then what is? The front page of the Sanic project's website gives six reasons for us to explore:

- Simple and lightweight

- Unopinionated and flexible

- Performant and scalable

- Production-ready
- Trusted by millions
- Community-driven

Simple and lightweight

The API is intentionally lightweight. This means that the most common properties and methods are easily accessible and that you do not need to spend a long time memorizing a particular call stack. Early on in the history of the project, there were discussions about adding certain features. But often, adding features can lead to bloat. The SCO decided that it was more important to focus on the specifics of providing a quality developer experience than on providing "bells and whistles" features.

For example, if my application is meant to be consumed by third-party applications, then why does it need CORS? There are so many differing needs for web applications that it was decided that these features are better left to plugins and developers. This leads to the next reason.

Unopinionated and flexible

Being unopinionated is a huge asset. It means that the developer decides if they want sessions or token authentication and if they want an ORM, raw SQL queries, a NoSQL database, a combination, or even no data store at all. That is not to say that these things can't be achieved in other frameworks. But there are certain design decisions to take into account. Rather than focusing on all these features, Sanic would rather provide you with the tools to implement the features you need, and nothing more. Tooling beats features. Borrowing from a popular proverb: *Sanic does not give you the fish, it teaches you how to fish*.

Performant and scalable

This is what Sanic is well-known for. With a performance-first approach toward development and implementations that include tools such as **uvloop** and **ujson**, Sanic tends to outperform other asynchronous Python frameworks. We will not spend too much time on benchmarks because I tend to feel they have limited use when it comes to comparing frameworks. Something more important regarding performance is the ability to build fast and scale fast. Sanic makes it simple to run multiple server instances from a single deployment. *Chapter 8, Running a Sanic Server*, will talk more about scaling. It's also important to note that Sanic is very well suited to building monolithic applications, microservices, and everything in-between because of the intentional flexibility of the API.

Production-ready

It is common for frameworks to ship with a development server. These development servers make the process of building simpler by including features such as auto-reload while you are working on a project. However, these servers don't tend to be ready for production environments. Sanic's web server is intentionally built to be the primary strategy for deployment in production systems. This leads to the next reason.

Trusted by millions

Sanic is installed in and powers many applications, both large and small. It is used in corporate-built web applications and personal web projects. It tends to be one of the most downloaded frameworks from PyPI. Between April 2019 and April 2020, there were 48 million downloads of Django. Sanic, in that same period, had about 44 million downloads. It is a project with high visibility and widespread adoption in a variety of use cases.

Community-driven

Since its move from an individual repository to a community organization in 2018, decision-making has become shared across the members of the community in what they call "lazy consensus." Here's what the SCO's website says:

> *In general, so long as nobody explicitly opposes a proposal or patch, it is recognized as having the support of the community. This is called lazy consensus; that is, those who have not stated their opinion explicitly have implicitly agreed to the implementation of the proposal.*

Source: `https://sanic.dev/en/org/scope.html#lazy-consensus`

Another important factor of the community is the ability of all members (whether a regular contributor or a first-time user) to enter the conversation and input valuable information into the conversation. As much as possible, Sanic attempts to be "of the community, by the community" to ensure its stability, feature set, and future.

The insistence on a community-first organization is meant to create a level of stability. All the work on the project is done by volunteers. It is a labor of love. With that said, projects driven by passion alone are at risk of becoming unmaintained if it rests upon the shoulders of a single person. This is exactly the scenario Sanic was trying to escape when the SCO was created. Being a project "of the community" means that multiple people are willing and capable to help carry the torch forward. Sanic achieves this with a rotating set of developers and balances long-term stability with staggered terms.

> **More on the SCO**
>
> If you want to learn more about the structure of the SCO and how you can get involved, check out the **Sanic Community Organization Policy E-manual** (**SCOPE**): `https://sanic.dev/en/org/scope.html#lazy-consensus`.

What drives code decisions?

Although not exactly formalized, there is an underlying set of principles that the architects and engineers of Sanic use when making coding decisions. Keeping in mind that this project is built by the hands of many people, from many backgrounds and experience levels, it is no surprise to learn that maintaining a set of consistent coding practices is itself a challenge.

I am not specifically talking about things such as formatting—tools such as **black**, **isort**, **flake8**, and **mypy** have abstracted that away. Rather, what should the code look like, how should it be organized, and what patterns should be followed?

The principles behind developing Sanic's code base are as follows:

- Performance
- Usability (unopinionated)
- Simplicity

Any line of code that is going to run during the execution of the request/response cycle will be highly scrutinized for its performance impacts. When faced with a question that puts two or more of these core philosophies in opposition, the performance consideration will almost always win. However, there are times when a slower alternative must be used. This will help not force developers into an awkward development pattern or add an undue level of complexity for developers. Part of the "speed" of Sanic is not just application performance, but also development performance.

When using Sanic, we can feel confident that there is a team of developers scrutinizing every line of code and its impact on performance, usability, and simplicity.

Let's imagine you are being asked to build an API by a project manager, who has a deadline in mind. To meet that goal, you want to get up and running as quickly as possible. However, you also want to make sure that you will have the freedom to iterate on the problems that face you without fear of being boxed into making bad decisions. One of the goals of this book is to help you identify useful patterns to adapt to help you get there.

Summary

It is helpful to understand the history and decisions behind Sanic to understand its feature set and implementation. Often, Sanic will be seen as an attempt to bring *async/await* style programming to a Flask app. While this may be a fair point of the original proof-of-concept, Sanic has developed upon a very divergent path, with the goal and impact of becoming a powerful tool designed for performance applications.

Due to this, Sanic is typically used by developers and teams that are looking to build a rich environment that addresses the unique—and *obvious*—design patterns that are required by their application's needs. The project intends to take away the difficult or cumbersome parts of building a web server and provide the tools to create performant and scalable web applications.

Now that we have learned about the background of Sanic, we should understand and appreciate the flexibility of using Sanic as a web framework. It is helpful to know the context in which Sanic was developed so that we can learn how to use it in our projects. The next step—beginning with *Chapter 2*, *Organizing a Project*—is to start learning about some of the foundational decisions we should make we're when starting any new web development project.

2
Organizing a Project

It is Day 0. You have a project in hand. You are fired up and ready to build a new web application. Ideas are swirling in your head, and your fingers are itching to start punching the keyboard. Time to sit down and start coding!

Or is it? It is tempting to start building an application as soon as the ideas about what we want to build begin to formulate in our heads. Before doing that, we should think about setting ourselves up for success. Having a solid foundation for the building will make the process much easier, reduce bugs, and result in a cleaner application.

The three foundations for beginning any Python web application project are as follows:

- Your IDE/coding editor

- An environment for running your development application

- A project application structure

These three elements take into account a lot of personal tastes. There are so many good tools and approaches. There is no way a single book could cover them all. If you are a more seasoned developer and already have a set of preferences, great, run with that and skip ahead to the next chapter.

In this chapter, we will explore a couple of modern options to get you up and running. The focus will be on foundation #2 (the environment) and foundation #3 (the application structure). We skip #1 and assume you are using a modern IDE of your own choosing. Popular choices in the Python world include VS Code, PyCharm, and Sublime Text. If you are not using one of these or something similar, go look them up and find one that works for you.

After we have set up our environment, we will explore some patterns to be implemented in Sanic that will help define your application architecture. This is not a software architecture book. I highly recommend you learn about approaches such as "Domain-Driven Design" and "Clean Architecture." This book is focused much more on the practical aspects and decisions of building a web application in Sanic, so feel free to adjust the patterns as you feel necessary.

In this chapter, we'll go through the following topics:

- Setting up an environment and directory
- Using blueprints effectively
- Wiring it all up
- Running our application

Technical requirements

Before we begin, we will assume that you have the following already set up on your computer:

- A modern Python installation (Python 3.8 or newer)
- A terminal (and basic knowledge of how to execute programs)
- An IDE (as discussed above)

Setting up an environment and directory

The first few steps that you take when starting any project have a monumental impact on the entirety of the project. Whether you are embarking on a multi-year project—or one that will be complete in a couple of hours—these early decisions will shape how you and others work on the project. But, even though these are important choices, do not fall into the trap of thinking that you need to find the *perfect* solution. There is no single "right way" to set up an environment or project directory. Remember our discussion from the previous chapter: we want to make the choices that fit the project at hand.

Environment

A good practice for Python development is to isolate its running environment from other projects. This is typically accomplished with virtual environments. In its most basic understanding, a virtual environment is a tool that allows you to install Python dependencies in isolation. This is important so that when we begin to develop our application, we have control of the requirements and dependencies in use. In its absence, we might mistakenly run our application and have requirements from other projects bleed into the application, thereby causing bugs and unintended behaviors.

The use of a virtual environment is so foundational in the Python development world that it has become the expected "norm" when creating a Python script or application. The first step you should always take when starting a new project is making a new virtual environment for it. The alternative to them is to run your application with your operating system's installation of Python. Do not do this. It may be fine for a while, but eventually, you will come across conflicting requirements, naming collisions, or other difficulties that all stem from a lack of isolation. The first step to becoming a better Python developer is to use virtual environments, if you are not doing so already.

It is also extremely helpful to acquaint yourself with the different tools that IDEs provide in hooking up to your virtual environment. These tools will often include things such as code completion and guide you as you start using features of your dependencies.

We do eventually want to run our application using containers. Being able to run our application inside a Docker container will greatly reduce the complexity associated with deploying our application down the road. This will be discussed further in *Chapter 9, Best Practices to Improve Your Web Applications*. However, I also believe that our application should be runnable (and therefore testable) from multiple environments. Even if we intend to use Docker down the road, we first need our application running locally without it. Debugging becomes much easier when our application does not rely upon an overly complex set of requirements just to run. Therefore, let's spend some time thinking about how to set up a virtual environment.

Many great tutorials and resources are available regarding how to use virtual environments. There are also many tools out there that are created to help manage the process. While I am a fan of the simple, tried and tested method of `virtualenv`, plus `virtuanenvwrapper`, many people are fans of `pipenv`, or `poetry`. These latter tools are meant to be a more "complete" encapsulation of your running environment. If they work for you, great. You are encouraged to spend some time to see what strikes a chord and resonates with your development pattern and needs.

We will leave virtual environments aside for now and briefly explore the usage of a relatively new pattern in Python. In Python 3.8, Python adopted a new pattern in PEP 582 that formalizes the inclusion of requirements in an isolated environment in a special __pypackages__ directory that lives inside the project. While the concept is similar to virtual environments, it works a little differently.

In order to implement __pypackages__, we are making it mandatory for our fictitious development team to use pdm. This is a relatively new tool that makes it super simple to adhere to some of the latest practices in modern Python development. If this approach interests you, take some time to read PEP 582 (https://www.python.org/dev/peps/pep-0582/) and look at pdm (https://pdm.fming.dev/).

You can get started by installing it with pip:

```
$ pip install --user pdm
```

Refer to the installation instructions on their website for more details: https://pdm.fming.dev/#installation. Pay particular attention to useful features such as shell completion and IDE integrations.

Now, let's proceed with setting up:

1. To get started, we create a new directory for our application and, from that directory, run the following and follow the prompts to set up a basic structure:

    ```
    $ mkdir booktracker
    $ cd booktracker
    $ pdm init
    ```

2. Now we will install Sanic:

    ```
    $ pdm add sanic
    ```

3. We now have access to Sanic. Just to confirm in our heads that we are indeed in an isolated environment, let's quickly jump into the Python REPL and check the location of Sanic using sanic.__file__:

    ```
    $ python
    >>> import sanic
    >>> sanic.__file__
    '/path/to/booktracker/__pypackages__/3.9/lib/sanic/__init__.py'
    ```

Sanic CLI

As discussed in *Chapter 8*, *Running a Sanic Server*, many considerations go into *how* to deploy and run Sanic. Unless we are specifically looking into one of these alternatives, you can assume in this book that we are running Sanic using the Sanic CLI. This will stand up our application using the integrated Sanic web server.

First, we will check to see what version we are running:

```
$ sanic -v
Sanic 21.3.4
```

And then we will check to see what options we can use with the CLI:

```
$ sanic -h
usage: sanic [-h] [-H HOST] [-p PORT] [-u UNIX] [--cert
CERT] [--key KEY] [-w WORKERS] [--debug] [--access-logs |
--no-access-logs] [-v] module


          Sanic
     Build Fast. Run Fast.


positional arguments:
  module                path to your Sanic app. Example: path.
to.server:app


optional arguments:
  -h, --help            show this help message and exit
  -H HOST, --host HOST  host address [default 127.0.0.1]
  -p PORT, --port PORT  port to serve on [default 8000]
  -u UNIX, --unix UNIX  location of unix socket
  --cert CERT           location of certificate for SSL
  --key KEY             location of keyfile for SSL.
  -w WORKERS, --workers WORKERS
                        number of worker processes [default 1]
  --debug
  --access-logs         display access logs
  --no-access-logs      no display access logs
  -v, --version         show program's version number and exit
```

Our standard form for running our applications right now will be as follows:

```
$ sanic src.server:app -p 7777 --debug --workers=2
```

What thought went into the decision behind using this command? Let's take a look.

Why src.server:app?

First, we are going to run this from the `./booktracker` directory. All of our code will be nested in an `src` directory.

Second, it is somewhat standard practice that our application creates a single `Sanic()` instance and assigns it to a variable called `app`:

```
app = Sanic("BookTracker")
```

If we were to place that in a file called `app.py`, then our module and variable would start to get confused:

```
from app import app
```

The preceding import statement is, well, ugly. It is beneficial to avoid naming conflicts between modules and the contents of that module as much as possible.

A bad example of this exists in the standard library. Have you ever done this one by accident?

```
>>> import datetime
>>> datetime(2021, 1, 1)
Traceback (most recent call last):
  File "<stdin>", line 1, in <module>
TypeError: 'module' object is not callable
```

Oops, we should have used `from datetime import datetime`. We want to minimize the replication of module names and properties, and to make our imports easy to remember and intuitive to look at.

Therefore, we will place our global `app` variable in a file called `server.py`. Sanic will look for our app instance when you pass in the `<module>:<variable>` form.

Why -p 7777?

We, of course, could choose any arbitrary port here. Many web servers will use port 8000 and that is the Sanic default if we just left it out completely. However, precisely because it is standard, we want to choose something else. Often, it is beneficial to choose a port that is less likely to collide with other ports that might be running on your machine. The more we can preserve common ports, the less likely we will run into collisions.

Why --debug?

While developing, having DEBUG mode enabled provides a more verbose output from Sanic, and an auto-reloading server. It can be helpful to see more logs, but make sure you turn this off in production.

The auto-reloading feature is particularly beneficial because you can start writing your app in one window, and have it running in a separate terminal session. Then, every time that you make a change and save the application, Sanic will restart the server, and your new code is immediately available for testing.

If you want auto-reloading but not all the extra verbosity, consider using --auto-reload instead.

Why --workers=2?

It is not an uncommon problem where someone begins to build an application and then realizes down the road that they have made a mistake by not preparing for horizontal scaling. Perhaps they added a global state that cannot be accessed outside of a single process:

```
sessions = set()

@app.route("/login")
async def login(request):
    new_session = await create_session(request)
    sessions.add(new_session)
```

Oops, now that person needs to go back and re-engineer the solution if they want to scale up the application. This could be a costly endeavor. Luckily, we are smarter than that.

By forcing our development pattern to include multiple workers from the beginning, it will help remind us as we are solving problems that our application *must* account for scaling. Even if our ultimate deployment does not use multiple Sanic workers per instance (and instead, for example, uses multiple Kubernetes pods with single worker instances; see *Chapter 9*, *Best Practices to Improve Your Web Applications*), this constant safeguard is a helpful way to keep the ultimate goal integral to the design process.

Directory structure

There are many different patterns you can follow for organizing a web application. Perhaps the simplest would be the single `server.py` file, where all of your logic exists together. For obvious reasons, this is not a practical solution for larger, real-world projects. So we will ignore that one.

What types of solutions are there? Perhaps we could use the "apps" structure that Django prefers, where discrete portions of our application are grouped into a single module. Or, perhaps you prefer to group by type, for example, by keeping all of your view controllers together. We make no judgments here about what is better for your needs, but we need to understand some consequences of our decisions.

When making a decision, you might want to learn some common practices. This might be a good opportunity to go and look up some of the following patterns:

- **Model View Controller** (**MVC**)

- **Model View ViewModel** (**MVVM**)

- **Domain-Driven Design** (**DDD**)

- **Clean Architecture** (**CA**)

Just to give you a flavor of the differences (or at least my interpretation of them), you might structure your project in one of the following ways:

You might use MVC:

```
./booktracker
├── controllers
│   ├── book.py
│   └── author.py
├── models
│   ├── book.py
│   └── author.py
├── views
```

```
|     ├── static
|     └── templates
└── services
```

Or you might use DDD:

```
./booktracker
└── domains
    ├── author
    |   ├── view.py
    |   └── model.py
    ├── book
    |   ├── view.py
    |   └── model.py
    └── universal
        └── middleware.py
```

In this book, we are going to adopt something that approximates to a hybrid approach. There is a time and place for applying these theoretical constructs. I urge you to learn them. The information is useful. But we are here to learn *how* to practically go about building an application with Sanic.

Here's the modified structure:

```
./booktracker
├── blueprints
|   ├── author
|   |   ├── view.py
|   |   └── model.py
|   └── book
|       ├── view.py
|       └── model.py
├── middleware
|   └── thing.py
├── common
|   ├── utilities
|   └── base
└── server.py
```

Let's break down each of these to see what they might look like and understand the thought process behind this application design.

./blueprints

This might strike you as odd since ultimately, this directory looks like it contains more than just blueprints. And, you would be right. Looking at the tree, you see that "blueprints" include both `view.py` and `model.py`. The goal of this directory is to separate your application into logical components, or domains. It functions much the same way as an `apps` directory might in a Django application. If you can isolate some construct or portion of your application as being a distinct entity, it should probably have a subfolder here.

A single module in this directory might contain models for validating incoming requests, utilities for fetching data from a database, and blueprints with attached route handlers. This keeps related code close together.

But why call it `blueprints`? Each subdirectory will contain much more than a single `Blueprint` object. The point is to reinforce the idea that everything in this directory resolves around one of these discrete components. The standard method for organization a so-called component in Sanic is the `Blueprint` method (which we will learn more about in the next section). Therefore, each subdirectory will have one, and only one, `Blueprint` object.

Another important rule is as follows: *nothing* inside the `./bluprints` directory will reference our Sanic application. This means that both `Sanic.get_app()` and `from server import app` are forbidden inside this directory.

It is generally helpful to think of a blueprint as corresponding to a portion of your API design pattern:

- `example.com/auth -> ./blueprints/auth`
- `example.com/cake -> ./blueprints/cake`
- `example.com/pie -> ./blueprints/pie`
- `example.com/user -> ./blueprints/user`

./middleware

This directory should contain any middleware that is meant to be global in scope:

```
@app.on_request
async def extract_user(request):
```

```
user = await get_user_from_request(request)
request.ctx.user = user
```

As discussed later in this chapter and in *Chapter 6, Operating Outside the Response Handler*, as well as in the Sanic user guide (`https://sanic.dev/en/guide/best-practices/blueprints.html#middleware`), middleware can be global or attached to blueprints. If you need to apply middleware to specific routes, perhaps blueprint-based middleware makes sense. In this case, you should nest them in the appropriate `./blueprints` directory and not here.

./common

This module is meant to be a place for storing class definitions and functions that will be used to build your application. It is for everything that will span your blueprints and be pervasive within your application.

> **Tip**
> Try to expand upon the directory structure here to meet your needs. However, try not to add too many top-level directories. If you start cluttering up your folders, think about how you might be able to nest directories inside one another. Usually, you will find that this leads to a cleaner architecture. There is also such a thing as going too far with nesting. For example, if you need to navigate ten levels deep in your application code, perhaps you should dial it back.

It's still Day 0. You still have a lot of great ideas in your head about what you want to build. And thanks to some thoughtful pre-planning, we now have an effective setup for building application locally. At this point, we should know how the application will run locally, and generally how the project will be organized. What we will learn next is the transition step from application structure to business logic.

Using blueprints effectively

If you already know what a blueprint is, imagine for a moment that you do not. As we are building out our application and trying to structure our code base in a logical and maintainable pattern, we realize that we need to constantly pass around our `app` object:

```
from some.location import app


@app.route("/my/stuff")
async def stuff_handler(...):
```

```
    . . .

@app.route("/my/profile")
async def profile_handler(...):

    . . .
```

This can become very tedious if we need to make changes to our endpoints. You can imagine a scenario where we would need to update a bunch of separate files to duplicate the same change over and over again.

Perhaps more frustratingly, we might end up in a scenario where we have circular imports:

```
# server.py
from user import *

app = Sanic(...)
# user.py
from server import app

@app.route("/user")
...
```

Blueprints solve both of these problems and allow us to abstract away some content so that the component can stand on its own. Returning to the preceding example, we take the common part of the endpoints (/my) and add it to the Blueprint definition:

```
from sanic import Blueprint

bp = Blueprint("MyInfo", url_prefix="/my")

@bp.route("/stuff")
async def stuff_handler(...):

    . . .

@bp.route("/profile")
async def profile_handler(...):

    . . .
```

In this example, we were able to group these routes together into a single blueprint. Importantly, this allows us to pull common parts of the URL path (/my) to the Blueprint, which gives us the flexibility to make changes in the future.

No matter how you decide to organize your file structure, you probably should always use blueprints. They make organization easier, and can even be nested. Personally, I will only ever use @app.route in the most simple of web applications. For any *real* projects, I always attach routes to blueprints.

Blueprint registration

Just creating our blueprints is not enough. Python would have no way to know they exist. We need to import our blueprints and attach them to our application. This is done through a simple registration method: app.blueprint():

```
# server.py
from user import bp as user_bp


app = Sanic(...)
app.blueprint(user_bp)
```

A common "gotcha" is misunderstanding what blueprint is doing. Something like this will not work as expected:

```
from sanic import Sanic, Blueprint


app = Sanic("MyApp")
bp = Blueprint("MyBrokenBp")


app.blueprint(bp)


@bp.route("/oops")
```

At the instant that we register a blueprint, everything that was attached to it will reattach to the application. This means that anything added to the blueprint *after* the call to app.blueprint() will not be applied. In the preceding example, /oops will not exist on the application. Therefore, you should try and register your blueprints as late as possible.

> **Tip**
>
> I think it is super convenient to always name blueprint variables bp. When I open a file, I automatically know what bp means. Some people may find it helpful to give their variable a more meaningful name: `user_bp` or `auth_bp`. For me, I would rather keep them consistent in the files I am always looking at, and just rename them at import: `from user import bp as user_bp`.

Blueprint versioning

A very powerful and common construct in API design is versioning. Let's imagine that we are developing our book API that will be consumed by customers. They have already created their integrations, and perhaps they have been using the API for some time already.

You have some new business requirements, or new features you want to support. The only way to accomplish that is to change how a particular endpoint works. However, this will break backward compatibility for users. This is a dilemma.

API designers often solve this problem by versioning their routes. Sanic makes this easy by adding a keyword argument to a route definition, or (perhaps more usefully) a blueprint.

You can learn more about versioning in the user guide (`https://sanic.dev/en/guide/advanced/versioning.html`) and we will discuss it in more depth in *Chapter 3, Routing and Intaking HTTP Requests*. For now, we will have to be content with knowing that our original API design needs a modification, and we will see how we can achieve that in the next section.

Grouping blueprints

As you begin to develop your applications, you might start to see similarities between blueprints. Just like we saw that we could pull common parts of routes out to `Blueprint`, we can pull common parts of `Blueprint` out into `BlueprintGroup`. This serves the same purpose:

```
from myinfo import bp as myinfo_bp
from somethingelse import bp as somethingelse_bp
from sanic import Blueprint

bp = Blueprint.group(myinfo_bp, somethingelse_bp, url_prefix="/api")
```

We have now added /api to the beginning of every route path defined inside myinfo and somethingelse.

By grouping blueprints, we are condensing our logic and becoming less repetitive. In the above example, by adding a prefix to the whole group, we no longer need to manage individual endpoints or even blueprints. We really need to keep the nesting possibilities in mind as we design the layout of our endpoints and our project structure.

In the last section, we mentioned using versions to provide an easy path to flexibly upgrade our API. Let's go back to our book tracking application and see what this might look like. If you recall, our application looked like this:

```
./booktracker
└── blueprints
    ├── author
    │   └── view.py
    └── book
        └── view.py
```

And we also have the view.py files:

```
# ./blueprints/book/view.py
bp = Blueprint("book", url_prefix="/book")
# ./blueprints/author/view.py
bp = Blueprint("author", url_prefix="/author")
```

Let's imagine the scenario where this API is already deployed and in use by customers when our new business requirements come in for a /v2/books route.

We add it to our existing architecture, and immediately it is starting to look ugly and messy:

```
└── blueprints
    ├── author
    │   └── view.py
    ├── book
    │   └── view.py
    └── book_v2
        └── view.py
```

Let's refactor this. We will not change `./blueprints/author` or `./blueprints/book`, just nest them a little deeper. That part of the application is already built and we do not want to touch it. However, now that we have learned from our mistake, we want to revise our strategy for `/v2` endpoints to look like this:

```
└── blueprints
    ├── v1
    │   ├── author
    │   │   └── view.py
    │   ├── book
    │   │   └── view.py
    │   └── group.py
    └── v2
        ├── book
        │   └── view.py
        └── group.py
```

We just created a new file, `group.py`:

```
# ./blueprints/v2/group.py
from .book.view import bp as book_bp
from sanic import Blueprint


group = Blueprint.group(book_bp, version=2)
```

Grouping blueprints is a powerful concept when building complex APIs. It allows us to nest blueprints as deep as we need to while providing us with both routing and organizational control. In this example, notice how we were able to assign `version=2` to the group. This means now that every route attached to a blueprint in this group will have a `/v2` path prefix.

Wiring it all up

As we have learned, creating a pragmatic directory structure leads to predictable and easy-to-navigate source code. Because it is predictable to us as developers, it is also predictable for computers to run. Perhaps we can use this to our advantage.

Earlier, we discussed one of the problems we often encounter when trying to expand our application from the single file structure: circular imports. We can solve this well with our blueprints, but it still leaves us wondering about what to do with things that we might want to attach at the application level (such as middleware, listeners, and signals). Let's take a look at those use cases now.

Controlled imports

It is generally preferred to break code up into modules using nested directories and files that help us both logically think about our code, but also navigate to it. This does not come without a cost. What happens when two modules are interdependent? This will cause a circular import exception, and our Python application will crash. We need to not only think about how to logically organize our code but also how different parts of the code can be imported and used in other locations.

Consider the following example. First, create a file called ./server.py like this:

```
app = Sanic(__file__)
```

Second, create a second file called ./services/db.py:

```
app = Sanic.get_app()

@app.before_server_start
async def setup_db_pool(app, _):
    ...
```

This example illustrates the problem. When we run our application, we need Sanic(__file__) to run before Sanic.get_app(). But, we need to import .services.db so that it can attach to our application. Which file evaluates first? Since the Python interpreter will run instructions sequentially, we need to make sure that we instantiate the Sanic() object before importing the db module.

This will work:

```
app = Sanic(__file__)

from .services.db import *
```

However, it sort of looks ugly and non-Pythonic. Indeed, if you run tools such as `flake8`, you will start to notice that your environment does not really like this pattern so much either. It breaks the normal practice of placing imports at the top of the file. Learn more about this anti-pattern here: `https://www.flake8rules.com/rules/E402.html`.

You may decide that you do not care, and that is perfectly okay. Remember, we are in this to find the solution that works for your application. Before we make a decision, however, let's look at some other alternatives.

We could have a single `startup` file that will be a controlled set of import ordering:

```
# ./startup.py
from .server import app
from .services.db import *
```

Now, instead of running `sanic server:app`, we want to point our server to the new `startup.py` file:

```
sanic startup:app
```

Let's keep looking for an alternative.

> **Tip**
>
> The `Sanic.get_app()` construct is a very useful pattern for gaining access to your app instance without having to pass it around by import. This is a very helpful step in the right direction, and you can learn more about it in the user guide: `https://sanic.dev/en/guide/basics/app.html#app-registry`.

Factory pattern

We are going to move our application creation into a factory pattern. You may be familiar with this if you come from Flask as many examples and tutorials use a similar construct. The main reason for doing this here is that we want to set up our application for good development practices in the future. It will also ultimately solve the circular import problem. Later on down the line in *Chapter 9, Best Practices to Improve Your Web Applications*, we will talk about testing. In the absence of a nice factory, testing will become much more difficult.

We need to create a new file, `./utilities/app_factory.py`, and redo our `./server.py` file:

```python
# ./utilities/app_factory.py
from typing import Optional, Sequence
from sanic import Sanic

from importlib import import_module

DEFAULT_BLUEPRINTS = [
    "src.blueprints.v1.book.view",
    "src.blueprints.v1.author.view",
    "src.blueprints.v2.group",
]


def create_app(
    init_blueprints: Optional[Sequence[str]] = None,
) -> Sanic:
    app = Sanic("BookTracker")

    if not init_blueprints:
        init_blueprints = DEFAULT_BLUEPRINTS

    for module_name in init_blueprints:
        module = import_module(module_name)
        app.blueprint(getattr(module, "bp"))

    return app
from .utilities.app_factory import create_app

app = create_app()
```

As you can see, our new factory will create the app instance, and attach some blueprints to it. We specifically are allowing for the factory to override the blueprints that it will use. Perhaps this is unnecessary and we could instead hardcode them all the time. However, I like the flexibility that this provides us, and find it helpful later on down the road when I want to start testing my application.

One problem that might jump out at you is that it requires our modules to have a global bp variable. While I mentioned that this is standard practice for me, it might not work in all scenarios.

Autodiscovery

The Sanic user guide gives us another idea in the *How to…* section. See https://sanic.dev/en/guide/how-to/autodiscovery.html. It suggests that we create an autodiscover utility that will handle some of the importing for us, and also have the benefit of automatically attaching blueprints. Remember how I said I like predictable folder structures? We are about to take advantage of this pattern.

Let's create ./utilities/autodiscovery.py:

```
# ./utilities/autodiscovery.py
from importlib import import_module
from inspect import getmembers
from types import ModuleType
from typing import Union


from sanic.blueprints import Blueprint



def autodiscover(app, *module_names: Union[str, ModuleType]) ->
None:
    mod = app.__module__
    blueprints = set()

    def _find_bps(module: ModuleType) -> None:
        nonlocal blueprints

        for _, member in getmembers(module):
            if isinstance(member, Blueprint):
                blueprints.add(member)
```

```
for module in module_names:
    if isinstance(module, str):
        module = import_module(module, mod)
    _find_bps(module)


for bp in blueprints:
    app.blueprint(bp)
```

This file closely matches what the user guide suggests (https://sanic.dev/en/guide/how-to/autodiscovery.html#utility.py). Noticeably absent from the code presented there is the idea of recursion. If you look up the function in the user guide, you will see that it includes the ability to recursively search through our source code looking for Blueprint instances. While convenient, in the application that we are building, we want the express control provided by having to declare every blueprint's location. Quoting Tim Peters, *The Zen of Python*, again:

Explicit is better than implicit.

What the autodiscover tool does is allow us to pass locations to modules and hands the task of importing them over to the application. After loading the module, it will inspect any blueprints. The last thing it will handle is automatically registering the discovered blueprints to our application instance.

Now, our server.py file looks like this:

```
from typing import Optional, Sequence
from sanic import Sanic

from .autodiscovery import autodiscover

DEFAULT_BLUEPRINTS = [
    "src.blueprints.v1.book.view",
    "src.blueprints.v1.author.view",
    "src.blueprints.v2.group",
]


def create_app(
    init_blueprints: Optional[Sequence[str]] = None,
```

```
) -> Sanic:
    app = Sanic("BookTracker")

    if not init_blueprints:
        init_blueprints = DEFAULT_BLUEPRINTS

    autodiscover(app, *init_blueprints)

    return app
```

> **Tip**
>
> In this example, we are using the import paths as strings. We could just as easily import the modules here and pass those objects since the `autodiscover` utility works with both module objects and strings. We prefer strings though since it will keep the annoying circular import exceptions away.

Another thing to keep in mind is that this autodiscover tool could be used for a module containing middleware or listeners. The given example is still fairly simplistic, and will not cover all use cases. How, for example, should we handle deeply nested blueprint groups? This is a great opportunity for you to experiment, and I highly encourage you to spend some time playing with the application structure and the autodiscover tool to figure out what works best for you.

Running our application

Now that we have laid our application foundations, we are almost ready to run our server. We are going to make one small change to `server.py` to include a small little utility to run at startup to show us what routes are registered:

```
from .utilities.app_factory import create_app
from sanic.log import logger

app = create_app()

@app.main_process_start
def display_routes(app, _):
```

```
logger.info("Registered routes:")
for route in app.router.routes:
    logger.info(f"> /{route.path}")
```

You can head over to the GitHub repository, https://github.com/
PacktPublishing/Python-Web-Development-with-Sanic/tree/main/
Chapter02, to see the full source code.

We can now start our application for the first time. Remember, this is going to be our pattern:

```
$ sanic src.server:app -p 7777 --debug --workers=2
```

We should see something like this:

```
[2021-05-30 11:34:54 +0300] [36571] [INFO] Goin' Fast @
http://127.0.0.1:7777
[2021-05-30 11:34:54 +0300] [36571] [INFO] Registered routes:
[2021-05-30 11:34:54 +0300] [36571] [INFO] > /v2/book
[2021-05-30 11:34:54 +0300] [36571] [INFO] > /book
[2021-05-30 11:34:54 +0300] [36571] [INFO] > /author
[2021-05-30 11:34:54 +0300] [36572] [INFO] Starting worker
[36572]
[2021-05-30 11:34:54 +0300] [36573] [INFO] Starting worker
[36573]
```

Hooray!

And now, for the tempting part. What does our code actually do? Head over to your favorite web browser and open http://127.0.0.1:7777/book. It might not be much to look at yet, but you should see some JSON data. Next, try going to /author and /v2/book. You should now see the content that we created above. Feel free to play around with these routes by adding to them. Every time you do, you should see your changes reflected in the web browser.

Our journey into web application development has officially begun.

Summary

We have looked at the important impact of some of the early decisions we make about setting up our environment and project organization. We can—and should—constantly adapt our environment and application to meet changing needs. We used pdm to leverage some of the newest tools to run our server in a well-defined and isolated environment.

In our example, we then started to build our application. Perhaps we were too hasty when we added our /book route because we quickly realized that we needed the endpoint to perform differently. Rather than breaking the application for existing users, we simply created a new group of blueprints that will be the beginning of /v2 of our API. By nesting and grouping blueprints, we are setting the application up for future flexibility and development maintainability. Going forward, let's stick to this pattern as much as possible.

We also examined a few alternative approaches for organizing our application logic. These early decisions will impact the import ordering and shape the look of the application. We decided to adopt a factory method that will help us in the future when we start to test the application.

With the basic application structure decided, in the next chapter, we will begin to explore the most important aspect of a web server and framework: handling the request/response cycle. We know that we will use blueprints, but it is time to dive in and look more closely at what we can do with Sanic routing and handlers. In this chapter, there was a taste of it with API versioning. In the next chapter, we will also look at routing more generally and try to understand some strategies for designing application logic within a web API.

Part 2:
Hands-On Sanic

This Part conducts a close inspection of the issues at play when building a web application and lays out potential approaches to take to solve the problems you may face. The goal is to not just identify issues and provide solutions but provide insight to help you make your own solutions. We will get our hands dirty learning how to use the tools that Sanic provides to meet your specific needs.

This Part contains the following chapters:

3
Routing and Intaking HTTP Requests

Back in *Chapter 1*, *Introduction to Sanic and Async Frameworks*, we looked at a raw HTTP request to see what kind of information it includes. In this chapter, we are going to take a closer look at the first line, which contains the HTTP method and the URI path. As we learned, the most basic function of a web framework is to translate a raw HTTP request into an actionable handler. Before we see how we can implement this, it is good to keep in mind what raw requests look like:

```
POST /path/to/endpoint HTTP/1.1
Host: localhost:7777
User-Agent: curl/7.76.1
Accept: */*
Content-Length: 14
Content-Type: application/json

{"foo": "bar"}
```

Looking at the request, we see the following:

- The first line (sometimes called the *start line*) contains three parts: the **HTTP method**, **request target**, and **HTTP protocol**.

- The second section contains zero or more lines of HTTP headers in `key: value` form, with each pair separated by a line break.

- Then, we have a blank line separating the head from the body.

- Lastly, we have the *optional* body.

The exact specification is covered by RFC 7230, 3: `https://datatracker.ietf.org/doc/html/rfc7230#section-3`

One of the goals of this book is to learn strategies to design API endpoints that will be easily consumable, keeping in mind the needs and limitations of the application we are building. The goal is to understand the first interactions that the server has with an incoming web request, and how to design our application around that. We will learn about how requests are structured, what choices Sanic makes for us and what choices it leaves to us, and other issues involved in turning an HTTP request into actionable code. Remember, the purpose of this book is not just to learn how to use a fancy new tool, but also to level up web development skills and knowledge. To become more informed developers, we seek to not only understand *how* to build with Sanic, but *why* we might build something in a particular way. We will learn to ask better questions and to make better decisions by understanding some of the mechanics. This does not mean we need to be experts on the HTTP protocol and specification. By being familiar, however, with what Sanic is doing with the raw request, we will ultimately be armed with a greater set of knowledge for building web applications.

In particular, we'll cover these topics:

- Understanding HTTP methods
- Paths, slashes, and why they matter
- Advanced path parameters
- API versioning
- Virtual hosts
- Serving static content

Technical requirements

In addition to what we have been building before, in this chapter, you should have the following tools at your disposal in order to be able to follow along with the examples:

- Docker Compose

- Curl

- You can access the source code for this chapter on GitHub: `https://github.com/PacktPublishing/Python-Web-Development-with-Sanic/tree/main/Chapter03`

Understanding HTTP methods

If you have built any kind of a website before, you probably have an understanding of the concept of **HTTP methods**, or at least the basic GET and POST methods. However, did you know that there are nine standard HTTP methods? In this section, we will learn about these different methods and how we can take advantage of them.

In the same way that an IP address or a web domain is a *place* on the internet, an HTTP method is an *action* on the internet. They are a collection of verbs in the language of the web. These HTTP methods have a shared understanding and meaning. Web applications will commonly use these methods in similar use cases. That does not mean that you must follow the conventions, or that your application will break if you venture away from the standards. We should learn the rules so that we know when it may be appropriate to break them. These standards exist to create a common language that web developers and consumers can use to communicate:

Method	Description	Has body?	Safe	Sanic support
CONNECT	Open two-way communication, like a tunnel to the resource	No	Yes	No
DELETE	Delete the resource	No (usually)	No	Yes
GET	Fetch the resource	No	Yes	Yes
HEAD	Fetch the metadata only for the resource	No	Yes	Yes
OPTIONS	Request permitted communication options	No	Yes	Yes
PATCH	Partially modify the resource	Yes	No	Yes
POST	Send data to the server	Yes	No	Yes
PUT	Create a new resource or update completely if existing	Yes	No	Yes
TRACE	Perform message loopback used for debugging	No	Yes	No

Table 3.1 – HTTP methods overview

When we talk about a method being *safe*, what we mean is that it should not be state-changing. That is not to say that a `GET` method cannot have side-effects. Of course, they can. For example, someone hitting the endpoint will trigger logs or some sort of a resource counter. These are technically what the industry might refer to as side-effects. *The important distinction here is that the user did not request the side-effects, so therefore cannot be held accountable for them.* RFC Section 2616, 9.1.1 (`https://datatracker.ietf.org/doc/html/rfc2616#section-9`). This means that from the perspective of the user accessing the resource, the determination of whether an endpoint is *safe* is a matter of intent. If the user intends to retrieve profile information, it is safe. If the user intends to update profile information, it is not safe.

While it is certainly helpful to try and stick to the descriptions of the methods in *Table 3.1*, undoubtedly you will come across use cases that do not fit nicely into these categories. When that happens, I encourage you to re-examine your application design. Sometimes the problem can be solved with new endpoint paths. Sometimes we need to create our own definitions. This is okay. I would, however, caution against changing a *safe* method to *unsafe*. Performing stateful changes with a `GET` request is considered poor form, and a *rookie mistake*.

After deciding what our HTTP methods should be, we will venture into the next section to learn about how we can implement them and attach them to routes.

Using HTTP methods on route handlers

We are finally ready to see what frameworks are all about! If you have used Flask in the past, this will look familiar. If not, what we are about to do is create a route definition that is a set of instructions to tell Sanic to send any incoming HTTP requests to our route handler. A route definition must have two parts: a URI path and one or more HTTP methods.

Matching on the URI path alone is not enough. HTTP methods are also used by Sanic to get your incoming request to the correct handler. Even when we implement the most basic form of a route definition, both parts must exist. Let's look at the simplest use case and see what default choices Sanic will make:

```
@app.route("/my/stuff")
async def stuff_handler(...):
    ...
```

In this example, we defined a route at /my/stuff. Usually, we inject the route() call with an optional methods argument to tell it what HTTP methods we want that handler to respond to. We have not here, so it will default to just GET. We have the option of telling the route that it should also handle other HTTP methods:

```
@app.route("/my/stuff", methods=["GET", "HEAD"])
async def stuff_handler(...):
    return text("Hello")
```

> **Important Note**
>
> We will look at the HEAD method a little later in this chapter. But it is important to know that a HEAD request should not have any response body. This is enforced by Sanic for us. Even though, technically, this endpoint is responding with the text Hello, Sanic removes the body from the response and only sends the metadata.

Now that we have a single endpoint set up with multiple methods, we can access it with both methods.

First, let's do so with a GET request (it should be noted that when using curl, if you do not specify a method, it will default to GET):

```
$ curl localhost:7777/my/stuff -i
HTTP/1.1 200 OK
content-length: 5
connection: keep-alive
content-type: text/plain; charset=utf-8

Hello
```

Then, with a HEAD request. You should take note that in the HEAD request, there is no body in the response. It is only headers that are returned:

```
$ curl localhost:7777/my/stuff -i --head
HTTP/1.1 200 OK
content-length: 5
connection: keep-alive
content-type: text/plain; charset=utf-8
```

For convenience, Sanic provides shortcut decorators to all of its supported HTTP methods on both the app instance and any blueprint instance:

```
@app.get("/")
def get_handler(...):
    ...


@app.post("/")
def post_handler(...):
    ...


@app.put("/")
def put_handler(...):
    ...


@app.patch("/")
def patch_handler(...):
    ...


@app.delete("/")
def delete_handler(...):
    ...


@app.head("/")
def head_handler(...):
    ...


@app.options("/")
def options_handler(...):
    ...
```

These decorators can also be stacked. The last example that we saw could also be written like this:

```
@app.head("/my/stuff")
@app.get("/my/stuff")
async def stuff_handler(...):
    return text("Hello")
```

This is the standard way routes are defined in Sanic, but not the only. Using decorators is convenient and highly encouraged, but you can also define the exact same endpoint like this:

```
async def stuff_handler(...):
    return text("Hello")


app.add_route(stuff_handler, "/my/stuff", methods=["GET",
"HEAD"])
```

Using the `add_route` method, we can define a new endpoint by passing the handler, along with all of the other arguments. It also becomes the basis for the third style: **Class-Based Views** (**CBV**), which we will review in detail in the later section called *Simplifying your endpoints with class-based views*. Here is a sneak peek of how the same endpoint would be written as a CBV:

```
from sanic.views import HTTPMethodView



class StuffHandler(HTTPMethodView):
    async def get(self, request):
        return text("Hello")

    async def head(self, request):
        return await self.get(request)


app.add_route(StuffHandler.as_view(), "/")
```

One more fundamental thing to know about HTTP methods is that you can access the incoming method on the HTTP request object. This is very helpful if you are handling multiple HTTP methods on the same handler, but need to treat them differently. Here is an example where we look at the HTTP method to change the behavior of the handler:

```
from sanic.response import text, empty
from sanic.constants import HTTPMethod

@app.options("/do/stuff")
@app.post("/do/stuff")
async def stuff_handler(request: Request):
    if request.method == HTTPMethod.OPTIONS:
```

```
        return empty()
    else:
        return text("Hello")
```

Before moving on to advanced method routing, there is some Sanic syntax we should mention. All of the examples here use the decorator syntax to define routes. This is by far the most common way to achieve this because of its convenience. There is, however, an alternative. All route definitions could be converted to functional definitions, as shown here:

```
@app.get("/foo")
async def handler_1(request: Request):
    ...


async def handler_2(request: Request):
    ...


app.add_route(handler_2, "/bar")
```

In some circumstances, this may be a more attractive pattern to use. We will see it again when we encounter CBVs later in this chapter.

Advanced method routing

Sanic does not support CONNECT and TRACE out of the box, two standard HTTP methods. But let's imagine that you want to build an HTTP proxy or some other system that needs to have the CONNECT method available in your route handler. Even though Sanic does not allow it out of the box, you have two potential approaches.

First, we could create a piece of middleware that is on the lookout for CONNECT and hijacks the request to provide a custom response. This *trick* of responding from middleware is a feature that allows you to halt the execution of the request/response lifecycle before the handlers take over and otherwise fail as a 404 Not Found:

```
async def connect_handler(request: Request):
    return text("connecting...")



@app.on_request
async def method_hijack(request: Request):
```

```
if request.method == "CONNECT":
    return await connect_handler(request)
```

You can see that a potential downside to this approach is that we need to implement our own routing system if we want to send different endpoints to different handlers.

A second approach might be to tell the Sanic router that CONNECT is a valid HTTP method. Once we do this, we can add it to a normal request handler:

```
app.router.ALLOWED_METHODS = [*app.router.ALLOWED_METHODS,
"CONNECT"]


@app.route("/", methods=["CONNECT"])
async def connect_handler(request: Request):
    return text("connecting...")
```

One important consideration for this strategy is that you will need to redefine app.router.ALLOWED_METHODS as early as possible before registering the new handlers. For this reason, it might be best for it to come directly after app = Sanic(...).

A side benefit that this strategy provides is the ability to create your own ecosystem of HTTP methods with your own definitions. This may not necessarily be advisable if you intend for your API to be used for public consumption. However, it may be useful, practical, or just plain fun for your own purposes. There may only be nine standard methods, but there are infinite possibilities. Do you want to create your own verbs? You are certainly free to do so:

```
ATTACK /path/to/the/dragon HTTP/1.1
```

Method safety and request body

As we have learned, there are generally two types of HTTP methods: **safe** and **unsafe**. The unsafe methods are POST, PUT, PATCH, and DELETE. These methods are generally understood to mean that they are state-changing. That is to say that by hitting these endpoints, the user is intending to change or alter the resource in some way.

The converse of this is safe methods: GET, HEAD, and OPTIONS. The purpose of these endpoints is to request information from the application, not to change state.

It is considered good practice to follow this practice: if an endpoint will make a change on the server, do not use GET.

Lining up with this division is the concept of the request body. Let's revisit the raw HTTP request one more time:

```
POST /path/to/endpoint HTTP/1.1
Host: localhost:7777
User-Agent: curl/7.76.1
Accept: */*
Content-Length: 14
Content-Type: application/json


{"foo": "bar"}
```

An HTTP request can optionally include a body. In the preceding example, the request body is the last line: {"foo": "bar"}.

It is important to note that Sanic will only take the time to read the message body for POST, PUT, and PATCH requests. It will stop reading the HTTP message after the headers if it is an HTTP request using any other HTTP method. This is a performance optimization since we generally do not expect there to be a message body on the *safe* HTTP requests.

You may have noticed this list did not include DELETE. Why? In general, the HTTP specification says that there *may* be a request body (https://datatracker.ietf.org/doc/html/rfc7231#section-4.3.5). Sanic assumes that it will not have one unless you tell it that it does. To do this, we simply set ignore_body=False:

```
@app.delete("/", ignore_body=False)
async delete_something(request: Request):
    await delete_something_using_request(request.body)
```

If we do not set ignore_body=False, and we send a body with our DELETE requests, Sanic will raise a warning in the logs to let us know that part of the HTTP message was not consumed. If you intend to use DELETE methods, you should be on the lookout for this since Sanic makes the assumption. It should also be noted that if you are in the habit of receiving GET requests with a body, you will also need to use ignore_body=False. However, I hope you have a very good reason for doing that since it breaks most web standards.

One helpful takeaway from this is that, out of the box, the following two endpoints are *not* equal:

```
@app.route("/one", methods=["GET"])
async def one(request: Request):
    return text("one")


@app.get("/two")
async def two(request: Request):
    return text("two")
```

Both /one and /two will behave similarly. Without further customization, however, the first request will spend time trying to read the request body even if it does not exist, while the second just assumes that it is the case that there is no body. While the performance difference will be small, it is generally preferred to use @app.get ("/two") over @app.route("/one", methods=["GET"]). The reason that these two endpoints differ is that they have different default values for ignore_body.

> **Important Note**
>
> If you are building a GraphQL application, then usually POST is used by the endpoint even for informational requests. This is because it is generally much more acceptable to pass a body on a POST request than a GET request. However, it is worth mentioning that we could consume a message body from a GET request if we really wanted to by setting ignore_body=False.

When deciding what method you should use, another factor to consider is **idempotency**. In short, idempotence means that you can perform the same action over and over again, and the result should be the same every time. The HTTP methods that are considered idempotent are GET, HEAD, PUT, DELETE, OPTIONS, and TRACE. Keep this in mind when designing your API.

RESTful API design

HTTP methods are often used in **RESTful API design**. There is a wealth of literature already written on building RESTful APIs, so we will not dive deeply into *what* it is, but more *how* we can practically implement it. We shall, however, first have a quick refresher of the basic premise.

Web API endpoints have a target. That target is *something* that either the user would like to fetch information about or manipulate by adding or changing it. Based upon a shared understanding, the HTTP method tells the server how you would like to interact with that target. The *target* is often called a *resource*, and we may use the terms interchangeably here.

To explain this concept, I like to think back to the adventure computer games I played as a child. My swashbuckling character would happen upon an object—let's say a rubber chicken. When I clicked on that object, a menu would appear with different verbs that told me what I could do with that object: pick up, look at, use, talk to, and so on. There was a target (the rubber chicken), and methods (the verbs or actions).

Putting this together with the HTTP methods we defined earlier, let's look at a concrete example. In our hypothetical situation, we will be building an API to manage a social media platform for people that love adventure computer games. Users need to be able to create a profile, view other profiles, and update their own. We might design the following endpoints:

METHOD	URI PATH	DESCRIPTION
GET	/profiles	A list of all of the member profiles
POST	/profiles	Create a new profile
GET	/profiles/<username>	Retrieve profile for a single user
PUT	/profiles/<username>	Remove the old profile and replace with a complete profile
PATCH	/profiles/<username>	Make a change to only a part of a profile
DELETE	/profiles/<username>	Remove a profile – but why would anyone want to remove their swashbuckling adventure game profile?

Table 3.2 – Example HTTP methods and endpoints

Before we go further, if you are unfamiliar with how routing works in Sanic (and what the <username> syntax means), you can get more information in the user guide at https://sanic.dev/en/guide/basics/routing.html, and we will also look at it more closely later in this chapter, in the section titled *Extracting information from the path*. Feel free to skip ahead and come back.

As you can see, there really are only two URI paths: /profiles and /profiles/<username>. Using the HTTP methods, however, we have been able to define six different interactions with our API! What might the profile blueprint look like?

```
from sanic import Blueprint, Request


bp = Blueprint("MemberProfiles", url_prefix="/profile")


@bp.get("")
```

```
async def fetch_all_profiles(request: Request):
    ...


@bp.post("")
async def create_new_profile(request: Request):
    ...


@bp.get("/<username>")
async def fetch_single_profile(request: Request, username:
str):
    ...


@bp.put("/<username>")
async def replace_profile(request: Request, username: str):
    ...


@bp.patch("/<username>")
async def update_profile(request: Request, username: str):
    ...


@bp.delete("/<username>")
async def delete_profile(request: Request, username: str):
    ...
```

Using HTTP methods to define our use cases seems helpful and having decorators that map them seems convenient. But, it seems like there is a lot of boilerplate there, and repetition. We will next look at CBVs and how we can simplify our code.

Simplifying your endpoints with CBVs

The previous example exposes a weakness with using functions and decorators alone to design your API. What happens when we want to add endpoint handlers for /profile/<user_id:uuid>? Or when we want to make some other change to the existing endpoint? We now have multiple places to make the same change, leading to a higher chance that we do not maintain parity among all our route definitions. This is a violation of the **don't repeat yourself (DRY)** principle and could lead to bugs. Maintaining these endpoints over the long term therefore might be more difficult than necessary.

This is one of the compelling reasons to use **CBVs**. This pattern will give us the opportunity to link together the first two endpoints and the last four endpoints so they become easier to manage. They are being grouped together because they share the same URI path. Instead of standalone functions, each HTTP method will be a functional method on a class and that class will be assigned a common URI path. A bit of code should make this easy to understand:

```python
from sanic import Blueprint, Request, HttpMethodView


bp = Blueprint("MemberProfiles", url_prefix="/profile")


class AllProfilesView(HttpMethodView):
    async def get(request: Request):
        """same as fetch_all_profiles() from before"""

    async def post(request: Request):
        """same as create_new_profile() from before"""


class SingleProfileView(HttpMethodView):
    async def get(request: Request, username: str):
        """same as fetch_single_profile() from before"""

    async def put(request: Request, username: str):
        """same as replace_profile() from before"""

    async def patch(request: Request, username: str):
        """same as update_profile() from before"""

    async def delete(request: Request, username: str):
        """same as delete_profile() from before"""


app.add_route(AllProfilesView.as_view(), "")
app.add_route(SingleProfileView.as_view(), "/<username>")
```

> **Important Note**
>
> Later in this book, we may see custom decorators used more and more often to add shared functionality. It is worth mentioning that we can also add them easily to CBVs, and I highly suggest you take a moment to refer to the user guide to see it in action: `https://sanic.dev/en/guide/advanced/class-based-views.html#path-parameters`.
>
> One thing to be on the lookout for when adding decorators to CBV methods is the `self` argument on instance methods. You may need to adjust your decorator or use `staticmethod` to get it to work as expected. The above-mentioned documentation explains how to do that.

Earlier, we saw how `add_route` could be used as an alternative approach to attaching a single function as a handler to a route definition. It looked like this:

```
async def handler(request: Request):
    ...


app.add_route(handler, "/path")
```

This pattern is one of the main ways to attach a CBV to a Sanic or blueprint instance. The caveat is that you need to call it using the class method `as_view()`. In our previous example, we saw what this looked like:

```
app.add_route(SingleProfileView.as_view(), "/<username>")
```

This could also be achieved by attaching the CBV when it is declared. This option will only work if you already have a known blueprint or application instance. We will rewrite `SingleProfileView` to take advantage of this alternative syntax:

```
from sanic import Sanic


app = Sanic.get_app()


class SingleProfileView(HttpMethodView, attach=app,
uri="/<username>"):
    async def get(request: Request, username: str):
```

```
        """same as fetch_single_profile() from before"""

    async def put(request: Request, username: str):
        """same as replace_profile() from before"""

    async def patch(request: Request, username: str):
        """same as update_profile() from before"""

    async def delete(request: Request, username: str):
        """same as delete_profile() from before"""
```

In this example, instead of using add_route, we can attach the CBV directly to the app instance and define the path in the class definition. How should you decide which style to use? Personally, I find this second version to be easier and cleaner. This is what most of my projects use. The big downside is that you cannot lazily create the CBV and attach it later since the application or blueprint needs to be known upfront. This can lead to issues with the ordering of your imports. But this is easily avoided using the create_app factory pattern introduced in *Chapter 9, Best Practices to Improve Your Web Applications*, in the Testing a full application section, and used throughout the examples in *Chapter 10, Implementing Common Use Cases with Sanic*, and *Chapter 11, A Complete Real-World Example*.

Blanket support for OPTIONS and HEAD

It is generally best practice to support the OPTIONS and HEAD methods on all of your endpoints, where it is appropriate. This could become tedious and include a lot of repetitive boilerplate. To achieve this with standard route definitions alone would require a lot of code duplication as seen in the following snippet. Here, we see that we need four route definitions where two would be sufficient. Now imagine if every endpoint needed to also have OPTIONS and HEAD!

```
@app.get("/path/to/something")
async def do_something(request: Request):
    ...


@app.post("/path/to/something")
async def do_something(request: Request):
    ...
```

```
@app.options("/path/to/something")
async def do_something_options(request: Request):
    ...


@app.head("/path/to/something")
async def do_something_head(request: Request):
    ...
```

We can use Sanic's router to our advantage to add on a handler for these requests to each and every route. The idea will be to loop over all of the routes defined in our application and dynamically add handlers for OPTIONS and HEAD if needed. Later on, in *Chapter 7, Dealing with Security Concerns*, we will use this strategy for creating our customized CORS policy. For now, however, all we need to keep in mind is that we want to be able to handle *any* request to a valid endpoint using one of these HTTP methods:

```
async def options_handler(request: Request):
    ...


async def head_handler(request: Request):
    ...


@app.before_server_start
def add_info_handlers(app: Sanic, _):
    app.router.reset()
    for group in app.router.groups.values():
        if "OPTIONS" not in group.methods:
            app.add_route(
                handler=options_handler,
                uri=group.uri,
                methods=["OPTIONS"],
                strict_slashes=group.strict,
            )
    app.router.finalize()
```

Let's take a closer look at this code.

First, we create route handlers: the functions that will do the work when the endpoint is hit. For now, they do not do anything. If you want to know what this endpoint *could* do, jump ahead to the CORS discussion in *Setting up an effective CORS policy* located in *Chapter 7, Dealing with Security Concerns*:

```
async def options_handler(request: Request):
    ...


async def head_handler(request: Request):
    ...
```

The next part needs to be done after we register all of our endpoints. In *Chapter 11, A Complete Real-World Example*, we accomplish this by running the code inside of a factory. Feel free to take a look ahead to the example there to be able to compare it with our current implementation.

In our current example, we do not have a factory and are adding the routes inside an event listener. Normally, this would not be possible since we cannot change our routes after the application is running. When a Sanic application starts up, one of the first things it does internally is to call app.router.finalize(). But, it will not let us call that method twice. Therefore, we need to run app.router.reset(), add our routes, and finally call app.router.finalize() after all of our dynamic route generation is complete. You can use this same strategy anywhere that you might want to dynamically add routes. Is this a good idea? In general, I would say that dynamically adding routes is a bad idea. Having changing endpoints might lead to unpredictability or weird bugs across a distributed application. However, the benefit we are gaining through dynamic route generation in this instance is great, and the risk is very low.

There are a few different properties that the Sanic router provides us with that we could loop over to see what routes are registered. The two most commonly used for public consumption are app.router.routes and app.router.groups. It is helpful to understand what they are and how they differ. We will put our discussion on OPTIONS and HEAD on pause for a brief moment to look at these two properties:

```
@app.before_server_start
def display(app: Sanic, _):
    for route in app.router.routes:
        print(route)


    for group in app.router.groups.values():
```

```
        print(group)


@app.patch("/two")
@app.post("/two")
def two_groups(request: Request):
    return text("index")



@app.route("/one", methods=["PATCH", "POST"])
def one_groups(request: Request):
    return text("index")
```

The first thing to notice is that one of them is producing `Route` objects and the other `RouteGroup` objects. The second obvious takeaway is that one is a list and the other a dictionary. But what are `Route` and `RouteGroup`?

In our console, we will see that there are three `Route` objects, but only two `RouteGroup` objects. This is because Sanic has grouped together similar-looking routes to more efficiently match them later. A `Route` is a single definition. Every time we call `@app.route`, we are creating a new `Route`. Here, we can see that they have been grouped by the URI path:

```
<Route: name=__main__.two_groups path=two>
<Route: name=__main__.two_groups path=two>
<Route: name=__main__.one_groups path=one>
<RouteGroup: path=two len=2>
<RouteGroup: path=one len=1>
```

Getting back to our discussion of automation, we are going to use `app.router.groups`. This is because we want to know which methods have and have not been assigned to a given path. The quickest way to figure that out is to look at the groups that Sanic has already provided for us. All we need to do is check if the group already includes a handler for the HTTP method (so we do not overwrite anything that already exists) and call `add_route`:

```
for group in app.router.groups.values():
    if "OPTIONS" not in group.methods:
        app.add_route(
            handler=options_handler,
```

```
            uri=group.uri,
            methods=["OPTIONS"],
            strict_slashes=group.strict,
        )
    if "GET" in group.methods and "HEAD" not in group.methods:
        app.add_route(
            handler=head_handler,
            uri=group.uri,
            methods=["HEAD"],
            strict_slashes=group.strict,
        )
```

Although we will not look at `options_handler` right now, we can look more closely at `head_handler`. A HEAD request is defined in RFC Section 7231 as being identical to a GET request: *The HEAD method is identical to GET except that the server MUST NOT send a message body in the response* (`https://datatracker.ietf.org/doc/html/rfc7231#section-4.3.2`).

This is pretty easy to implement in Sanic. Really, what we want to do is retrieve the response from the GET handler for the same endpoint, but only return the metadata, *not* the request body. We will use `functools.partial` to pass the GET handler to `head_handler`. Then, all it needs to do is run `get_handler` and return the response. As we saw earlier in the chapter, Sanic will do the work for us of removing the body before it sends the response to the client:

```
from functools import partial


for group in app.router.groups.values():
    if "GET" in group.methods and "HEAD" not in group.methods:
        get_route = group.methods_index["GET"]
        app.add_route(
            handler=partial(
                head_handler,
                get_handler=get_route.handler
            ),
            uri=group.uri,
            methods=["HEAD"],
            strict_slashes=group.strict,
            name=f"{get_route.name}_head",
```

```
        )

async def head_handler(request: Request, get_handler, *args,
**kwargs):
    return await get_handler(request: Request, *args, **kwargs)
```

> **Important Note**
>
> In the preceding example, we added `name=f"{get_route.name}_`
> `head"` to our `add_route` method. This is because all routes in Sanic
> get a name. If you do not manually supply one, then Sanic will attempt
> to generate a name for you using `handler.__name__`. In this case,
> we are passing a `partial` function as a route handler, and Sanic does not
> know how to generate a name for that since partial functions in Python have
> no `__name__` property.

Now that we know how to use HTTP methods to our advantage, we will look at the next big area in routing: **paths**.

Paths, slashes, and why they matter

Way back in the stone age when the internet was invented, if you navigated to a URL, you were literally being delivered a file that existed on a computer somewhere. If you asked for `/path/to/something.html`, the server would look in the `/path/to` directory for a file called `something.html`. If that file existed, it would send it to you.

While this does still exist, times have certainly changed for many applications. The internet is still largely based upon this premise, but often a generated document is sent instead of a static document. It is helpful to still keep this mental model in your head though. Thinking that a path on your API should lead to a resource of some kind will keep you away from certain API design flaws. Let's look at an example:

```
/path/to/create_something   << BAD
/path/to/something          << GOOD
```

Your URI paths should use nouns, not verbs. If we want to perform an action and tell the server to do something, we should manipulate the HTTP method as we learned, not the path of the URI. Going down this path—believe me, I've done it—will lead to some messy-looking applications. Very likely you will wake up one day and look at a mess of disjointed and incoherent paths and ask yourself: what have I done? There might, however, be a time and place for this, so we will revisit it shortly.

Knowing that our paths should contain nouns, the obvious next question is whether they should be singular or plural. I do not think there is consensus on the internet about what is right here. Many people always use plurals, many people always use the singular form, and some wild hooligans decide to mix and match. While the decision may seem small, it is nevertheless important to establish consistency. Picking a system and applying consistency is more important than the actual decision.

With that out of the way, I will give you my opinion. Use plural nouns. Why? It makes for very nice nesting of paths, which could translate nicely to the nesting of blueprints:

```
/users       << to get all users
/users/123  << to get user ID 123
```

I do encourage you to use singular nouns if it makes sense to you. But if you do, you must do it everywhere. As long as you stay consistent and logical about your choice, your API will feel polished. Mixing plural and singular paths will make your API feel haphazard and amateurish. A really nice resource that explains how to *consistently* break the two rules I just laid out (use nouns, use plurals) is found here: `https://restfulapi.net/resource-naming/`. Again, it is important and helpful for us to not only learn the *rules* or the *right way* to do something but to also learn when to break them, or when to make our own set of rules. Sometimes following standards makes sense, sometimes not. This is how we go from someone that merely is capable of making a web application, to someone that knows how to design and build one. The difference is expertise.

It is also encouraged when designing paths to favor hyphens (-) over spaces, capitalization, or underscores. This increases the human readability of your API. Consider the difference between these:

```
/users/AdamHopkins        << BAD
/users/adam_hopkins       << BAD
/users/adam%20hopkins     << BAD
/users/adam-hopkins       << GOOD
```

Most people would consider the last option to be the easiest to read.

Strict slashes

Because of the legacy paradigm where endpoints were equivalent to the file structure of a server, the trailing slash in a path took on a specific meaning. It is widely accepted that paths with and without trailing slashes are not the same and are not interchangeable.

If you navigated to /characters, you might expect to receive a list of all the characters in our fictitious social media application. However, /characters/ technically means *show me a list of everything that is in the* characters *directory*. Because this could be confusing, you are encouraged to avoid using trailing slashes.

On the other hand, it is also widely accepted that these *are* the same thing. In fact, a lot of browsers (and websites) treat them the same. I'll show you how you can test this for yourself.

Open your web browser and go to https://sanic.readthedocs.io/en/stable/.

Now open a second tab and go to https://sanic.readthedocs.io/en/stable.

It is the same page. In fact, it seems that this web server breaks the rule that I just mentioned and prefers the trailing slash to not having it at all. So, where does this leave us, and what should we implement? It is really up to you to determine, so let's see how we can control it in Sanic.

If you do nothing, Sanic will drop the trailing slash for you. Sanic does, however, provide you with the ability to control whether that trailing slash should have meaning or not by setting the strict_slashes argument. Consider an application set up with and without trailing slashes, and with and without strict_slashes:

```
@app.route("/characters")
@app.route("/characters/")
@app.route("/characters", strict_slashes=True)
@app.route("/characters/", strict_slashes=True)
async def handler(request: Request):
    ...
```

These definitions will fail. Why? When Sanic sees a trailing slash on a path definition it will remove it, *unless* strict_slashes=True. Therefore, the first and second routes are considered identical. Furthermore, the third route is also the same, therefore causing a conflict.

While the generally accepted rule is that a trailing slash *should* have meaning, this is not the case for a trailing slash that is the only part of a path. RFC 7230, Section 2.7.3 states that an empty path ("") is the same thing as a single slash path ("/") (https://datatracker.ietf.org/doc/html/rfc7230#section-2.7.3).

I put together a deeper discussion about how Sanic handles the possible scenarios of trailing slashes. If this is something you are considering using, I suggest you take a look here: https://community.sanicframework.org/t/route-paths-how-do-they-work/825.

If you were to ask me my opinion, I would say do not use them. It is much more forgiving to allow /characters and /characters/ to have the same meaning. Therefore, I personally would define the route as follows:

```
@app.route("/characters")
async def handler(request: Request):
    ...
```

One of the purposes of crafting good endpoint paths is to properly convey information from the user to the server. It is therefore natural that you will need to access certain information in the path in your route handlers. Next, we will look at the methodology provided by Sanic for extracting information from the path.

Extracting information from the path

The last thing we need to consider in this section is extracting usable information from our request. The first place we often look is the URI path. Sanic provides a simple syntax for extracting parameters from the path:

```
@app.get("/characters/<name>")
async def profile(request: Request, name: str):
    print text(f"Hello {name}")
```

We have declared the second segment in our path to contain a variable. The Sanic router extracts that and injects it as an argument in our handler. It is important to note that if we do nothing else, that injection will be a str type value.

Sanic also provides an easy mechanism for converting the type. Suppose we want to retrieve a single message from a message feed, query it in the database, and return the message. In this case, our call to the database requires message_id to be an int:

```
@app.get("/messages/<message_id:int>")
async def message_details(request: Request, message_id: int):
    ...
```

This route definition will tell Sanic to convert the second segment into an `int` before injecting it. It is also important to note that if the value is something that cannot be cast as an `int`, it will raise a `404 Not Found`. Therefore, the parameter type does more than just typecasting. It is also involved in route handling.

You can refer to the next section and the user guide to learn what all of the allowed parameter types are: `https://sanic.dev/en/guide/basics/routing.html#path-parameters`.

Besides extracting information from the path itself, the two other places we may want to look for user data are the query parameters and the request body. Query parameters are the part of the URL that comes after a ?:

`/characters?count=10&name=george`

How should we decide whether information should be passed in the path, the query arguments, or as a part of form data or JSON body? Best practices dictate that information should be accessed as follows:

- **Path parameters**: Information to describe *what* the resource is we are looking for

- **Query parameters**: Information that can be used to filter, search, or sort the response

- **Request body**: Everything else

It is a good habit to get into very early on in your application development to learn where different usable bits of information can come from. *Chapter 4, Ingesting HTTP Data*, dives much further into passing data through query parameters and the request body. Just as valuable is of course the HTTP path itself. We just looked at how important crafting purposeful paths might be. Next, we will take a deeper look at extracting data from the HTTP path.

Advanced path parameters

In the last section, we learned the basics of extracting information from a dynamic URL path to something we can code with. This is truly a fundamental feature of all web frameworks. It is also extremely common among many frameworks to allow you to specify what that path parameter should be. We learned that `/messages/<message_id:int>` would match `/messages/123` but not `/messages/abc`. We also learned about the convenience that Sanic provides in converting the match path segment to an integer.

But what about more complex types? Or what if we need to modify the matched value before using it in our application? In this section, we will explore a couple of helpful patterns to achieve these goals.

Custom parameter matching

Out of the box, Sanic provides eight path parameter types that can be matched:

- `str`: Matches any valid string
- `slug`: Matches standard path slugs
- `int`: Matches any integer
- `float`: Matches any number
- `alpha`: Matches only alphabet characters
- `path`: Matches any expandable path
- `ymd`: Matches `YYYY-MM-DD`
- `uuid`: Matches a `UUID`

Each of these provides a type that corresponds to the matched parameter. For example, if you have the path `/report/<report_date:ymd>`, the `date` object in your handler will be a `datetime.date` instance:

```
from datetime import date


@app.get("/report/<report_date:ymd>")
async def get_report(request: Request, report_date: date):
    assert isinstance(report_date, date)
```

This is a very helpful pattern because it accomplishes two things for us. First, it makes sure that the incoming request is in the correct format. A request that is `/report/20210101` would receive a `404 Not Found` response. Second, when we go to work with that `report_date` instance in our handler, it has already been cast into a usable data type: `date`.

What happens when we need routing for types outside of the standard types? Sanic does of course allow us to achieve the first part by defining a custom regular expression for a path segment. Let's imagine that we have an endpoint that we want to match on a valid IPv4 address: `/ip/1.2.3.4`.

The simplest approach here would be to find a relevant regular expression and add it to our path segment definition:

```
IP_ADDRESS_PATTERN = (
    r"(?:(?:25[0-5]|2[0-4][0-9]|[01]?[0-9][0-9]?)\.){3}"
    r"(?:25[0-5]|2[0-4][0-9]|[01]?[0-9][0-9]?)"
)


@app.get(f"/<ip:{IP_ADDRESS_PATTERN}>")
async def get_ip_details(request: Request, ip: str):
    return text(f"type={type(ip)} {ip=}")
```

Now, when we access our endpoint, we should have a valid match:

```
$ curl localhost:7777/1.2.3.4
type=<class 'str'> ip='1.2.3.4'
```

Using regular expression matching also allows us to narrowly define an endpoint between a limited number of options:

```
@app.get("/icecream/<flavor:vanilla|chocolate>")
async def get_flavor(request: Request, flavor: str):
    return text(f"You chose {flavor}")
```

We now have routing based upon our two available choices:

```
$ curl localhost:7777/icecream/mint
⬜ 404 — Not Found
==================
Requested URL /icecream/mint not found


$ curl localhost:7777/icecream/vanilla
You chose vanilla
```

While regular expression matching is incredibly helpful sometimes, the problem is that the output is still `str`. Going back to our first IPv4 example, we would need to manually cast the matched value into `ipaddress.IPv4Address` if we wanted an instance of that class to work with.

While this might not seem like a big deal if you have one or two handlers, if you have a dozen endpoints that need a dynamic IP address as a path parameter, it could become cumbersome. Sanic's solution to this is custom pattern matching. We can tell Sanic that we want to create our own parameter type. To do this we need three things:

- A short descriptor that we will use to name our type

- A function that will return the value we want or raise `ValueError` if there is no match

- A fallback regular expression that also matches our value

In the IP address example, we will do the following:

1. We will name the parameter `ipv4`.

2. We can use the standard library's `ipaddress.ip_address` constructor.

3. We already have our fallback regular expression from earlier. We can proceed to register the custom parameter type:

```
import ipaddress


app.router.register_pattern(
    "ipv4",
    ipaddress.ip_address,
    IP_ADDRESS_PATTERN,
)



@app.get("/<ip:ipv4>")
async def get_ip_details(request: Request, ip: ipaddress.
IPv4Address):
    return text(f"type={type(ip)} {ip=}")
```

Now, we have a more usable object in the handler (`ipaddress.IPv4Address`), and we also have a very easy to reuse path parameter (`<ip:ipv4>`).

What about our second example with ice cream flavors? Instead of having a `str` type, what if we wanted to have an Enum or some other custom model? There is, unfortunately, no function in Python's standard library for parsing ice cream flavors (maybe someone should build that), so we will need to create our own:

1. To start, we will create our model using an Enum. Why an Enum? It is a fantastic tool to keep our code nice and consistent. If our environment is set up right—which it is because we took care in *Chapter 2, Organizing a Project*, to use good tools—we have a single place where we can maintain our flavors with code completion:

    ```
    from enum import Enum, auto


    class Flavor(Enum):
        VANILLA = auto()
        CHOCOLATE = auto()
    ```

2. Next, we need a regular expression that we can later use in our route definition for matching incoming requests:

    ```
    flavor_pattern = "|".join(
        f.lower() for f in Flavor.__members__.keys()
    )
    ```

    The resulting pattern should be `vanilla|chocolate`.

3. We also need to create a function that will act as our parser. Its job is to either return our target type or raise `ValueError`:

    ```
    def parse_flavor(flavor: str) -> Flavor:
    try:
        return Flavor[flavor.upper()]
    except KeyError:
        raise ValueError(f"Invalid ice cream flavor:
    {flavor}")
    ```

4. We can now proceed to register that pattern with Sanic. Just like the IP address example before, we have the name of our parameter type, a function to check the match, and a fallback regular expression:

    ```
    app.router.register_pattern(
        "ice_cream_flavor",
        parse_flavor,
    ```

```
        flavor_pattern,
    )
```

5. With our pattern now registered, we can proceed to use it in all of our ice cream endpoints:

```
@app.get("/icecream/<flavor:ice_cream_flavor>")
async def get_flavor(request: Request, flavor: Flavor):
    return text(f"You chose {flavor}")
```

When we access the endpoint now, we should have an Enum instance, but still only accept requests that match one of our two defined flavors. Yum!

```
$ curl localhost:7777/icecream/mint
404 — Not Found
===============
Requested URL /icecream/mint not found


$ curl localhost:7777/icecream/vanilla
You chose Flavor.VANILLA
```

The key to this example is having a good parse function. In our example, we know that if a bad flavor is entered into the Enum constructor, it will raise KeyError. This is a problem. If our application cannot match mint, it will throw KeyError and the application will respond with 500 Internal Server Error. This is not what we want. By catching the exception and casting it to ValueError, Sanic is able to understand that this is expected, and it should respond with 404 Not Found.

Modifying matched parameter values

As we have learned, using path parameter types is extremely helpful in building our API to respond to intended requests and ignoring bad paths. As much as possible, it is best practice to be as specific as your endpoint needs to get the right data in. We just explored how we might also use parameter types to recast the matched value to a more useful data type. But what if we are not concerned about changing the type of the value, but the actual value itself?

Returning to our character profile application example, imagine that we have some URLs that include **slugs**. If you are not familiar with a slug, it is basically a string that uses lower case letters and hyphens to make human-friendly content in URL paths. We saw an example of this earlier: /users/adam-hopkins.

In our hypothetical application, we need to build an endpoint that returns details about a character instance:

1. First, we will create a model for what the character object will look like:

    ```
    @dataclass
    class Character:
        name: str
        super_powers: List[str]
        favorite_foods: List[str]
    ```

2. We want to be able to return specific details about our character. For example, the endpoint /characters/george/name should return George. So, our next task is to define our route:

    ```
    @app.get("/characters/<name:alpha>/<attr:slug>")
    async def character_property(request: Request, name: str,
    attr: str):
        character = await get_character(name)
        return json(getattr(character, attr))
    ```

3. It is a fairly simple route. It searches for the character and then returns the requested attribute. Let's check it out in action:

    ```
    $ curl localhost:7777/characters/george/name
    "George"
    ```

4. Now, let's try getting George's superpowers:

    ```
    $ curl localhost:7777/characters/george/super-powers
    500 — Internal Server Error
    ================================
    'Character' object has no attribute 'super-powers'
    ```

Uh oh, what happened? The property we are trying to access is Character. super_powers. But our endpoint accepts slugs (because they are easier for people to read). So we need to convert the attribute. Just like in the previous section, where we *could* cast our value inside of the handler, it becomes more difficult to scale that solution. We *could* run attr.replace("-", "_") inside of our handler, and perhaps this is a viable solution. It does make for extra code inside the handlers.

Luckily, we have an alternative. This is a good use case for middleware where we need to convert all slugs (for example, this-is-a-slug) to snake case (for example, this_is_snake_case) so that they can be used programmatically down the road. By converting the slugs, we can look for super_powers instead of super-powers.

5. Let's make that middleware:

```
@app.on_request
def convert_slugs(request: Request):
    request.match_info = {
        key: value.replace("-", "_")
        for key, value in request.match_info.items()
    }
```

What this will do is modify the Request instance before it gets executed by the route handler. For our use case, this means that every value that is matched will be converted from a slug to snake case. Note that we are *not* returning anything in this function. If we do, Sanic will think that we are trying to halt the request/response cycle by providing an early return. This is not the intention. All we want to do is modify Request.

6. Let's test that endpoint again:

```
$ curl localhost:7777/characters/george/super-powers
["whistling","hand stands"]
```

7. Middleware is not the only solution to this problem though. Sanic makes use of signals to dispatch events that your application can listen to. Instead of the preceding middleware, we could do something similar to this with signals:

```
@app.signal("http.routing.after")
def convert_slugs(request: Request, route: Route,
handler, kwargs):
    request.match_info = {
        key: value.replace("-", "_")
        for key, value in kwargs.items()
    }
```

As you can see, it is a very similar implementation. Perhaps the biggest difference to us as developers is that the signal provides us with some more arguments to work with. Although, to be honest, `route`, `handler`, and `kwargs` are all properties that could be accessed from the `Request` instance. Middleware and signals are discussed in greater depth in *Chapter 6*, *Operating Outside the Response Handler*. For now, just know that these are two methods for altering the request/response cycle outside of the route handler. Later on, we will learn more about the differences between them and when it might be preferable to choose one or the other.

In the next section we will learn about a common practice of prepending paths with a version number, and how Sanic makes this practice easy.

API versioning

Back in *Chapter 2*, *Organizing a Project*, we discussed how you could implement API versioning using blueprints. If you recall, it was simply a matter of adding a keyword value to the blueprint definition.

Given the following blueprint definition, we get the URL path `/v1/characters`:

```
bp = Blueprint("characters", version=1, url_prefix="/
characters")


@bp.get("")
async def get_all_characters(...):
    ...
```

That `version` keyword argument is available at the route level as well. If the version is defined in multiple places (for example, on the route and also the blueprint), priority is given to the narrowest scope. Let's look at an example of different places where the version can be defined, and see what the result is. We will define it at the route level, the blueprint level, and the blueprint group level:

```
bp = Blueprint("Characters")
bp_v2 = Blueprint("CharactersV2", version=2)
group = Blueprint.group(bp, bp_v2, version=3)


@bp.get("", version=1)
async def version_1(...):
    ...
```

```
@bp_v2.get("")
async def version_2(...):

    ...


@bp.get("")
async def version_3(...):

    ...


app.blueprint(group, url_prefix="/characters")
```

We now have the following routes. Take a closer look at the example to see how we manipulate the blueprints and the `version` argument to control the handler that each path is delivered to:

- `/v1/characters <Route: name=main.Characters.version_1 path=v1/characters>`

- `/v3/characters <Route: name=main.Characters.version_3 path=v3/characters>`

- `/v2/characters <Route: name=main.CharactersV2.version_2 path=v2/characters>`

Adding versions to endpoint paths is fairly simple. But why should we do it? It is a good practice because it keeps your API flexible but also consistent and stable for your users. By allowing endpoints to be versioned, you maintain the ability to make changes to them and still allow legacy requests to not be denied. It is incredibly beneficial as, over time, you transition your API to add, remove, or enhance features.

Even if the only consumer of your API is your own website, it is still a good practice to version your API so that you have an easier path towards upgrades without potentially causing application regressions.

It is a common practice to "lock in" features with a version. This is a form of creating what is known as an API contract. Think of an API contract as a promise by the developer that the API will continue to work. In other words, once you put an API into usage—and especially if you publish documentation—you are creating a promise to the user that the API will continue to function as is. You are free to add new features, but any breaking changes that are not backwards compatible violate that contract. Therefore, when you do need to add breaking changes, versions might be the right trick in your tool bag to accomplish your goal.

Here is an example. We're building out our database of character profiles. The first version of our API has an endpoint to create a new profile and it looks something like this:

```
@bp.post("")
async def create_character_profile(request: Request):
    async create_character(name=request.json["name"], ...)
    ...
```

This endpoint is built upon the assumption that the incoming JSON body will be fairly simple like this:

```
{
    "name": "Alice"
}
```

What happens when we want to handle some more complex use cases?

```
{
    "meta": {
        "pseudonuym": "The Fantastic Coder",
        "real_name": "Alice"
    },
    "superpowers": [
        {
            "skill": "Blazing fast typing skills"
        }
    ]
}
```

It might start getting complicated, messy, and overall difficult to maintain our route handler if we put too much logic into it. As a general practice, I like to keep my route handlers very concise. If I see my code creeping up to 50 lines of code inside a view handler, I know there is probably some refactoring that needs to be done. Ideally, I like to keep them to about 20 lines or less.

One way we can keep our code clean is to split these use cases. Version 1 of the API will still be able to create characters using the simpler data structure, and version 2 has the capability of the more complex structure.

Should all of my routes bump versions?

Often, you will have a need to increase a version on a single endpoint, but not all of them. This raises the question: what version do I use on the unchanged endpoints? Ultimately, this is going to be a question that can only be dictated by the application. It might be helpful to keep in mind how the API is being used.

Very often, you will see APIs bumping versions when there is a complete break or some major overhaul in the API structure. This could accompany a new technology stack, or a new API structure or design pattern. An example of this is when GitHub changed its API from v3 to v4. The older version of their API (v3) is RESTful, similar to what we discussed earlier in this chapter. The newer version (v4) is based upon GraphQL (see *Chapter 10, Implementing Common Use Cases with Sanic*, for more on GraphQL). This is a complete redesign of the API. Because v3 and v4 are completely incompatible, they changed the version number.

In GitHub's case, it was clear all endpoints needed to change as it was effectively a brand new API. Drastic changes like this are not the only catalyst for version changing, however. What if we are only changing compatibility on a smaller portion of our API and keeping the rest intact?

Some people may find that it makes sense to implement the new version number on all of their endpoints. One way to accomplish this is to add multiple route definitions to an endpoint:

```
v1 = Blueprint("v1", version=1)
v2 = Blueprint("v2", version=2)

@v1.route(...)
@v2.route(...)
async def unchanged_route(...):
    ...
```

The downside of this approach is that could become very cumbersome to maintain. If you needed to add a new route definition to *every* handler when you want to change a version, you might be discouraged from adding versions in the first place. Take this into consideration.

How about nesting blueprints? How about a function that dynamically adds routes at startup? Can you think of a solution? We have already seen various tools and strategies earlier in this book that might help us out. This might be a good time to put the book down and jump into your code editor on your computer. I encourage you to play around with versions and nesting to see what is and is not possible.

Remember `app.router.routes` and `app.router.groups`? Try adding a single handler to multiple blueprints. Or try adding the same blueprints to different groups. I challenge you to come up with a pattern to have the same handler on different versions without multiple definitions like the preceding example. Start with this, and see what you can come up with, without doubling up the route definition as before:

```
v1 = Blueprint("v1", version=1)
v2 = Blueprint("v2", version=2)

@v1.route(...)
async def unchanged_route(...):
    ...
```

Here is a handy snippet you can use while developing to see which paths are defined:

```
from sanic.log import import logger

@app.before_server_start
def display(app: Sanic, _):
    routes = sorted(app.router.routes, key=lambda route: route.
uri)
    for route in routes:
        logger.debug(f"{route.uri} [{route.name}]")
```

Getting back to our question: should all of my routes bump versions? Some people will say yes, but it seems artificially complex to bump the version of all routes when only one has changed. By all means, if it makes sense, bump everything simultaneously.

If we only want to bump the routes that are changing, it causes another problem. What should we bump it to? Many people will tell you that versions should *only* ever be integers: v1, v2, v99, and so on. I find this limiting, and it really makes the following set of endpoints feel unnatural:

- `/v1/characters`
- `/v1/characters/puppets`
- `/v1/characters/super_heroes`
- `/v1/characters/historical`
- `/v2/characters`

While I am not discounting this approach, it does seem like there *should* be a v2 for all of the routes, even if they did not change. We are trying to avoid that. Why not use minor versions like semantic versioning? It seems more natural and accepting to have a single /v1.1 endpoint than a single /v2. Again, this is going to be a matter of what works for your application needs, and what is reasonable given the types of users that will be consuming your API. Should you decide that semantic versioning style will work for your application needs, you can add it by using a float for the version argument as seen here:

```
@bp.post("", version=1.1)
async def create_character_profile_enhanced(request: Request):
    async create_character_enhanced(data=request.json)
```

> **Important Note**
>
> Semantic versioning is an important concept in software development, but beyond the scope here. In brief, the concept is to create a version by declaring a major, minor, and patch number, which are connected by a period, for example, 1.2.3. Generally speaking, semantic versioning states that an increment of the major version corresponds to a backwards-incompatible change, the minor version to a new feature, and the patch version to a bug fix. If you are unfamiliar with it, I suggest taking some time to read through the documentation for it since it is widely used throughout software development: https://semver.org/.

> **Tip**
>
> It is highly recommended that you use version with your endpoints if you intend for there to be third-party integration with your API. If the API is only meant to be used by your own application, perhaps this is less important. Nevertheless, it may still be a useful pattern. Therefore, I recommend using version=1 for new projects or version=2 for projects that are replacing an existing API even if the legacy application did not have a version scheme.

Version prefixing

The standard way to use versions in Sanic is version=<int> or version=<float>. The version will *always* be inserted into your path at the very beginning. It does not matter how deeply nested and how many layers of url_prefix you have. Even a deeply nested route definition can have a single version and it will be the first segment in the path: /v1/some/deeply/nested/path/to/handler.

This does, however, impose a problem when you are trying to build multiple layers on your application. What if you want to have some HTML pages and an API and keep them separate based upon their path? Consider the following paths that we might like to have in our application:

- `/page/profile.html`
- `/api/v1/characters/<name>`

Notice how the versioned API route starts with `/api`? This is impossible to control only with URIs and blueprint URI prefixes since Sanic *always* puts the version before the rest of the path. However, Sanic provides a `version_prefix` argument in all of the same places that `version` can be used. The default value is `/v`, but feel free to update it as needed. In the following example, we can nest our entire API design in a single blueprint group to automatically add `/api` to the front of every endpoint:

```
group = Blueprint.group(bp1, bp2, bp3, version_prefix="/api/v")
```

> **Tip**
> The same path parameters are available here. You could, for example, do something like this: `version_prefix=/<section>/v`. Just make sure you remember that `section` will now be an injected keyword argument in every route handler.

You should now have a good grasp of how and when to use versions. They are a powerful tool in making your API more professional and maintainable since they allow for more flexible development patterns. Next, we will explore another tool for creating flexibility and reusability in your application code: virtual hosts.

Virtual hosts

Some applications can be accessed from multiple domains. This gives the benefit of having a single application deployment to manage, but the ability to service multiple domains. In our example, we will imagine that we completed the computer adventure game social media site. The API is truly amazing.

It is so incredible in fact that both Alice and Bob have approached us about the opportunity to be resellers and to *white label* our application, or reuse the API for their own social media sites. This is a somewhat common practice in the internet world where one provider builds an application and other providers simply point their domain to the same application and operate as if it is their own. To achieve this, we need to have distinct URLs:

- `mine.com`
- `alice.com`
- `bob.com`

All of these domains will be set up with their DNS records pointing to our application. This can work without any further changes inside the application. But what if we need to know which domain a request is serving, and do something slightly different for each one? This information should be available to us in the request headers. It should simply be a matter of checking the headers:

```
@bp.route("")
async def do_something(request: Request):
    if request.headers["host"] == "alice.com":
        await do_something_for_alice(request)
    elif request.headers["host"] == "bob.com":
        await do_something_for_bob(request)
    else:
        await do_something_for_me(request)
```

This example may seem small and simple, but you can probably imagine how the complexity could increase. Remember earlier I stated how I like to keep the lines of code per handler to a minimum? This is certainly a use case where you can imagine the handlers could get very lengthy.

Essentially, what we are doing in this endpoint is host-based routing. Depending upon the incoming request host, we are routing the endpoint to a different location.

Sanic already does that for us. All we need to do is break the logic into separate route handlers and give each one a `host` argument. This achieves the routing that we need but keeps it out of our response handlers:

```
@bp.route("", host="alice.com")
async def do_something_for_alice(request: Request):
    await do_something_for_alice(request: Request)
```

```
@bp.route("", host="bob.com")
async def do_something_for_bob(request: Request):
    await do_something_for_bob(request: Request)


@bp.route("", host="mine.com")
async def do_something_for_me(request: Request):
    await do_something_for_me(request: Request)
```

If you find yourself in this situation, you do not need to define a host for every endpoint, only the endpoints where you would want to have host-based routing. Following this pattern, we can reuse the same application across multiple domains, and still have some endpoints capable of distinguishing between them, and others ignorant to the fact that multiple domains are reaching them.

One thing that is important to keep in mind: if you create an endpoint that has host-level routing, then all routes on that same path must also have it. You cannot, for example, do the following. Notice how the third route does *not* define the host argument.

The following example will *not* work, and will raise an exception at startup:

```
@bp.route("", host="alice.com")
async def do_something_for_alice(request: Request)::
    await do_something_for_alice(request: Request)


@bp.route("", host="bob.com")
async def do_something_for_bob(request: Request):
    await do_something_for_bob(request: Request)


@bp.route("")
async def do_something_for_me(request: Request):
    await do_something_for_me(request: Request)
```

To solve this, make sure that all routes that could be grouped together have a host value. This way they can be distinguished. If one of them has a host, they all need to have one.

We have generally now discussed all of the considerations to make when routing web requests to our response handlers. But, we have not yet looked at how Sanic delivers requests to static content (that is, actual files on your web server that you want to send such as images and style sheets). Next, we will discuss some options both with and without using Sanic.

Serving static content

So far, all of our discussion in this chapter has been about dynamically generating content for responses. We did, however, discuss that passing files that exist inside of a directory structure is a valid use case that Sanic supports. This is because most web applications have the need to serve some static content. The most common use cases would be for delivering JavaScript files, images, and style sheets to be rendered by the browser. Now, we are going to dive into static content to see how that works, and we can deliver this type of content. After learning how Sanic does it, we will see another very common pattern to serve the content outside of Sanic with a proxy.

Serving static content from Sanic

Our `app` instance has a method on it called `app.static()`. That method requires two arguments:

- A URI path for our application
- A path to tell Sanic where it can access that resource

That second argument can either be a single file or a directory. If it is a directory, everything inside of it will be accessible, like the old school web servers we talked about at the beginning of the chapter.

This is very helpful if you plan to serve all of your web assets. What if you have a folder structure like this?

```
.
├── server.py
└── assets
    ├── index.html
    ├── css
    │   └── styles.css
    ├── img
    │   └── logo.png
    └── js
        └── bundle.js
```

We can use Sanic to serve all of those assets and make them accessible like this:

```
app.static("/static", "./assets")
```

Those assets are now accessible:

```
$ curl localhost:7777/static/css/styles.css
```

Serving static content with Nginx

Now that we have seen how to serve static files with Sanic, a good next question is: should you?

Sanic is very fast at creating the sort of dynamic endpoints that are required by most web APIs. It even does a pretty good job serving static content, keeps all of your endpoint logic in one application, and allows for manipulating those endpoints or renaming files. As we discussed in *Chapter 1, Introduction to Sanic and Async Frameworks*, Sanic applications are also meant to be fast to build.

There is however a potentially faster method for delivering static content. For a single-page application that is meant to be consumed by a browser that requests data through your API, one of your biggest stumbling blocks will be reducing the time to your first page render. This means that you must package up all of your JavaScript, CSS, image, or other files as quickly as possible in the browser to reduce rendering latency.

For this reason, you might want to consider using a proxy layer such as Nginx in front of Sanic. The purpose of the proxy would be to:

1. Send any requests to the API through toSanic

2. Handle serving static content itself.

You may want to consider this option especially if you intend to serve a lot of static content. Nginx has a caching engine built in to be able to deliver static content much faster than any Python application could.

Chapter 8, Running a Sanic Server, discusses deployment strategies and considerations to make when deciding whether to use tools such as Nginx and Docker. For now, we will use Docker Compose to really quickly and easily stand up Nginx:

1. We need to make our `docker-compose.yml` manifest:

    ```
    version: "3"

    services:
      client:
        image: nginx:alpine
        ports:
    ```

```
      - 8888:80
    volumes:
      - ./nginx/default.conf:/etc/nginx/conf.d/default.
conf
      - ./static:/var/www
```

If you are not familiar with Docker Compose or how to install and run it, you should be able to find a wealth of tutorials and information online.

This simple setup we are going for in our example will require that you set the path for ./static in our docker-compose.yml file to whatever directory you have your static assets in.

> **Tip**
>
> This is intentionally a super simple implementation. You should make sure that a real Nginx deployment includes things such as TLS encryption and proxy secrets. Check out the user guide for more details and a helpful walk-through: https://sanic.dev/en/guide/deployment/nginx. html#nginx-configuration.

2. Next, we will create the ./nginx/default.conf file needed to control Nginx:

```
upstream example.com {
    keepalive 100;
    server 1.2.3.4:8000;
}

server {
    server_name example.com;
    root /var/www;

    location / {
        try_files $uri @sanic;
    }
    location @sanic {
        proxy_pass http://$server_name;
        proxy_set_header Host $host;
        proxy_set_header X-Forwarded-Proto $scheme;
        proxy_set_header X-Real-IP $remote_addr;
```

```
            proxy_set_header X-Forwarded-For $proxy_add_x_
    forwarded_for;
        }
        location ~* \.(jpg|jpeg|png|gif|ico|css|js|txt)$ {
            expires max;
            log_not_found off;
            access_log off;
        }
    }
```

We start it using the following command:

```
$ docker-compose up
```

The most important thing to change here is the server address. You should change
1.2.3.4:8000 to whatever address and port your application can be accessed
at. Keep in mind that this will *NOT* be 127.0.0.1 or localhost. Since Nginx
will be running inside of a Docker container, that local address will point to
the container itself, and not your computer's local network address. Instead, for
development purposes, you should consider setting it to your local IP address.

3. You will need to make sure that Sanic knows to serve on that network address. Do
    you remember how we said we are running Sanic back in *Chapter 2, Organizing a
    Project*? It looked like this:

    ```
    $ sanic server:app -p 7777 --debug --workers=2
    ```

    For this example, we will change that to this:

    ```
    $ sanic server:app -H 0.0.0.0 -p 7777 --debug --workers=2
    ```

    My local IP address is 192.168.1.7, therefore I will set the upstream block in
    my Nginx configuration to server 192.168.1.7:7777;.

4. You should now be able to access any static files in your ./static directory. I
    have a file called foo.txt. I am using the -i flag with curl to be able to see the
    headers. The important headers to see are Expires and Cache-Control. These
    help your browser to cache the file instead of re-requesting it:

    ```
    $ curl localhost:8888/foo.txt -i
    HTTP/1.1 200 OK
    Server: nginx/1.21.0
    Date: Tue, 15 Jun 2021 18:42:20 GMT
    Content-Type: text/plain
    ```

```
Content-Length: 9
Last-Modified: Tue, 15 Jun 2021 18:39:01 GMT
Connection: keep-alive
ETag: "60c8f3c5-9"
Expires: Thu, 31 Dec 2037 23:55:55 GMT
Cache-Control: max-age=315360000
Accept-Ranges: bytes

hello...
```

If you try to send a request to a file that does not exist, Nginx will send that route on to your Sanic application. This setup is just the tip of the iceberg when it comes to proxying and Nginx. It is, however, very common for Python web applications to use a strategy like this. As mentioned earlier, we will dig deeper into this topic when we discuss deployment options in *Chapter 8, Running a Sanic Server*.

Streaming static content

It is also worth reiterating that the Sanic server is built and intended to be a frontline server. That means that it can certainly stand as your point of ingress without a proxy server in front of it, including serving static content. The decision about whether to proxy or not—at least as it relates to delivering static files—is likely a question of how much traffic and how many files your application may need to deliver.

Another important factor to consider is whether your application needs to stream files. Streaming will be discussed in depth in *Chapter 5, Building Response Handlers*. Let's create a really simple web page to stream a video and see what that might look like:

1. First, the HTML. Store this as index.html:

    ```html
    <html>
        <head>
            <title>Sample Stream</title>
        </head>
        <body>
            <video width="1280" height="720" controls>
                <source src="/mp4" type="video/mp4" />
            </video>
        </body>
    </html>
    ```

2. Next, find an mp4 file that you want to stream. It can be any video file. If you do not have one, you can download a sample file for free from a website such as this: `https://samplelib.com/sample-mp4.html`.

3. We will now create a small Sanic app to stream that video:

    ```python
    from sanic import Sanic, response


    @app.route("/mp4")
    async def handler_file_stream(request: Request):
        return await response.file_stream("/path/to/sample.mp4")


    app.static("/index.html", "/path/to/index.html")


    @app.route("/")
    def redirect(request: Request):
        return response.redirect("/index.html")
    ```

4. Run the server as normal and visit it in your web browser: `http://localhost:7777`.

You should notice that the root URI (/) redirected you to /index.html. Using app. static, the application tells Sanic that it should accept any requests to /index.html and serve back the static content that is located on the server at /path/to/index. html. This should be your delivered content from above. Hopefully, you have a play button, and you can now stream your video to your browser. Enjoy!

Summary

This chapter covered a lot of material on taking an HTTP request and turning it into something usable. At the core of a web framework is its ability to translate a raw request into an actionable handler. We learned about how Sanic does this and how we can use HTTP methods, good API design principles, paths, path parameter extraction, and static content to build useful applications. As we learned earlier in this book, a little bit of upfront planning goes a long way. Before putting too much code together, it is really helpful to think about the tools HTTP offers, and how Sanic allows us to take advantage of those features.

If we did a good job in *Chapter 2, Organizing a Project*, of setting up directories, it should be very easy for us to loosely mirror that structure and nest blueprints to match our intended API design.

There are some key takeaways from this chapter. You should purposely, and thoughtfully, design your API endpoint paths—using nouns—that point to an intended target or resource. Then, HTTP methods should be used as the verbs that tell your application and users *what* to do with that target or resource. Finally, you should extract helpful information from those paths to be used in your handlers.

We mainly focused our attention on the first line of the raw HTTP request: the HTTP method and URI path. In the next chapter, we will dive into extracting more data from the request, including the headers and the request body.

4
Ingesting HTTP Data

The next building block in application development involves **data**. Without data, the web has little utility. I do not mean to get too philosophical here, but it is axiomatic that the purpose of the internet is the facilitation of the transfer of data and knowledge from one location to another. Therefore, it is critical to our development as web professionals to learn how data can be transferred not just *from* our applications (which we deal with in *Chapter 5*, *Building Response Handlers*), but also *to* our applications (which is the purpose of this chapter). The simplest applications we can build simply provide data. However, to become interactive web applications participating in the global exchange of knowledge, even simple applications must be capable of extracting data from web requests.

A web application that receives no data is like a *screencast*. Viewers can come to watch the presentation, but the presenter has no *personal* connection to the people watching. During the COVID-19 global pandemic, I was fortunate enough to still be able to participate in several Python conventions. Much applause is due to the volunteers who pushed forward to present the community with a continuation of the sharing and learning atmosphere that exists with technology conferences. However, I would be remiss to point out that, as a presenter, I had zero connection to my audience. It was not until after the presentation was done that I even knew how many people watched my content.

This model can be useful to disperse information to those who need to intake information from it. However, the transaction is entirely one-sided. My presentations could not be adjusted based upon cues from the audience, and even during chat or question and answer sessions, there was an interpersonal experience that was missing. In much the same way, a web application that receives no data operates under a similar principle. The server has no knowledge about who is listening and cannot alter its behavior or content based upon user input. These types of applications are purely for the dissemination of data and resources only.

Generally, web APIs of this kind only have GET methods since they exist entirely to spit back information. They can be useful for relaying information about the weather, flight details, or other centralized repositories of information that many people might want to access.

To build a truly interactive API, we need it to operate not like a screencast, but more like a video chat. Both sides of the conversation will participate in the passing of information back and forth. And it is this bidirectional communication that we will explore in this chapter.

If you recall from our earlier discussions, there are three main sections in the raw HTTP request: **the first line**, **the HTTP headers**, and **the body**. So far, we have focused on intaking HTTP requests as they relate to the HTTP method and the path: that is, the information that appears on the first line of the HTTP request.

In this chapter, we will learn how to get data from the client from all three sections. Data can be passed to a web server in query arguments, headers, and, of course, the body itself. Therefore, in this chapter, we will explore the following:

- Reading cookies and headers
- Reading forms, query arguments, files, JSON, and more
- Validating data

Technical requirements

In this chapter, you should have, at your disposal, the same tools available as in the previous chapters in order to follow along with the examples (such as an IDE, modern Python, and curl). You can access the source code for this chapter on GitHub at https://github.com/PacktPublishing/Python-Web-Development-with-Sanic/tree/main/Chapter04.

Reading cookies and headers

As we have learned from the earlier chapters of this book, when an HTTP client sends a request to a web server, it includes one or more headers that are in a key/value pair. These headers are meant to be part of a meta-conversation between the client and the server. And since an HTTP connection is a two-sided transaction with both a request and a response, we must bear in mind that there is a distinction between request headers and response headers.

This chapter focuses only on HTTP requests. Therefore, we will only be covering material related to request headers. This is worth pointing out because there are some headers that are commonly found in both the request and the response. One such example is **Content-Type**, which can be used by both HTTP requests and HTTP responses. So, keep this in mind when we talk about Content-Type in this section it relates to HTTP requests only. There is a time and a place for discussing response headers. Feel free to skip ahead, or read this section in conjunction with *Chapter 5*, *Building Response Handlers*, where we will discuss the other side of the same coin.

Headers are flexible

HTTP headers are not magic. There is no predefined, finite list of header names. Furthermore, deviating from what is considered *standard* will have no impact on your application. *Remember when we discussed HTTP methods, and we said that you could invent your own methods?* Well, you have that control and ability to create your own *headers* as well.

This practice is both commonplace and encouraged. *Are you familiar with Cloudflare?* In short, **Cloudflare** is a popular tool used as a proxy for web applications. We will discuss proxies further in *Chapter 8*, *Running a Sanic Server*. The idea is simple: Cloudflare runs a web server, a request comes into their server, they do *something* to it, and then bundle that up and send the request on to your server. When they do that, they include their own set of non-standard headers. For example, they will forward the request to you with `CF-Connection-IP` and `CF-IPCountry` headers to give you some helpful information about the IP address and the location of its origin.

Let's imagine that we are building an API to be used by a farmer's market. They want to set up a web API that will help coordinate among the various participants in the market: farmers, restaurant owners, and consumers. The first endpoint we build will be used to provide information about the market stalls for a given day:

```
@app.get("/stalls/<market_date:ymd>")
async def market_stalls(request: Request, market_date: date):
```

```
info = await fetch_stall_info(market_date)
return json({"stalls": info})
```

The response content from this endpoint does not require authentication (we will discuss this further later), but it really ought to be tailored to each of the types of users. A farmer might want to know how many stalls are available. Consumers and restaurant owners might be more interested in knowing the kinds of products that will be available instead. Therefore, we have identified at least two different use cases for the same endpoint.

One option might be to split this single endpoint into two: */stalls/<market_ date:ymd>/ availability* and */stalls/<market_date:ymd>/products.*

However, this does add some complexity to the overall API design. Furthermore, `availability` and `products`, as used in this context, are not really resources in and of themselves. Giving them their own endpoint sort of muddies the water of the current structure of our API.

What we are really saying is that we have a single resource—the collection of market stalls for a given day of the year—and we simply want to present those resources in different ways based upon the participant type. It really is only one endpoint with two different ways of displaying the same information.

Perhaps instead of two different endpoints, a second option might be to use query parameters (more on those later in the *Query arguments* section). That would look like this: `/stalls/<market_date:ymd>?participant=farmer`, and `/stalls/<market_date:ymd>?participant=consumer`. This also sort of breaks the paradigm of query parameters—or at least the way I like to use them—which are, typically, meant to be used for filtering and sorting results.

Instead, we will opt for creating a custom header for our use case: `Participant-Type: farmer`. We will also create an `Enum` to help us validate and limit the acceptable participants:

```
from enum import Enum, auto
class ParticipantType(Enum):
    UNKNOWN = auto()
    FARMER = auto()
    RESTAURANT = auto()
    CONSUMER = auto()


@app.get("/stalls/<market_date:ymd>")
```

```
async def market_stalls(request: Request, market_date: date):
    header = request.headers.get("participant-type", "unknown")
    try:
        paticipant_type = ParticipantType[header.upper()]
    except KeyError:
        paticipant_type = ParticipantType.UNKNOWN
    info = await fetch_stall_info(market_date, paticipant_type)

    return json(
        {
            "meta": {
                "market_date": market_date.isoformat(),
                "paticipant_type": paticipant_type.name.
lower(),
            },
            "stalls": info,
        }
    )
```

When the request comes in, the handler will try and read the header, expecting there to be a valid ParticipantType object. If there is no Participant-Type header, or if the passed value is of an unknown type, we will simply fall back to ParticipantType.UNKNOWN.

> **Important Note**
>
> As you can see in the preceding example, request.headers.
> get("participant-type") is in lowercase. This really does not matter. It could be uppercase, lowercase, or a mix. All headers will be read as case-insensitive keys. So, even though the request.headers object is a dict, it is a special kind of dictionary that does not care about the case. It is only a convention to use lowercase letters when retrieving headers from Sanic. Feel free to do what makes sense to you. However, I would caution you to try and stay consistent throughout a project. It can be confusing if, sometimes, you see headers.get("Content-Type"), and other times, you see headers.get("content-type").

> **Tip**
>
> Enums are great. You should really use them everywhere you can. While using them for validation, as we are doing here, might not be their most obvious use case, they are super helpful when needing to pass around some types of constants. Imagine having to remember deep inside the bowels of your application: *is it restaurant-owner, restaurant_owner, or restaurant?* Using enums can help reduce bugs, can provide a single place to maintain and update, and can provide you with code completion if your IDE supports it. In this book, you will see me use enums in a variety of ways. Next to `asyncio`, the standard library `enum` package might just be one of my favorites.

Getting back to our example, we will now try and hit our endpoint with a few different examples to see how it responds with different headers:

1. We will access the information with a known type pretending to be a farmer:

```
$ curl localhost:7777/stalls/2021-06-24 -H "Participant-
Type: farmer"
{
  "meta": {
    "market_date": "2021-06-24",
    "paticipant_type": "farmer"
  },
  "stalls": [...]
}
```

2. Now, we will leave out the header to see how the endpoint will respond to the absence of any type:

```
$ curl localhost:7777/stalls/2021-06-24
{
  "meta": {
    "market_date": "2021-06-24",
    "paticipant_type": "unknown"
  },
  "stalls": [...]
}
```

3. Finally, we will hit the endpoint with a type we have not anticipated:

```
$ curl localhost:7777/stalls/2021-06-24 -H "Participant-
Type: organizer"
{
  "meta": {
    "market_date": "2021-06-24",
    "paticipant_type": "unknown"
  },
  "stalls": [...]
}
```

We have successfully implemented a custom HTTP header that can be used by our endpoint to decide how it will display and customize the output. We might be getting ahead of ourselves since we will cover middleware in *Chapter 6, Operating Outside the Response Handler, but what if we want to reuse the* Participant-Type *header on other endpoints?* Here is a quick demonstration to make this universal to our entire application:

```
@app.on_request
async def determine_participant_type(request: Request):
    header = request.headers.get("participant-type", "unknown")
    try:
        paticipant_type = ParticipantType[header.upper()]
    except KeyError:
        paticipant_type = ParticipantType.UNKNOWN

    request.ctx.paticipant_type = paticipant_type


@app.get("/stalls/<market_date:ymd>")
async def market_stalls(request: Request, market_date: date):
    info = await fetch_stall_info(market_date, request.ctx.
paticipant_type)

    return json(
        {
            "meta": {
                "market_date": market_date.isoformat(),
```

```
                    "paticipant_type": request.ctx.paticipant_type.
    name.lower(),
                },
            "stalls": info,
        }
    )
```

By evaluating the header inside middleware, we can now place `participant_type` inside the request object for easy access.

The last thing that I would like to point out about this development example is the mindset toward testability. Notice how we identified three different potential uses of the endpoint: a known type, a lack of a type, and an unknown type. We will talk about testing in *Chapter 9, Best Practices to Improve Your Web Applications*. However, as we continue working through this book, it is good to be reminded not only of how to work with Sanic, but the types of things we should be thinking about when we uncover a problem. Thinking ahead about how the application could be used will help us to understand the types of use cases we might want to test for and, therefore, the types of use cases that our application needs to handle.

> **Tip**
>
> Also, it is worth pointing out that the `request.ctx` object is there for you to attach any information you want to it. This is really powerful to help pass information around and to abstract some logic to middleware, as shown earlier. Keep in mind that this only lasts as long as the request lasts. After there is a response, anything on that `request.ctx` object will be disposed of. There is also a similar context for the entire lifespan of the application and the lifespan of a single client connection. These are `app.ctx` and `request.conn_info.ctx`, respectively. Please refer to *Chapter 6, Operating Outside the Response Handler*, for more information regarding these `ctx` objects.

Even though it is entirely possible to create your own set of headers—and indeed, I highly encourage it—there does exist a set of common headers that are standard among clients and servers. In the next sections, we will explore what some of those are.

Common headers

There is a set of predefined standard headers in *section 5 of RFC 7231*: `https://datatracker.ietf.org/doc/html/rfc7231#section-5`. If you are so inclined, put this book down and go read that section. We'll be waiting for you. If not, let's try and pull out some highlights and several of the more important request headers that you should probably know about.

Authentication headers

One of the primary mechanisms for authenticating web requests is through the use of headers. The other basic method is with cookies (which, technically, is also a header, but more on that later, in the *Getting information from cookies (yum!)* section). While there are certainly different types of authentication schemes (such as basic auth, JWT, and session tokens, to name a few), generally, they share the same construct: the use of the `Authorization` header.

You might have just noticed something peculiar. We are talking about *authentication*—at least, that's what the title of this section is called. But we just said that the primary *authentication* header is called *Authorization. How can this be?*

We will cover this in more detail in *Chapter 7, Dealing with Security Concerns*, when we discuss access control more thoroughly, but it is worth mentioning the distinction and the fundamental questions that these two related concepts are trying to answer:

- **Authentication**: *Do I know who this person is?*

- **Authorization**: *Should I let them in?*

I like to keep these two concepts of authentication and authorization straight in my head by imagining a gatekeeper at a medieval castle. The gatekeeper's job is to protect access to the castle and is one of the first lines of defense. This is done in two phases. The first function of the gatekeeper is to identify a person approaching the castle and seeking access to enter. In medieval Europe, they might have used colorful flags or banners for this purpose.

The gatekeeper must decide whether the identity of the person approaching is authentic: do they believe that the person is who they claim to be? If their identity does not appear to be authentic, this means that they are *Unauthorized* to even move on to the second phase. After positively identifying the person—which means that their identity has been *authenticated*—the gatekeeper's next task is to determine whether this person should be allowed entry. If not, their entry to the castle is *Forbidden*.

A failure to authenticate results in a *401 Unauthorized* error message, and a failure of authorization is a *403 Forbidden* error message. It is an unfortunate quirk in the history of the internet that these terms have been muddled up and that they developed as they did. They are confusing and inconsistent. But if you think of both authentication and authorization as a consecutive, linear process, it might be easier to consider the resulting error message as it relates to the subsequent step of the process. Therefore, a failure of authentication means that the user is unauthorized to even present itself for authorization consideration.

So, even though the HTTP header used for authentication is called *Authorization*, and even though its failure should lead to an *Unauthorized* response, we are still exclusively talking about authentication and answering the question of *do I know who this person is?* Or, as the medieval gatekeeper might ask, *does this person's flag and identity appear authentic?*

Because Sanic does not take a stance in terms of how you should build your application, we have a lot of freedom in choosing how we want to consume the *Authorization* request header. Three main strategies come to mind here:

- **Decorators**
- **Middleware**
- **Blueprints**

Let's look at each of these individually.

Decorators

First, let's look at an example using decorators:

```
from functools import wraps
from sanic.exceptions import Unauthorized


def authenticated(handler=None):
    def decorator(f):
        @wraps(f)
        async def decorated_function(request, *args, **kwargs):
            auth_header = request.headers.get("authorization")
            is_authenticated = await check_authentication(auth_
header)

            if is_authenticated:
```

```
            return await f(request, *args, **kwargs)
        else:
            raise Unauthorized("who are you?")

    return decorated_function

return decorator(handler) if handler else decorator


@app.route("/")
@authenticated
async def handler(request):
    return json({"status": "authenticated"})
```

The core of this example is inner `decorated_function`. Essentially, this is saying that before running our actual handler (which is f), run `check_authentication`. This allows us the opportunity to execute code inside the route but before we get to the actual defined handler.

This decorator pattern is extremely common in Sanic—not only for running checks but also for injecting arguments into our handler. If you are not using some form of decorator in your application, you are leaving some real power on the table. It is a useful way to duplicate logic across endpoints, and I highly recommend you get familiar and comfortable with using them. There is a very helpful starter example that can be found in the *Sanic User Guide* at `https://sanic.dev/en/guide/best-practices/decorators.html`.

> **Tip**
>
> Notice `handler=None` and the last return line:
>
> ```
> def authenticated(handler=None):
>
>     ...
>
>     return decorator(handler) if handler else decorator
> ```
>
> The reason we do this is because we are allowing our decorator to be used in one of two ways: either via `@authenticated` or `@authenticated()`. You will have to decide which one is (or whether both are) appropriate for your needs.

Middleware

Now that we have seen how this works with decorators, how can we achieve the same logic with middleware? In the next example, we will try and achieve the same functionality that the decorator example provided, except we will be using middleware:

```
@app.on_request
async def do_check_authentication(request: Request):
    is_authenticated = await check_authentication(auth_header)
    if not is_authenticated:
        raise Unauthorized("who are you?")
```

The downside of this method is that we have just locked up our *entire* API! *What about our* /stalls/<market_date:ymd> *endpoint or even the endpoints that are meant for logging in?* One way to fix this is to check whether the request has a matched Route instance (it should unless we are responding to a *404 Not Found* error), and if it does, make sure it is not one of the exempt routes. Here, we can see an example of how to do that by cross-referencing the name of the matched route with an express list of exempt endpoints:

```
@app.on_request
async def do_check_authentication(request: Request):
    if request.route and request.route.name not in (
        "MyApp.login",
        "MyApp.market_stalls",
    ):
        is_authenticated = await check_authentication(auth_
header)
        if not is_authenticated:
            raise Unauthorized("who are you?")
```

This time, in the middleware, we are taking a look at the route's name to check whether it is one of the routes that we know should be safe.

> **Important Note**
>
> As a quick aside—since we have not seen it before—all routes will have a *name*. You can, of course, name them manually:
>
> ```
> @app.route(..., name="hello_world")
> ```
>
> More likely than not, we can just let Sanic name our routes. By default, it will use the handler function's name, and then append it to our application name (and any blueprints) with dot notation. That is why we see `MyApp.login` and `MyApp.market_stalls`. They are presuming that our application is called `MyApp`, and the handlers for our exempt endpoints are `login` and `market_stalls`, respectively.

Hang on a minute?! Do you want me to keep a list of names of exempt endpoints? That sounds like a nightmare to maintain! True. If you are only handling two items, such as this simple use case, it is probably manageable enough. But once we begin really building out an application, this might start to get super unwieldy. Feel free to decide which of the two patterns makes more sense. Using decorators is clearer and far more explicit.

However, it does lead to more code repetition. The middleware alternative is simpler to implement and easier to audit to ensure that we are not forgetting to protect any routes. Its downside is that it hides some functionality and would be harder to maintain if the list of *safe* endpoints grows. If you are in any doubt about which makes sense for your needs, I would suggest more explicit authentication decorators. However, this does indicate that there are usually different ways to tackle the same problems. Coming back to the point of *Chapter 1, Introduction to Sanic and Async Frameworks*, if one of these solutions appears more *obviously* correct to you, then that is likely to be the one that you should use.

Blueprints

And this is where our third solution comes in: our friend blueprints again. This time, we are going to continue using middleware, but we are only going to apply middleware to the blueprints that contain the protected endpoints:

```
protected = Blueprint("Protected")


@protected.route("/")
async def handler(request):
    return json({"status": "authenticated"})


@protected.on_request
async def do_check_authentication(request: Request):
```

```
        auth_header = request.headers.get("authorization")
        is_authenticated = await check_authentication(auth_header)
        if not is_authenticated:
            raise Unauthorized("who are you?")
```

Since we are placing the middleware on the `protected` blueprint, it will only run on the routes that are attached to it. This leaves everything else open.

Context headers

These headers provide you with some information about the web browser where the request originated. Generally, they are useful in analytics and logging to provide some information about how your application is being used. Let's examine some of the more common contextual headers:

- **Referer**: This header contains the name of the page that directed the user to the current request. It is really helpful if you want to know from which page on your application an API request came from. If your API is not meant to be used by a browser, perhaps it is less important. Yes, it is misspelled. The internet is not perfect.

  And now for a bit of trivia knowledge: RFC 1945 was published in 1996 as a specification of the HTTP/1.0 protocol. The team that published it included none other than Tim Berners-Lee (that is, the inventor of the World Wide Web). *Section 10.13* introduced the `Referer` header, but it was inadvertently misspelled in the specification! Subsequent specifications and implementations have adopted this misspelling, and it has stuck with us for almost 30 years. If nothing else, it is a certain warning about the use of spellcheck: `https://datatracker.ietf.org/doc/html/rfc1945#section-10.13`.

- **Origin**: This header is similar to `Referer`. While the `Referer` header will generally include the full path of where the request originated, the `Origin` header is just a URL in the form of `<scheme>://<hostname>:<port>` without the path. We will look at how we can use it to protect our application from CORS attacks in *Chapter 7, Dealing with Security Concerns*.

- **User-Agent**: This header is almost always sent by every HTTP client. It identifies the type of application that is accessing your API. Commonly, it is a browser, but it could also be `curl`, a Python library, or a tool such as Postman or Insomnia.

- **Host**: Back in *Chapter 3, Routing and Intaking HTTP Requests*, we learned how to do host-based routing with *virtual hosts*. This is accomplished by reading the `Host` header. While `Origin` is the domain where the request is coming from, `Host` is where it is going to. Usually, we know this information ahead of time. Except, sometimes, we either have a dynamic host (such as a wildcard subdomain) or multiple domains pointing to one application.

- **Forwarded headers**: This encompasses both `Forwarded` headers and a bunch of `X-Forwarded-*` headers. Generally, when you see a header that starts with `X-`, it means it is a header that has come into common practice and usage, but its implementation is not necessarily standard.

What are these headers? They contain details about the web request and are used by intermediary proxies (such as Nginx or Cloudflare) to pass along relevant details about the request. The most common is `X-Forwarded-For`. This is a list of all of the IP addresses from the originating request to the current server that handled the request (this is not the same as a traceroute). This is incredibly helpful and important when trying to identify a request by an IP address.

> **Important Note**
>
> As with *all* headers and input data, you should *never* assume that incoming user data is accurate and harmless. It is very simple for someone to spoof headers. As always, we need to be cautious when reading headers and not just take them at face value.

Sanic extracts header data for us

Sanic will automatically extract information about the request from headers and place them onto easily accessible attributes in the `Request` object. This makes them very useful to access when needed. Here is a reference for some of the common attributes that you might encounter:

Request property	HTTP header used to generate
request.accept	Accept
request.forwarded	Forwarded
request.host	Host
request.id	X-Request-ID (can be configured)
request.remote_addr	Forwarded or X-Forwarded-For (depends on additional configuration)
request.token	Authorization

Table 4.1 – Extracted header data

> Tip
>
> Sometimes, it might be confusing to know when to use `request.ip` and when to use `request.remote_addr`. The former property will always be set and will always return the IP address of the client that is connecting to it. This might not actually be what you want. If your application is behind a proxy server, and you need to rely on `X-Forwarded-For`, then, most likely, the attribute you want is `request.remote_addr`.

Headers as multi-dict

Headers are stored in Sanic as a **multi-dict**. This is a special data type that will operate both as a one-to-one key-to-value dictionary and a one-to-many key-to-value dictionary. To illustrate the point, here is what both of these dictionaries will typically look like:

```
one_to_one = {
    "fruit": "apples"
}


one_to_many = {
    "Fruit": ["apples", "bananas"]
}
```

The header object in Sanic functions as both of these simultaneously. Moreover, it regards the keys as case-insensitive. *Did you notice, in the last example, that the keys are different cases?* Using standard dictionaries, the following would be false:

```
"fruit" in one_to_one and "fruit" in one_to_many
```

However, because the HTTP specification allows for HTTP headers to be case-insensitive, the Sanic header object is also case-insensitive. *But how does it handle the issue between one-to-one and one-to-many?*

Again, the HTTP specification allows for multiple identical headers to be concatenated without overriding one another. Sanic opts for this special data type to be standards-compliant. If you do nothing special and merely treat the header object as a regular Python `dict` in your application, it will work just fine. You might not ever even notice that it is not a regular dictionary. However, you will only ever access the first value passed to it for each header. If you need to support multiple values for the same header, you can access the full *list* of values.

Consider the following example:

```
@app.route("/")
async def handler(request):
    return json(
        {
            "fruit_brackets": request.headers["fruit"],
            "fruit_get": request.headers.get("fruit"),
            "fruit_getone": request.headers.getone("fruit"),
            "fruit_getall": request.headers.getall("fruit"),
        }
    )
```

Now, let's hit this endpoint with multiple `Fruit` headers:

```
$ curl localhost:7777/ -H "Fruit: apples" -H "Fruit: Bananas"
{
  "fruit_brackets": "apples",
  "fruit_get": "apples",
  "fruit_getone": "apples",
  "fruit_getall": [
    "apples",
    "Bananas"
  ]
}
```

Using either square brackets or the `.get()` method provides us with `apples` because that was the first `Fruit` header that was sent. A more explicit usage would be to use `.getone()`. Alternatively, we can use `.getall()` to return the full list of `Fruit` header values. Again, the case does *not* matter for header keys. For values, however, it does. Notice how, in our example, `Fruit` became `fruit`, but `Bananas` did not change its case at all.

Getting information from cookies (yum!)

Building a web application without cookies is like ending a meal without cookies. Sure, it could be done. *But why would you want to?* Given a choice, pick the cookies.

Jokes aside, cookies are, of course, an extremely important topic to consider. They are the backbone of many of the rich user experiences of web applications. Additionally, cookies are inherently full of potential security pitfalls. Generally, the security issues are more of a concern when we talk about setting cookies (*Chapter 5*, *Building Response Handlers*) and securing our web applications (*Chapter 7*, *Dealing with Security Concerns*). Here, we are mainly interested in how to access cookies so that we can read the data from them.

A web cookie is a specialized HTTP header: `Cookie`. This header contains a structured set of data defined by *RFC 6265*, *Section 5.4*: `https://tools.ietf.org/html/ rfc6265#section-5.4`. The incoming cookie from a request is treated, in Sanic, like a regular dictionary:

1. In order to get an explicit look at how cookies are structured, set up a debug handler as follows:

    ```
    @app.route("/cookies")
    async def cookies(request):
        return json(request.cookies)
    ```

2. Now, we will send some cookies using `curl`:

    ```
    $ curl localhost:7777/cookie -H "Cookie: name=value;
    name2=value2; name3=value3"
    {
      "name": "value",
      "name2": "value2",
      "name3": "value3"
    }
    ```

As you can see, the data is just a simple key/value dictionary. Therefore, accessing cookies should be very straightforward. Like other forms of data, it is, of course, advisable to treat them with suspicion. These values are not immune to tampering and can easily be spoofed. Nonetheless, they are an important part of the web, especially if your application needs to support a frontend UI.

While using cookies will be an invaluable source of data for your applications, the primary method for users to pass information will come in other forms. Next up, we look at the other methods of passing data from a web client to a web server.

Reading forms, query arguments, files, JSON, and more

Now that we know about pulling input from the path and the headers, we will turn our attention to more classic types of passing input values. Typically, we think of request data as being those bits of information that come from the request body. However, before we turn to the request body, we still have one more item in the first line of the HTTP request to examine: **Query arguments**.

Query arguments

As a reminder, the first line of an HTTP request looks like this:

```
GET /stalls/2021-07-01?type=fruit HTTP/1.1
```

If you have previous web experience, you might know that a URL can have a section of arbitrary parameters separated from the rest of the path by a question mark (?). These are known as query arguments (or parameters), follow in the form of key=value, and are concatenated with an ampersand (&). Sometimes, they are called *parameters*, and sometimes, they are called *arguments*. Here, we will call them arguments since this is what Sanic opts for to be able to distinguish them from path parameters.

Query arguments are very simple to use, and we can gain access to them on our request instance:

```
@app.route("/")
async def handler(request: Request):
    print(request.args)
    return text(request.args.get("fruit"))
```

We can send query arguments to our endpoint using curl, as follows:

```
$ curl localhost:7777\?fruit=apples
apples
```

> **Important Note**
> You might have noticed that my curl command included \ ? instead of just ?. This is a necessary pattern in some command-line applications since ?, by itself, could have a different meaning. It just as well could have been wrapped in quotes: curl "localhost:7777?fruit=apples", but I prefer to remove the quotes and opt for character escaping.

The usage seems simple enough, right? Well, not so fast. The obvious next question is *what happens when the key is repeated? Or, what happens when we want to pass an array of data?*

There is *no* single standard way to pass array data on the internet inside query arguments. Several methods do exist:

- `?fruit[]=apples&fruit[]=bananas`
- `?fruit=apples,bananas`
- `?fruit=[apples,bananas]`
- `?fruit=apples&fruit=bananas`

The first three approaches were rejected by Sanic, which has, instead, opted to implement the fourth option. A quick look at the three rejected models will explain why the chosen model makes sense, and how we can use it going forward:

- *First*, `fruit[]` is a strange construct that is not obvious to newcomers. In fact, it is a hijacking and alteration of the key. Yuck, no thank you.

- *Second*, `fruit=apples,bananas` *seems nice, but what if we wanted to just pass an* `apples,bananas` *string and not actually separate them?* Hmm, this does not seem possible. Pass.

- *Third*, `fruit=[apples,bananas]` seems better, but it is again somewhat awkward and nonintuitive. Additionally, it suffers the same ambiguity problem. *Is* `apples,bananas` *a single string or two items?*

Furthermore, the second and third options suffer another problem in terms of how to handle duplicate keys. *Should they take the first? The last? Merge? Error?* Again, there is no consensus and different servers will handle this differently.

The most reasonable approach seems to be the fourth, which can handle all of these problems. Keep it simple: we have a key and a value, nothing more. If there are duplicate keys, we treat them as a list append. There are no surprising losses of data, no errors, and data integrity is maintained.

In our last example, we printed the value of `request.args` to the console. Here is the output:

```
{'fruit': ['apples']}
[INFO] [127.0.0.1:53842]: GET http://
localhost:7777/?fruit=apples   200 6
```

Wait?! A list? I thought it was a single value: apples. At least that is what the response gave us. Query arguments are a special dictionary that contains lists. They have a unique .get() method that will only fetch the first value from that list. If you want all of the elements, use .getlist():

```
@app.route("/")
async def handler(request: Request):
    return json(
        {
            "fruit_brackets": request.args["fruit"],
            "fruit_get": request.args.get("fruit"),
            "fruit_getlist": request.args.getlist("fruit"),
        }
    )
```

Now, when we hit this endpoint, we can see what the values are:

```
$ curl localhost:7777\?fruit=apples\&fruit=bananas
{
  "fruit_brackets": ["apples","bananas"],
  "fruit_get": "apples",
  "fruit_getlist": ["apples","bananas"]
}
```

Another point worth mentioning is that request.args is not the only way to look at these key/value pairs. Additionally, we have request.query_args, which is just a list of tuples of all the pairs that were passed. The preceding request would look something like this:

```
request.query_args == [('fruit', 'apples'), ('fruit',
'bananas')]
```

Of course, a data structure such as this can easily be turned into a standard dictionary if desired. Just be careful because you will lose out on duplicate key data and be left only with the last of each duplicated key:

```
>>> dict([('fruit', 'apples'), ('fruit', 'bananas')])
{'fruit': 'bananas'}
```

Next, we will look at the forms and files that can be accessed and used in a similar way to query arguments.

Forms and files

By learning how we can pull data from the query arguments, we have also, inadvertently, learned how to get both form data and uploaded file data! That is because query arguments, forms, and files all operate identically. To prove this, we will set up a couple of endpoints, just as we have done before, and see what happens:

```python
@app.post("/form")
async def form_handler(request: Request):
    return json(request.form)
```

```python
@app.post("/files")
async def file_handler(request: Request):
    return json(request.files)
```

Next, we will test the form handler:

```
$ curl localhost:7777/form -F 'fruit=apples'
{"fruit":["apples"]}
```

```
$ curl localhost:7777/form -F 'fruit=apples' -F 'fruit=bananas'
{"fruit":["apples","bananas"]}
```

Just like before, we can see that it looks like a dict object with a list. Well, that's because it is. However, it will still behave in the same way as request.args. We can use the .get() method for the first item, and .getlist() for all of them in a list:

```python
assert request.form.get("fruit") == "apples"
assert request.form.getlist("fruit") == ["apples","bananas"]
```

And, of course, we will see the same result with files:

```
$ curl localhost:7777/files -F 'po=@/tmp/purchase_order.txt'
{
  "po": [
    ["text\/plain","product,qty\napples,99\n","purchase_order.
txt"]
  ]
}
```

We might want to take a closer look at this one to see what it is doing.

When you upload a file to Sanic, it will convert the file into a `File` object. The `File` object is really just a named tuple containing the basic information about the file. If we execute `print(request.files.get("po"))`, we should see an object that looks like this:

```
File(
    type='text/plain',
    body=b'product,qty\napples,99\n',
    name='purchase_order.txt'
)
```

> Tip
>
> *Are you familiar with named tuples?* They are a really great tool for modeling concise objects. I highly recommend using them since they behave as tuples but with the convenience of having dot notation to access specific properties. They are great to use instead of dictionaries, as long as you do not need to modify their contents. This is why Sanic uses them here as file objects. It is a convenient small structure that is easy for us, as developers, to work with while keeping some safety around the data so that it is not accidentally corrupted.

Consuming JSON data

Arguably, the most important type of request data is JSON. Modern web applications have embraced and clung to serializing and transmitting data with JSON because of its simplicity. It supports basic types of scalar values, is easy for humans to read, and is easy to implement and widely supported in many programming languages. It is no wonder that it is the default methodology.

Therefore, it should come as no surprise that Sanic makes it very easy:

```
@app.post("/")
async def handler(request: Request):
    return json(request.json)
```

Our request JSON is converted into a Python dictionary:

```
$ curl localhost:7777 -d '{"foo": "bar"}'
{"foo":"bar"}
```

Now we have seen all of the typical ways to access data in a single request. Next, we will learn about how data can alternatively be streamed to Sanic in multiple chunks.

Getting streaming data

The term *streaming* has become somewhat of a buzzword. Many people, even outside the tech industry, use it all the time. The word—and, more specifically, the actual technological concept that it represents—has become an important part of society as the consumption of media content continues its march to the cloud. *What exactly is streaming?* For those who are not entirely clear about what this term means, we will spend a brief moment attempting to understand it before moving on.

Streaming is the act of sending data in multiple, consecutive chunks from one side of an open connection to the other. One of the core foundations of the HTTP model is that there is a request followed by a response after a connection has been established between the client and the server. The client sends a complete HTTP request message and then waits for the server to send back a complete HTTP response message. It looks like this:

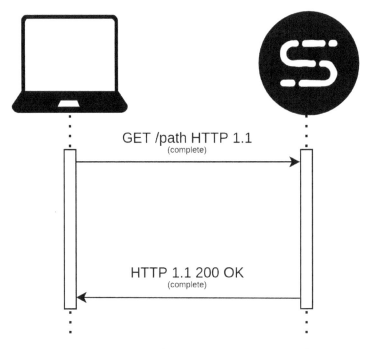

Figure 4.1 – A normal HTTP request/response cycle

I like to think of these as finite transactions. Both the request and the response have definite and known endpoints. Up until now, these finite requests are what we have been looking at. A request comes in, the server does something to process it, and a response goes out. It is important to note that both the request and the response are sent as a whole within a single block.

One header that we did not discuss earlier is the `Content-Length` header. This header can be found on both requests and responses. The actual specification regarding when it *should* be sent versus when it *must* be sent is really beyond the scope of this discussion. Sanic will take care of providing this for us when necessary. I bring it up here because this header is exactly what it purports to be: the length of the content in an HTTP message. This tells the recipient that there is a message of a certain length that is being transmitted. And it is important here because the known length of a message cannot necessarily be computed when the request headers are sent.

What happens if there is a large amount of data to be sent that might overwhelm a single connection or if the data being sent is not 100% available when the connection opens? Streaming is a methodology for one side of the connection to tell the other side it is transmitting some bytes but that it is not yet complete. The connection should be held open so that more data can be sent. The way that this interaction happens is by replacing the `Content-Length` header with a `Transfer-Encoding: chunked` header. In this way, one side of the connection can tell the other that it should continue to receive data until notified that the stream of data is closed.

When most laypeople hear the term *streaming*, their immediate thought goes to streaming media, such as movies or music. They might describe the concept as the consumption of the media before it is fully downloaded. And this is correct. Streaming is the sending of data in multiple *chunks* instead of sending it all at once. This is extremely efficient and can reduce the overall resource overhead. When supported, it allows the receiving side to start processing that data if it desires instead of blocking and waiting for it to be completed. So, when you go to watch your favorite movie, you can start watching it without having to wait for the entire file to download.

However, streaming does not only apply to media, and it is not only done by a server. There are two basic flavors we are concerned with: request streaming and response streaming. Here is what those flows look like:

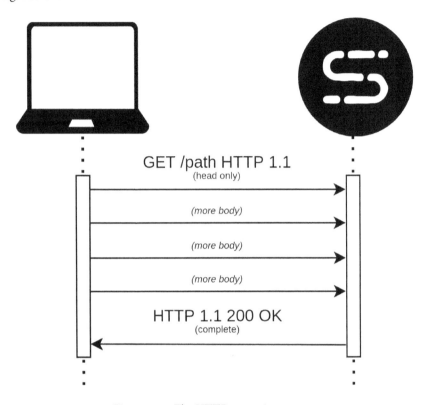

Figure 4.2 – The HTTP streaming request

In *Figure 4.2*, we can see what a streaming request looks like. Once the HTTP connection is opened, the client starts sending data. But it does not send the message all at once. Instead, it breaks the message up into chunks, sending each chunk of bytes on its own:

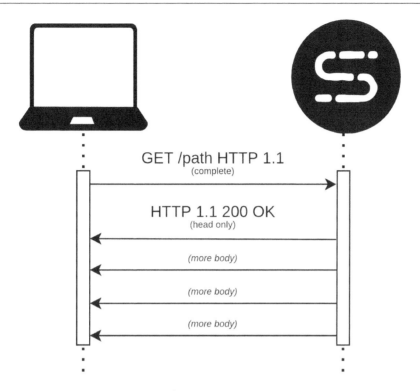

Figure 4.3 – The HTTP streaming response

Essentially, the streaming response in *Figure 4.3* is the reverse of the streaming request. The request is sent in its entirety, but the server decides to send the response in chunks until it is complete. When someone talks about streaming media, they are referring to the response stream. We will discuss this option in more detail in *Chapter 5, Building Response Handlers*, when we talk about different types of responses.

Currently, we are concerned with learning about request streaming, as depicted in *Figure 4.2*. It should definitely be noted that between the two, this is by far the lesser utilized feature. When you search for *streaming HTTP* on the internet, you are likely to find less information on it. Nonetheless, it can be a powerful tool in the right situation.

So, first, we will ask: when should we think about using request streaming? One potential use case is if a client wants to *warm up* the HTTP connection. Let's say you are building a stock trading platform. The latency between the frontend UI and the backend server is critically important. The difference in milliseconds has potential financial impacts. Your task is to get the data from the frontend as quickly as possible. The solution is to initiate the POST request as soon as the user clicks on the input box. Simultaneously, the frontend UI opens the HTTP connection with a Transfer-Encoding: chunked header, signaling that more data is coming.

So, while the user is typing in their values, we have already performed the operations and suffered any overhead that is related to the opening of a connection. Now the server is on alert waiting for data to come as soon as the user hits the *Enter* button.

What might this endpoint look like? Take a look at the following:

```python
async def read_full_body(request: Request):
    result = ""
    while True:
        body = await request.stream.read()
        if body is None:
            break
        result += body.decode("utf-8")
    return result


@app.post("/transaction", stream=True)
async def transaction(request: Request):
    body = await read_full_body(request)
    data = ujson.loads(body)
    await do_transaction(data)
    return text("Transaction recorded", status=201)
```

Let's point out a few important parts, one at a time:

1. We need to tell Sanic that we are going to be streaming the request here. There are two options: passing `stream=True` in the route definition or using the `@stream` decorator. They work in the same way, so it is more a matter of personal choice:

    ```python
    from sanic.views import stream


    @app.post("/transaction", stream=True)
    async def transaction(request: Request):
        ...


    # OR


    @app.post("/transaction")
    @stream
    ```

```
async def transaction(request: Request):
    ...
```

2. There should be some sort of a loop that continues to read from the stream until it is complete. *How do we know it is complete?* There will be an empty read from the stream. If you skip the `if body is None` line, you could end up crashing your server as it gets stuck in an infinite loop.

    The data when read is a `bytes` string, so you might want to convert that into a regular `str` value, as we do here:

    ```
    result = ""
    while True:
        body = await request.stream.read()
        if body is None:
            break
        result += body.decode("utf-8")
    ```

    It is important to note that, in this example, we are reading the body by completely draining it before continuing with processing the request. Another alternative might be to take those bytes and write them to something else that can consume and act upon them immediately. In just a moment, we will see an example that does this.

3. You need to decode the body yourself. In regular requests, if you send JSON data, Sanic will decode it for you. However, here, all we have are the raw bytes (converted into a string). If we need further processing, we should do it ourselves. In our example, we use `ujson.loads`, which ships with Sanic as a speedy way to convert the JSON into a Python `dict` type.

Our example works because we are expecting a single *delayed* input from the client. Another important place where you might use this is with file uploads. If you are expecting large file uploads, you might want to start reading and writing the bytes as soon as they are received.

The following shows an example of how to do that:

```
@app.post("/upload")
@stream
async def upload(request: Request):
    filename = await request.stream.read()
    async with aiofiles.open(filename.decode("utf-8"),
mode="w") as f:
        while True:
```

```
        body = await request.stream.read()
        if body is None:
            break
        await f.write(body.decode("utf-8"))
    return text("Done", status=201)
```

Notice that the loop here looks very similar to the last one. The concept is the same: loop until there is nothing left to read. The difference is that instead of writing the data to a local variable, we are using the `aiofiles` library to asynchronously write bytes to the file.

Why would you want to do this? Well, the biggest reason would be efficiency and memory utilization. If you use the regular `request.files` accessor to read file data, then you are effectively reading the entire contents before doing anything with them. This could be a lot of memory usage if there are large files at play. By reading and writing in chunks, we keep the buffers small.

This chapter has focused entirely upon different methods of reading data. We know we can access it from the body, files, form data, streams, and query arguments. All of these mechanisms on their own lack one critical component: validation.

Validating data

Next, we are going to take our first glimpse at the security-related topics in this book. We cover additional concepts later, in *Chapter 7, Dealing with Security Concerns*. However, this is not a security book. Unfortunately, there is too much material to be able to cover it all in this book. There are too many risks and too many potential mitigation measures for our one chapter dedicated to security. Therefore, instead, we will touch upon the concepts generally for those who are unfamiliar with them, and then show several ways to combat the issues in Sanic.

The first of those topics is data validation. If you have been around the web, you will understand what I am saying, and the *why* will be obvious to you. You are concerned with **SQL injection attacks** or **XSS attacks**. You know the potential threats posed by blindly accepting data and acting upon it. I trust that you already know this is a big *no-no* and are here to learn *how* to implement standard practices in Sanic. If the concept of data validation is completely foreign to you, I suggest you spend time searching other online materials regarding the security issues posed by attacks such as those mentioned earlier.

Web API security is not a singular approach. Data validation is only a small part of a much larger plan that you will need to protect your applications, resources, and users. In this section, our primary focus will be on the most common scenario in modern web applications: making sure that JSON data conforms to expectations. These techniques alone will not make your application secure from attacks. Please refer to *Chapter 7, Dealing with Security Concerns*, for more on this. Here, our goal is far more modest: when we expect a number, we get a number, and when we expect a UUID, we get a UUID.

If you recall in *Chapter 3, Routing and Intaking HTTP Requests*, we actually had our first brush with data validation. We were trying to make sure that the data received was from a known list of ice cream flavors. We are going to expand this concept here. There are *many* libraries out there that can do this for us. Some popular choices include `marshmallow`, `attrs`, and `pydantic`. Before we try and leverage an existing package, we are going to try to build our own validation library using Python's dataclasses.

It is good to remember why we are doing this. As we know, Sanic tries hard to not make decisions for developers. Data validation is one of the most critical components of an application, and it can vary wildly from one use case to the next. Therefore, the core Sanic project does not have a single method for doing this and leaves the choice to you: the developer.

Of course, there are a number of plugins out there that add validation, but we are going to take a crack at building one ourselves that will fit our needs. Ultimately, I hope this inspires some ideas in your own projects for you to take these principles and apply them to your own unique situations. This next section will deviate from Sanic and is more about Python programming in general. Ultimately, however, I think it is illustrative to see how Sanic tries to get out of your way to allow you to implement your own solutions and business logic, and only involve itself where needed.

With that said, let's get started.

Step 1—getting started and making a decorator

The first thing we need to do is create the framework that we are going to work in. To accomplish our goals, we are going to rely heavily upon decorators. This is a fantastic approach because it allows us to create per route definitions but also easily repeat our logic across the application, as needed. What we are after is something that looks like this:

```
@app.post("/stalls")
@validate
async def book_a_stall(request: Request, body: BookStallBody):
    ...
```

This looks like a super-clean interface. *What will this achieve?* Take a look at the following:

- **No repetition**: Rather than explicitly telling the `validate` function what to do, we are going to use some Python tricks to read `body: BookStallBody` from the handler signature.

- **Dependency injection**: Our `validate` function will need to inject a `body` argument. This means that we should have a clean data structure with exactly the information that we want and cast the types of data we expect them to be. If something is missing, it should raise an exception and cause a failure response.

- **Type annotations**: By annotating the `body` argument, we will have helpful features from `mypy` and our IDE to make sure our code is clean, consistent, and bug-free.

To begin, we want to create a decorator that will be capable of being callable or not callable. This will give us both `@validate` and `@validate()`, which will make our experience more flexible and easier as we expand our usage. We already saw an example of this in *Chapter 3, Routing and Intaking HTTP Requests*. Let's see what a minimal decorator looks like:

```
def validate(wrapped=None):
    def decorator(handler):
        @wraps(handler)
        async def decorated_function(request, *args, **kwargs):
            return await handler(request, *args, **kwargs)

        return decorated_function

    return decorator if wrapped is None else decorator(wrapped)
```

With this in place, we have a minimally viable decorator. Of course, it does not do anything useful yet, but we can begin to build our validation logic on top of this.

Step 2—reading the handler signature

The next thing we want to do is determine which parts of the request we want to validate. We are going to start with the JSON body. In our target implementation, we want to control this using the handler signature. However, we do have an alternative approach. For example, we could try for this:

```
@app.post("/stalls")
@validate(model=BookStallBody, location="body")
```

```
async def book_a_stall(request: Request, body: BookStallBody):
    ...
```

Arguably, this is a *much* easier decorator to build. In this version, we are explicitly telling the `validate` function that we want it to look in the `Request` body and validate against a model called `BookStallBody`. However, if we also want type annotations, we end up with duplicated code since we need to put the model in a typed function argument. We are not going to let the difficulty scare us away! After all, we know this decorator will be used all over our application. Building a better version upfront will help us down the road as we reuse and expand the implementation.

So, how do we get the model and the location information? We are going to use Python's `typing` module that comes with the standard library. We need to be very careful here. When dealing with decorators, we need to remember that different layers get executed at different times. Since we are evaluating the handler, we only want to do this *once*. If we set this up wrong, we might end up executing the setup code on *every single request*! We will try to avoid that.

This is where we are at now:

```
import typing


def validate(wrapped=None):
    def decorator(handler):
        annotations = typing.get_type_hints(handler)
        body_model = None

        for param_name, annotation in annotations.items():
            if param_name == "body":
                body_model = annotation


        # Remainder of decorator skipped
```

We inspect the handler and loop over the parameters that are defined inside it. If there is a parameter that is called body, then we grab its annotation and save it for later use.

One potential downside to this approach is that we are boxing ourselves in by *only* allowing our validations to be on a parameter called body. *What if we have a URL that has a dynamic path variable such as* /path/to/<body>? *Or do we just simply want to call the variable something else?* Let's make the decorator slightly more flexible by introducing body_arg:

```
def validate(wrapped=None, body_arg="body"):
    def decorator(handler):
        annotations = typing.get_type_hints(handler)
        body_model = None

        for param_name, annotation in annotations.items():
            if param_name == body_arg:
                body_model = annotation

        @wraps(handler)
        async def decorated_function(request, *args, **kwargs):
            nonlocal body_model

            return await handler(request, *args, **kwargs)

        return decorated_function

    return decorator if wrapped is None else decorator(wrapped)
```

So, by moving the name of the body argument to body_arg, we have the flexibility to rename it if we want to.

Step 3—modeling

The next critical piece is our model. This could be a place where we add in a prebuilt library—for example, one of the packages mentioned earlier. Of course, I suggest you take a look at them.

There are many devoted contributors that have spent a lot of time building, testing, and supporting these packages who will cover far more use cases than our simple example. However, since we are still learning here, we will continue building our own validation logic on top of the data classes.

Let's create a basic payload that we might expect in our endpoint:

```python
from dataclasses import dataclass
from enum import Enum, auto


class ProductType(Enum):
    def _generate_next_value_(name, *_):
        return name.lower()

    FRUIT = auto()
    VEGETABLES = auto()
    FISH = auto()
    MEAT = auto()


class ValidatorModel:
    def __post_init__(self):
        ...


@dataclass
class Product(ValidatorModel):
    name: str
    product_type: ProductType


@dataclass
class BookStallBody(ValidatorModel):
    name: str
    vendor_id: UUID
    description: str
    employees: int
    products: List[Product]
```

Okay, so there is not too much new here. We are defining some models using Python's data classes. I encourage you to go look them up if you are unfamiliar with them. In brief, they are type-annotated data structures that will be super easy for us to work with.

One problem with them is that the type annotations are *not* enforced at runtime. Even though we say that `BookStallBody.vendor_id` is a UUID type, Python will happily inject a Boolean or another kind of value there. This is where the `ValidatorModel` class comes in. We are going to add some simple logic to `dataclass` to make sure it is populating with the correct data type.

Another nice trick added to this simple structure is that we are defining `ProductType` as Enum. By defining `_generate_next_value_`, we are forcing the values of each Enum value to be a lowercase string value of the key. For example, consider the following:

```
assert ProductType.FRUIT.value == "fruit"
```

> **Tip**
>
> Whenever your application is dealing with an ID of any kind, you should try and avoid passing the sequential ID record to it from your database. Many common databases increment the row number every time you insert a record. If your API relies on that ID, you are inadvertently broadcasting information about the state of your application to the world. Stick with UUIDs or another form that will add some obscurity for client-facing applications. Do not let your database IDs leave your server.

Step 4—model hydration

Ultimately, we want to be able to send a JSON request to our endpoint that looks like this:

```
{
    "name": "Adam's Fruit Stand",
    "vendor_id": "b716337f-98a9-4426-8809-2b52fbb807b3",
    "employees": 1,
    "description": "The best fruit you've ever tasted",
    "products": [
        {
            "name": "bananas",
            "product_type": "fruit"
        }
    ]
}
```

Therefore, our goal is to turn this nested structure into Python objects. Data classes can get us part of the way there. What is missing is the specific typecasting and nesting. The following is what our `ValidatorModel` class will provide for us:

```python
class ValidatorModel:
    def __post_init__(self):
        for fld in fields(self.__class__):
            existing = getattr(self, fld.name)
            hydrated = self._hydrate(fld.type, existing)

            if hydrated:
                setattr(self, fld.name, hydrated)
            elif type(existing) is not fld.type:
                setattr(self, fld.name, fld.type(existing))

    def _hydrate(self, field_type, value):
        args = get_args(field_type)
        check_type = field_type

        if args:
            check_type = args[0]

        if is_dataclass(check_type):
            if isinstance(value, list):
                return [self._hydrate(check_type, item) for
    item in value]
            elif isinstance(value, dict):
                return field_type(**value)

        return None
```

It might appear as though there is a lot going on here, but it is really quite simple. After a model instance is created, we loop through all of its fields. There are now two options: either the field annotation is another dataclass, or it's something else. If it is something else, then we just want to ensure that we cast it to the new type.

If we are dealing with a dataclass, then we have two more options we need to determine. Either it is a single item or a list of items. If it is a list, then we simply need to make sure we loop over all of the values and try to hydrate each individual item.

Admittedly, this will not cover all use cases. However, since we are creating our own solution, we only care that it covers the cases we need and that it is relatively simple to maintain if we need to add more complexity in the future.

This solution will do that for us.

Step 5—performing validations

Now that our models are capable of handling nested logic and converting all of our values into their desired types, we need to hook it back up to our decorator.

This is where we currently stand:

```
def validate(wrapped=None, body_arg="body"):
    def decorator(handler):
        annotations = get_type_hints(handler)
        body_model = None

        for param_name, annotation in annotations.items():
            if param_name == body_arg:
                body_model = annotation

        @wraps(handler)
        async def decorated_function(request, *args, **kwargs):

            if body_model:
                kwargs[body_arg] = do_validation(body_model,
    request.json)

            return await handler(request, *args, **kwargs)

        return decorated_function

    return decorator if wrapped is None else decorator(wrapped)
```

The important changes are the following lines:

```
if body_model:
    kwargs[body_arg] = do_validation(body_model, request.json)
```

This will convert our raw JSON request data into usable (and well-annotated) data structures. If there is a failure within a data type, an exception should be raised. *So, what would that look like?* Take a look at the following:

```
from sanic.exceptions import SanicException


class ValidationError(SanicException):
    status_code = 400


def do_validation(model, data):
    try:
        instance = model(**data)
    except (ValueError, TypeError) as e:
        raise ValidationError(
            f"There was a problem validating {model} "
            f"with the raw data: {data}.\n"
            f"The encountered exception: {e}"
        ) from e
    return instance
```

If our data class models cannot cast a value into the expected type, then it should raise `ValueError` or `TypeError`. We want to catch either one of them and convert it into our own `ValidationError` for two reasons. First, by subclassing `SanicException`, we can give the `status_code` exception, and when that exception is raised, Sanic will automatically know to return a `400` response. In *Chapter 9, Best Practices to Improve Your Web Applications*, we will discuss exception handling in more detail, which is another important consideration. For now, just know that Sanic will give us some exception handling out of the box in both debug mode and regular mode.

Taking it to the next level with third-party packages

The input validation from the preceding section was admittedly a bit thin. It works well for our very limited use cases, but lacks some of the richness that can be achieved from a proper package. In the future, if your projects require some customized validation logic, then, by all means, use what was started to launch your project.

However, we are going to switch our mode here. Instead of using plain vanilla data classes and our custom ValidatorModel class, we are going to use a third-party package. We will keep the rest of what we built, so we are not entirely grabbing an off-the-shelf solution.

Let's see what it would be like if we used pydantic.

Validation with pydantic

pydantic is a popular package for creating models in Python. Generally, it plays very nicely with type annotations and even has a drop-in replacement for dataclasses. Therefore, we can take our previous example, change the dataclass import line, and remove ValidatorModel, and we have upgraded our capabilities:

1. We change our models to use the pydantic dataclass:

    ```
    from pydantic.dataclasses import dataclass
    ```

2. Remove the ValidatorModel class since it is no longer required:

    ```
    @dataclass
    class Product:
        name: str
        product_type: ProductType


    @dataclass
    class BookStallBody:
        name: str
        vendor_id: UUID
        description: str
        employees: int
        products: List[Product]


    @dataclass
    class PaginationQuery:
        limit: int = field(default=0)
        offset: int = field(default=0)
    ```

3. The only other change is to make sure that do_validation will raise the appropriate error message (there is more on exception handling in *Chapter 6, Operating Outside the Response Handler*):

```
def do_validation(model, data):
    try:
        instance = model(**data)
    except PydanticValidationError as e:
        raise ValidationError(
            f"There was a problem validating {model} "
            f"with the raw data: {data}.\n"
            f"The encountered exception: {e}"
        ) from e
    return instance
```

It is an almost identical solution. Please take a look at the full example in the GitHub repository. Now we have the full power of a proper library to handle much more complicated validation logic. Perhaps we should build out our decorator just a bit more to handle other types of input validation:

1. First, here is a model of what our expected query parameters will look like:

```
from dataclasses import field


@dataclass
class PaginationQuery:
    limit: int = field(default=0)
    offset: int = field(default=0)
```

2. Then, we extend the decorator to handle both the body and query parameters:

```
def validate(
    wrapped=None,
    body_arg="body",
    query_arg="query",
):
    def decorator(handler):
        annotations = get_type_hints(handler)
        body_model = None
```

```
            query_model = None


        for param_name, annotation in annotations.
items():
            if param_name == body_arg:
                body_model = annotation
            elif param_name == query_arg:
                query_model = annotation
```

3. Now that the variable models and argument names have been defined, we can put them to use when the request is executed:

```
        @wraps(handler)
        async def decorated_function(request: Request,
*args, **kwargs):
            if body_model:
                kwargs[body_arg] = do_validation(body_
model, request.json)
            if query_model:
                kwargs[query_arg] = do_validation(query_
model, dict(request.query_args))

            return await handler(request, *args,
**kwargs)

        return decorated_function

    return decorator if wrapped is None else
decorator(wrapped)
```

4. With this decorator, we can act on both the body argument and the query argument. The implementation between them looks remarkably similar. Now we can reuse our decorator in other situations:

```
@app.get("/stalls/<market_date:ymd>")
@validate
async def check_stalls(
    request: Request,
    query: PaginationQuery,
    market_date: date,
```

```
    ):
        ...
```

Now it is time for a little experiment. We started by validating against the request JSON. This was validated and injected as the `body` argument. Then, we saw that it was super easy to extend this to the `query` arguments with the `query` argument.

Your challenge now is to put the book down and see whether you can make a similar implementation for both regular forms and file upload validation. Take a look at the previous approaches, and also reference the `request.files` and `request.form` objects we talked about earlier in the book.

Summary

It is a fairly safe assumption that all web APIs require some input from users at some point. Even APIs that are read-only often might allow for filtering, searching, or paginating data. Therefore, to become proficient at building web applications in general, and Sanic applications specifically, you must learn the data tools at your disposal.

In this chapter, we covered a great deal of material. We learned how to pull data from headers, cookies, and the request body. When using headers, form data, query arguments, and file data, we discovered that these objects could operate as regular dictionaries or dictionaries of lists to be both compliant with HTTP standards and also usable for most regular use cases. Additionally, we saw that the request body itself could be sent as a single chunk or in multiple chunks.

However, perhaps the biggest takeaway is that reading data cannot and does not take a single path. As a reminder, Sanic provides the tools to build the most obvious solution for your needs. While many other projects could fill a similar discussion with the minutiae of how to implement form data retrieval in their specific API, much of our focus was on how to build solutions *with* Sanic, not *from* Sanic. It is a framework that tries not to get in the way.

For example, we learned that it was super simple to add both custom and *off-the-shelf* validation logic. Sanic did not tell us how to do it. Instead, it provided some conveniences to help make our business logic easier to build. Decorator logic gave us the flexibility to have reusable code across the application. Exception definitions can automatically catch and handle responses. Building applications with Sanic is much more about building well-structured Python applications.

Once the information has been gathered and validated, it is time to do something with it. This is the purpose of the next chapter, where we explore how to handle and, ultimately, respond to web requests.

5
Building Response Handlers

Up until this point, our applications have largely been reactive. We have worked on different parts of web applications to learn how to manage the incoming HTTP request. If we imagine the HTTP request/response cycle as a conversation, so far, we have been only listening. Our applications have been built to hear what the incoming client has to say.

Now, it is our turn to talk. In this chapter, we will begin to explore different facets of the HTTP response. Just as we began our learning of the HTTP request by looking at a raw request object, we will look at the raw response. It looks nearly identical and, by now, should be familiar. We will go on to explore some of the powerful tools that Sanic has to offer. Of course, there are mechanisms for JSON and HTML responses, which are probably the most popular types of content to be delivered on the web today. However, Sanic has an advantage by being an async framework: it is super easy to implement server-driven responses such as **websockets**, **Server-Sent Events (SSE)**, and streaming responses. In this chapter, we will also explore these main topics:

- Examining the HTTP response structure
- Rendering HTML content
- Serializing JSON content
- Streaming data

- Server-sent events for push communication
- Websockets for two-way communication
- Setting up response headers and cookies

Technical requirements

Some of our examples are going to start getting a little longer than what we have previously seen. For the sake of convenience, you might want to keep the GitHub repository handy as you read through this chapter. You can find it at `https://github.com/PacktPublishing/Web-Development-with-Sanic/tree/main/chapters/05`.

Examining the HTTP response structure

Back in *Chapter 3*, *Routing and Intaking HTTP Requests*, we looked at the structure of the HTTP request. When a web server is ready to send back a response, the format is very similar to what we have already seen. The HTTP response will look something like this:

```
HTTP 1.1 200 OK
Content-Length: 13
Connection: keep-alive
Content-Type: text/plain; charset=utf-8

Hello, world.
```

What we see is the following:

- The first line contains the HTTP protocol used, a status code, and a status description.
- Response headers are in the `key: value` format and are separated by a line break.
- There is a blank row.
- There is a response body.

We are not looking at this here because we need to know to build a web application. After all, building these response objects to a valid HTTP specification is precisely one of the reasons that we use web frameworks. Without them, building these blobs would be tedious and error-prone. Instead, it is helpful for us to review and understand what is happening so that we can increase our grasp of HTTP and web application development.

A lot of the structure is duplicative of what we have already learned.

The HTTP response status

If you compare the HTTP request and response objects, perhaps the most identifying difference is the first line. While the first request line had three distinct parts, it is easier to think of the response as only having two: the HTTP protocol in use and the response status.

Earlier in this book, we discussed the HTTP protocol (please refer to *Chapter 3, Routing and Intaking HTTP Requests*), so we will skip it here and focus on the response status. The response status is meant to be both a computer-friendly and human-friendly tool to let the client know what happened to the request. *Did it succeed? Was the request wrong? Did the server make a mistake?* These questions, and more, are answered by the HTTP response status.

If you have built a website in the past, you likely have a basic understanding of different response codes. Even people who have never built an application have surely, at some point, landed on a web page that said `404 Not Found` or `500 Internal Server Error`. These are response statuses. HTTP response statuses consist of a number and a description. The meanings of these numbers and the specific descriptions associated with them are defined in *Section 6* of *RFC 7231*: `https://datatracker.ietf.org/doc/html/rfc7231#section-6`.

To clarify, if you see the terms *response status*, *status code*, or *response code*, they are all describing the same thing. In general, I prefer to use *response status* to describe the general concept and *status code* when talking about the numeric value of the status. However, they are fairly interchangeable, and this book uses the terms interchangeably as well.

The three most common statuses are as follows:

- `200 OK`
- `404 Not Found`
- `500 Internal Server Error`

In general, Sanic will attempt to respond with the most appropriate status. If there is an unhandled error, you will likely get a `500` status. If the path does not exist, it will be a `404` status. And, if the server can respond properly, Sanic defaults to using `200`. Let's dig a little deeper to see how the statuses are organized.

Response groupings

The standard responses are grouped in series of 100s, as follows:

- **100s**: *Informational* – these are provisional responses with information about how the client should proceed.

- **200s**: *Successful* – these are responses that indicate the request was processed as expected.

- **300s**: *Redirection* – these are responses that indicate the client must take further action.

- **400s**: *Client error* – there are responses where it appears the client made a mistake in trying to access or proceed with some resource.

- **500s**: *Server error* – these are responses where the server made a mistake, and a response, as expected, could not be generated.

Beyond the big three responses, there are some other important responses that you should be familiar with:

Code	Description	Use case
201	Created	The endpoint successfully created a new resource; often, the response will include the new data and/or an ID that can be used to look it up.
202	Accepted	The application has taken the request and pushed it to a queue or background process for further operation.
204	No content	There is no body; this is typical in an OPTIONS request.
301	Moved permanently	The target resource is now located at a new permanent URI.
302	Found	The target resource is temporarily located at a different URI.
400	Bad request	The server is refusing to respond to the request because the client did something improperly.
401	Unauthorized	The request lacks valid authentication credentials to be allowed access.
403	Forbidden	The request is authenticated, but the server does not recognize valid authorization to proceed with the response.

Table 5.1 – Common status codes

> **Important Note**
>
> Another important error code that you should be familiar with is 502. This is particularly important if you are running a proxy server of some kind in front of Sanic. If you receive a 502 Bade Gateway error, the likely problem is that your Sanic server has crashed and is no longer serving requests. You should check your logs to check why it is not running any longer.

A response through exceptions

Many of Sanic's built-in exceptions are associated with a specific status code. This means that we can raise an exception, and Sanic will automatically catch that exception and provide an appropriate response with the proper status code. This makes it super convenient and simple to respond to.

For example, let's imagine that we are building a music player application. One of our endpoints allows users that are logged in to view their playlist. However, it is protected behind authentication, and only users with whom the playlist has been shared are able to access it. It looks something like this:

```
from sanic.exceptions import NotFound

@app.get("/playlist/<playlist_name:str>")
async def show_playlist(request, playlist_name: str):
    can_view = async check_if_current_user_can_see_playlist(
        request,
        playlist_name
    )
    if not can_view:
        raise NotFound("Oops, that page does not exist")
    ...
```

By raising NotFound, Sanic will automatically know that it should return a 404 Not Found response:

```
$ curl localhost:7777/playlist/adams-awesome-music -i
HTTP/1.1 404 Not Found
content-length: 83
connection: keep-alive
content-type: application/json
```

```
{"description":"Not Found","status":404,"message":"Oops, that
page does not exist"}
```

Additionally, we could extend this concept with our own custom exception handlers:

```
from sanic.exceptions import SanicException


class NotAcceptable(SanicException):
    status_code = 406
    quiet = True



@app.post("/")
async def handler(request):
    if "foobar" not in request.headers:
        raise NotAcceptable("You must supply a Foobar header")
    return text("OK")
```

In this example, by subclassing `SanicException`, we can associate the exception with a status code. Additionally, we set a class property of `quiet=True`. This is not necessary but might be desirable. What it means is that the exception and its traceback (the details about the type and location of an exception) will not appear in your logging. This is a particular feature of `SanicException`. It is helpful for exceptions that might be expected (but otherwise uncaught) in the regular course of your application.

Custom status

Just as we saw with HTTP methods, it is possible to make up your own status codes as long as they have three digits. I am not suggesting this is a *good* idea—merely pointing out that it is possible—and Sanic will let you do it, even though you probably should not. Creating your own status codes might confuse browsers or clients that are using your application. Throwing caution to the wind, we will try it by performing the following steps anyway, just to demonstrate that Sanic allows us to do it:

1. Add a new status type to an otherwise *private* variable (remember, it's just Python, so we can hack it if we want):

    ```
    from sanic.headers import _HTTP1_STATUSLINES


    _HTTP1_STATUSLINES[999] = b"HTTP/1.1 999 ROCK ON\r\n"
    ```

```
@app.get("/rockon")
async def handler(request):
    return empty(status=999)
```

Nice! Now let's see what happens.

2. Check the HTTP return, making sure to use -i so that we see the raw response:

```
$ curl localhost:7777/rockon -i
HTTP/1.1 999 ROCK ON
content-length: 0
connection: keep-alive
content-type: None
```

3. To wrap up, here is a fun little experiment and a quirk of the HTTP specification. Enter the following route into your application:

```
@app.get("/coffee")
async def handler(request):
    return text("Coffee?", status=418)
```

4. Now, query it using curl so that you can see the response (don't forget -i):

```
$ curl localhost:7777/coffee -i
```

Did you observe the output? To see that it is actually a part of the real HTTP specification, take a look at https://datatracker.ietf.org/doc/html/rfc2324#section-2.3.2 and https://datatracker.ietf.org/doc/html/rfc7168#section-2.3.3. For more RFC humor, I would also suggest taking a look at the protocol specification for IP over Avian Carriers: https://datatracker.ietf.org/doc/html/rfc2549. These documents are a reminder that programming should be fun.

Headers

The second part of the HTTP response is the same as the second part of the HTTP request: headers are arranged, one per line, in a key: value format. As before, the keys are case-insensitive and can be repeated more than once in the response.

One interesting thing to bear in mind is that when a web server responds with an informational status (series 100), it does *not* include headers. These responses are generally used only in the context of *upgrading* an HTTP connection to a websocket connection. Since this is a responsibility of the framework, we can safely ignore this, and just file it away as good information to have.

Generally, headers are pretty simple to use in Sanic. We will dig into them deeper later on, but for now, we need to keep in mind that we can simply pass a dictionary with values:

1. Add a `headers` argument with a dictionary of values into any response function. Here, we use `empty` because we are not sending a `body` response, just the headers:

    ```
    @app.get("/")
    async def handler(request):
        return empty(headers={"the-square-root-of-four":
    "two"})
    ```

2. Let's see what the response looks like using `curl`. Make sure that you use `-i` so that we can see the headers:

    ```
    $ curl localhost:7777/ -i
    HTTP/1.1 204 No Content
    the-square-root-of-four: two
    connection: keep-alive
    ```

    The astute mathematician looking at my example will notice that I got it only partially correct: `two` is not the only value. *How can we have duplicate header keys?* Since Python's regular dictionaries will not allow us to duplicate keys, we can use a special data type that Sanic offers us to do the job.

3. Using the same response as before, insert a `Header` object with two of the same keys, as shown in the following snippet:

    ```
    from sanic.compat import Header
    @app.get("/")
    async def handler(request):
        return empty(
            headers=Header(
                [
                    ("the-square-root-of-four", "positive
    two"),
                    ("the-square-root-of-four", "negative
    ```

```
    two"),
                    ]
            )
        )
```

4. Hopefully, we will now see our more mathematically correct response headers; the same key appears twice but with a different value each time:

```
$ curl localhost:7777/ -i
HTTP/1.1 204 No Content
the-square-root-of-four: positive two
the-square-root-of-four: negative two
connection: keep-alive
```

Usually, it will suffice to just use a regular `dict` object for assigning headers. Often, the need will not arise where you require two duplicate keys in the headers. However, since it is allowed by the HTTP protocol, you should know that it is possible if the need arose in the course of your application development.

The response body

The last part of the HTTP response is the body. It is arguably the most important part of this whole business we call HTTP. We can realistically state that the HTTP response body is what the whole driving force of the web is after: the sharing of content.

The remainder of this chapter will focus on some of the different and more popular ways that we can structure data in the HTTP response body. Whether it is HTML, JSON, or raw bytes, what we are about to dive into will be part of the cornerstone of every web application you build. First up is HTML content; here, we will explore methodologies for sending static HTML content and generating dynamic HTML content.

Rendering HTML content

The foundation of the web is HTML. It is the media that enables browsers to function; therefore, it is fundamental that a web server is capable of delivering HTML content. Whether building a traditional page-based application or a single-page application, HTML delivery will be necessary. In *Chapter 3, Routing and Intaking HTTP Requests*, we discussed how we could route web requests to our static files. If you have static HTML files, then this is a great option. *But what if you need to generate dynamic HTML for your application?*

Since there are numerous ways that this could be accomplished, we will take a look at some of the general patterns that could be used with Sanic.

Delivering HTML files

Generally, serving HTML content is a simple operation. We need to send back a response to the client with HTML text and a header that tells the recipient that the document should be treated as HTML. Ultimately, the raw HTTP response is going to look like this:

```
HTTP/1.1 200 OK
content-length: 70
connection: keep-alive
content-type: text/html; charset=utf-8


<!DOCTYPE html><html lang="en"><meta charset="UTF-
8"><title>Hello</title><div>Hi!</div>
```

Notice the critical HTTP response header: `content-type: text/html; charset=utf-8`. Sanic has a convenient response function that will appropriately set the response headers for HTML content:

```
from sanic import html, HTTPResponse


@app.route("/")
async def handler(request) -> HTTPResponse:
    return html(
        '<!DOCTYPE html><html lang="en"><meta charset="UTF-
8"><title>Hello</title><div>Hi!</div>'
    )
```

> **Note**
>
> As a quick side note, while the preceding example might be valid HTML, not all the following examples will be. Getting 100% accurate HTML semantics is not the aim of this book, so we might break a few rules.

Now let's imagine that we are building a music player application. The first thing that needs to happen when someone lands on our website is to log in. If that person already has an active session, we want them to navigate to the **What's new** page. In *Chapter 6, Operating Outside the Response Handler*, and *Chapter 7, Dealing with Security Concerns*, we will look at how to use middleware and integrate it with authentication. For now, we will assume our application has already dealt with authentication and authorization. It has stored those values as `request.ctx.user`:

```
@app.route("/")
async def handler(request) -> HTTPResponse:
    path = "/path/to/whatsnew.html" if request.ctx.user else "/
path/to/login.html"
    with open(path, "r") as f:
        doc = f.read()
    return html(doc)
```

Have you noticed a pattern so far? All we really need to do to generate HTML content with Sanic is basic string building! So, if we can inject values into a string with string interpolation, then we have dynamic HTML. Here's a simple demonstration:

```
@app.route("/<name:str>")
async def handler(request, name: str) -> HTTPResponse:
    return html(f"<div>Hi {name}</div>")
```

Instead of using `curl`, let's see what it looks like in a browser this time:

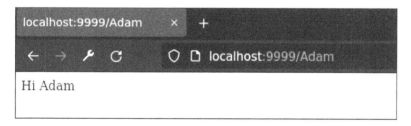

Figure 5.1 – A browser screenshot of the interpolated HTML

String interpolation of HTML is just a fancy way of saying templating.

Basic templating

In the past, I have presented at a couple of Python web conferences. While preparing my talks, I looked for tools that would make it super simple to generate a slide presentation. Since I am most comfortable working in my text editor, I was particularly interested in solutions that would translate markdown into slides. I found a tool called `remark.js`. If you want to learn more about remark, please visit `https://remarkjs.com/`.

In order to render slides from markdown, all I needed was an HTML file and some markdown text:

```
<!-- Boilerplate HTML here -->
    <textarea id="source">


class: center, middle


# Title


---


# Agenda


1. Introduction
2. Deep-dive
3. ...


---


# Introduction


    </textarea>
    <script src="https://remarkjs.com/downloads/remark-latest.
min.js">
<!-- Boilerplate HTML and here -->
```

This was super simple and exactly what I was looking for. However, there was a problem because my IDE did not know that the text inside `<textarea>` was markdown. Therefore, I had no syntax highlighting. Bummer!

The solution was quite simple really. I just needed a way to inject my markdown into the HTML file and serve that.

Here is a quick fix for the HTML:

```html
<!-- Boilerplate HTML here -->
    <textarea id="source">
__SLIDES__
    </textarea>
    <script src="https://remarkjs.com/downloads/remark-latest.
min.js">
<!-- Boilerplate HTML and here -->
```

Voila! An HTML template. Now, let's *render* it:

```python
from pathlib import Path


PRESENTATION = Path(__file__).parent / "presentation"


@app.get("/")
def index(_):
    with open(PRESENTATION / "index.html", "r") as f:
        doc = f.read()

    with open(PRESENTATION / "slides.md", "r") as f:
        slides = f.read()

    return html(doc.replace("__SLIDES__", slides))
```

Just like that, we have built a templating engine. The basic idea of any templating engine is that there is some protocol for telling the application how to convert and inject dynamic content. Python does this with its multiple forms of string interpolations. In my super-simple solution engine, all I needed to do was replace the __SLIDES__ value. I am sure you can start to formulate ideas regarding how you could build your own simple engine.

In fact, maybe you should try that now. Here is an HTML template:

```html
<p>
    Hi, my name is <strong>__NAME__</strong>.
</p>
<p>
```

```
    I am <em>__AGE__</em> years old.
</p>
```

Let's get started:

```
def render(template: str, context: Dict[str, Any]) -> str:
    ...


@app.get("/hello")
async def hello(request) -> HTTPResponse:
    return html(
        render("hello.html", {"name": "Adam", "age": 38})
    )
```

Now it is your turn to fill in the rest by building the rendering agent. Try to build a
render function to work with *any* variable names, not just name and age. We want
this to be reusable in more than just one location.

Using a templating engine

Of course, you do not always need to make your own templating engine. There are many
great choices that have already been built. Some popular template engines in Python are
Genshi, Mako, and *Jinja2*. But remember, all we really need to do is build a string. So, any
tools that you have that can do this will work. These packages can be thought of as fancy
versions of the Python format function.

They take strings and inject data into them to generate a bigger string. Any Python
templating tool that you pick up will work with Sanic. Specifically, regarding Jinja2,
there are some Sanic plugins already out there that make the interactions between Sanic
and Jinja2 super simple. Feel free to check them out in your own time. On a basic level,
templating with Jinja2 can be as lightweight as follows:

```
from jinja2 import Template


template = Template("<b>Hello {{name}}</b>")


@app.get("/<name>")
async def handler(request, name):
    return html(template.render(name=name))
```

And now you can see the result:

```
$ curl localhost:7777/Adam
<b>Hello Adam</b>
```

To move our templates out of Python and into their own HTML files, we can use Jinja2's `Environment` construct:

1. Create some HTML using the Jinja2 syntax. This will be saved as `index.html` in a `templates` directory. You can see the structure that we have used in the GitHub repository:

    ```html
    <!DOCTYPE html>
    <html>
        <head>
            <title>Adam's Top Songs</title>
        </head>
        <body>
            <h1>Adam's Top Songs</h1>
            <ul>
                {% for song in songs %}
                    <li>{{song}}</li>
                {% endfor %}
            </ul>
        </body>
    </html>
    ```

2. Now, set up `Environment` and attach it to our application context so that it is easily available throughout our application:

    ```python
    from pathlib import Path
    from jinja2.loaders import FileSystemLoader
    from jinja2 import Environment


    @app.before_server_start
    def setup_template_env(app, _):
        app.ctx.env = Environment(
            loader=FileSystemLoader(Path(__file__).parent /
    "templates"),
    ```

```
        autoescape=True,
    )
```

3. Finally, grab the template by its filename in our route handler, and inject some content into it:

```
@app.get("/")
async def handler(request):
    template = request.app.ctx.env.get_template("index.
html")
    output = template.render(
        songs=[
            "Stairway to Heaven",
            "Kashmir",
            "All along the Watchtower",
            "Black Hole Sun",
            "Under the Bridge",
        ]
    )
    return html(output)
```

With that all done, we should be able to visit our application in a web browser and see the rendered HTML.

> **Tip**
>
> When building with Sanic, you might have noticed how handy it is to have `auto_reload` enabled. Every time you hit the **Save** button, the application restarts and is available for you to test immediately. *Wouldn't it be great if the same were true when building HTML files?* There is a tool that does this, called **LiveReload**. Essentially, it injects some JavaScript into your HTML to make it listen to commands to refresh the page. When building that slide presentation, which I mentioned earlier, I made a LiveReload server so that I could keep the browser open alongside my IDE while I typed. Every time I hit **Save**, my browser refreshed, and I could see the rendered content without having to lift my fingers off the keyboard. If you are interested in more details regarding this topic, check out *Chapter 11, A Complete Real-World Example*.

Serializing JSON content

Next to HTML content, JSON is one of the most common forms of data transferred on the web. If you are building a **Single-Page Application** (**SPA**), (also known as a **Progressive Web Application** or **PWA**), it is likely that your backend server *only* or *mostly* returns JSON content. A common build pattern for a modern web application is to build a frontend user interface with a JavaScript framework powered by a backend server that feeds the frontend with dynamic JSON documents.

Choosing a serializer

Of course, the Python standard library ships with a JSON package that makes serializing Python objects to JSON strings (and the reverse) very simple. However, it is not the most performant implementation. In fact, it is quite slow. Many third-party packages have popped up to attempt to fix this problem. We will explore two of the common packages that are often used with Sanic.

When talking about response serialization, what we care about is the operation of the dumps() method. Each of these projects provides an interface with this method. To select a serializer, what we need to do is to set the dumps() method in one of two locations: at the response level or application-wide. We will learn how to do both shortly.

UJSON

UltraJSON (also known as **ujson**) is an alternative JSON implementation that is written in **C**. Because of its emphasis on performance, it was adopted as the default JSON tool for Sanic. If you do nothing else, this is the package that Sanic will use.

It includes some helpful encoder options such as encode_html_chars, ensure_ascii, and escape_forward_slashes. Consider the following example:

```
return json(
    {
        "homepage": request.app.url_for(
            "index",
            _external=True,
            _server="example.com",
        )
    },
)
```

When we access this endpoint, `ujson` will, by default, escape our slashes:

```
$ curl localhost:7777
{"homepage":"http:\/\/example.com\/index.html"}
```

We can use `functools.partial` to change the behavior:

```python
dumps = partial(ujson.dumps, escape_forward_slashes=False)


@app.get("/")
async def handler(request):
    return json(
        {
            "homepage": request.app.url_for(
                "index",
                _external=True,
                _server="example.com",
            )
        },
        dumps=dumps,
    )
```

By using the `dumps` keyword argument, we have told Sanic to use a different serializer. The result should be what we want:

```
$ curl localhost:7777
{"homepage":"http://example.com/index.html"}
```

If you do not want to use `ujson` in your projects, then you can force Sanic to skip the installation of `ujson`:

```
$ export SANIC_NO_UJSON=true
$ pip install --no-binary :all: sanic
```

While `ujson` is a great project that adds some much-needed performance to JSON string manipulation in Python, it *might* not actually be the fastest. Next, we will look at another relatively new package that attempts to bring performance to JSON manipulation.

orjson

A newer player to the game is **orjson**. It is written in **Rust** and claims to be the fastest alternative according to benchmarks. For this reason, many people like to swap out ujson for orjson.

An interesting thing to note about orjson is that it has built-in support for serializing common Python objects such as datetime.datetime and uuid.UUID. Since these are both very common when building web applications, it is very convenient to not have to think about how to handle these object types. Also, it should be noted that while the standard library and ujson return an str value, orjson returns a bytes string.

We can easily tell Sanic to use orjson everywhere:

```
import orjson

app = Sanic(__name__, dumps=orjson.dumps)
```

Using the dumps argument across an entire application can be very convenient. It is especially helpful when you realize that you might need to handle the serialization of more complicated custom objects. In the next section, we will learn how this sort of serialization can be accomplished.

Serializing custom objects

In the last two sections, you might have noticed there are two ways to override the default dumps method. The first is by changing a single response:

```
return json(..., dumps=orjson.dumps)
```

The second will apply to all routes globally:

```
Sanic(..., dumps=orjson.dumps)
```

Feel free to mix and match both the handler-specific method and the application-wide method to meet your application needs.

We briefly looked at two alternative packages. Of course, there are others. *So, how should you decide which package to use?* When deciding on an implementation, often, one of the biggest considerations is how to handle custom non-scalar objects. That is to say, how do we want objects that do not have an obvious and built-in mapping to JSON types (such as strings, integers, floats, Booleans, lists, and dictionaries) to behave when rendered to JSON? To make this point clear, consider the following example.

Let's say we have a `Thing` object. It looks like this:

```
class Thing:
    ...


data = {"thing": Thing()}
```

If we do nothing, serializing a `Thing` object will not be so straightforward, and JSON tools will usually throw an error because they do not know what to do with it. Without resorting to manual intervention, we can rely upon each of the tools' methodology to explicitly provide instructions when coming across a `Thing` object. We will consider each alternative to see how we can reduce `Thing` to a JSON-accessible object.

Perhaps, the simplest is `ujson`. Besides its performance, this happens to be one of my favorite features. If an object has a `__json__` method, `ujson` will call it when converting the object into JSON:

```
class Thing:
    def __json__(self):
        return json.dumps("something")


ujson.dumps(data)
```

Because of this functionality, when I am working on a project, one of the things that I often do is identify some base models for my objects and include a `__json__` method. *But what about the other tools?*

`orjson` allows us to pass a `default` function to the serializer. If it does not know how to render an object, it will call this. While `ujson` opts to handle this on the object/model, `orjson` opts to handle it in each individual serializer. Also, it should be noted that as of **version 4.2**, `ujson` supports the `default` keyword approach like `orjson`. The sky is really the limit in terms of the complexity you want to add. Since I am a fan of using `__json__` methods on my custom objects, we can achieve the same functionality with `orjson` as follows:

```
def default(obj):
    if hasattr(obj, "__json__"):
        return json.loads(obj.__json__())
    raise TypeError


orjson.dumps(data, default=default)
```

This might get a bit repetitive if you are constantly redefining the serializer method in the response handlers. Instead, maybe it is worth using the standard library to help. We can create a `partial` function with the `default` argument already populated:

```
from functools import partial

odumps = partial(orjson.dumps, default=default)

odumps(data)
```

The most cumbersome implementation is the standard library that requires you to pass a custom encoder (`CustomEncoder`) class. It is very similar to the `orjson` method, albeit with a little more boilerplate needed:

```
class CustomEncoder(json.JSONEncoder):
    def default(self, obj):
        return default(obj)

json.dumps(data, cls=CustomEncoder)
```

On seeing the preceding example, you should now be able to add the `__json__` approach to `CustomEncoder`.

No matter the project, you are very likely to come up against this issue. Having a standard and consistent way to handle non-scalar objects is important. Assess how you plan to build and look for meaningful patterns. Generally, I find this to be more of an important decision than raw performance. The incremental performance changes from one package to the next are likely not going to be as impactful as making a decision based upon how your application will be built and maintained.

For example, what if you need to render an integer that is larger than 64 bits? Both `ujson` and `orjson` have limitations where they will raise exceptions and not be capable of handling your data. However, the standard library implementation does have this capacity. As we stated earlier, make the right decisions that are the most obvious for your needs. But let's turn to some common practices and see what we might be able to learn from them.

Best practices

There are a number of common practices for what *typical* JSON responses look like. Of course, the content and your organization are something that will be determined by your application's needs. However, there is one common question you will find that is often discussed on developer forums: *How should I format an array of objects?*

Coming back to our earlier example, let's imagine that we are still building our music app. Now, we want to build an endpoint that lists out all of the available songs. Each individual song *object* will look something like this:

```
{
    "name": "Kashmir",
    "song_uid": "75b723e3-1132-4f73-931b-78bbaf2a7c04",
    "released": "1975-02-24",
    "runtime": 581
}
```

How should we organize an array of songs? There are two schools of thought: only using top-level objects and using whatever structure fits your data best. What we are talking about is the difference between the following code snippets:

```
{
    "songs": [
        {...},
        {...}
    ]
}
```

Consider this second snippet:

```
[
    {...},
    {...}
]
```

Why is there even a debate? And why do some people only strictly use top-level objects? There was a JSON security flaw in browsers uncovered in 2006 that would allow attackers to execute code based upon the second option, the top-level JSON array. For this reason, many people suggested that using the first structure was more secure.

While this is no longer a concern since the impacted browsers are long out of date, I still like the top-level object pattern. It still provides one critical benefit over the second option: flexibility without compromising on compatibility.

If our array of objects is nested inside a top-level object, then we can easily modify our endpoints in the future to add new keys to the top level without impacting anyone using that endpoint. One pattern that I like to include is having a `meta` object that includes some of the details of the query and contains pagination information:

```
{
    "meta": {
        "search_term": "Led Zeppelin",
        "results": 74,
        "limit": 2,
        "offset": 0
    },
    "songs": [
        {...},
        {...}
    ]
}
```

Therefore, I suggest that when given the choice, you should nest your objects like this. Some people also like to nest single objects:

```
{
    "song": {
        "name": "Kashmir",
        "song_uid": "75b723e3-1132-4f73-931b-78bbaf2a7c04",
        "released": "1975-02-24",
        "runtime": 581
    }
}
```

The argument goes that the same principle applies. The endpoint is more easily extensible if the objects are nested. However, this argument seems less convincing and practical when dealing with single objects. Generally, any change in the endpoint would be related to the object itself. So, perhaps this is a use case for versioning, which we explored in *Chapter 3, Routing and Intaking HTTP Requests*.

No matter your decision on how to structure the data, sending information about our songs in JSON format is still just a structural decision that will be dictated by the constraints of the application being built. Now, we want to move on to the next step: actually sending the song itself. Let's see how we can do that next.

Streaming data

When introducing the concept of streaming in *Chapter 4*, *Ingesting HTTP Data*, I said that request streaming was probably the less popular of the two types. I do not have any empirical data to confirm this, but it seems readily apparent to me that when most people hear the term *streaming*—whether they are a developer or a layperson—the implication is that there is a consumption of some form of media from *the cloud*.

In this section, what we are looking to achieve is to learn how we can accomplish this. *How exactly does this work?* When building a streaming response, Sanic will add the same `Transfer Encoding: chunked` header that we saw with streaming requests. This is the indication to the client that the server is about to send incomplete data. Therefore, it should leave the connection open.

Once this happens, it is time for the server to send data at its discretion. *What is a chunk of data?* It follows a protocol whereby the server sends the number of bytes it is about to send (in hexadecimal format), followed by a `\r\n` line break, followed by some bytes, and then followed by another `\r\n` line break:

```
1a\r\n
Now I'm free, free-falling\r\n"
```

When the server is done, it needs to send a `0` length chunk:

```
0\r\n
\r\n
```

As you can probably guess, Sanic will take care of much of the plumbing in setting up the headers, determining chunk sizes, and adding the appropriate line breaks. Our job is to control the business logic. Let's see what a super simple implementation looks like, and then we can build from there:

```
@app.get("/")
async def handler(request: Request):
    resp = await request.respond()
    await resp.send(b"Now I'm free, free-falling")
    await resp.eof()
```

When we were consuming streaming requests, we needed to use the `stream` keyword argument or decorator. For responses, the simplest method is to generate the response upfront: `resp = await request.respond()`. In our example, `resp` is a `<class 'sanic.response.HTTPResponse'>` type object.

Once we have a `response` object, we can write to it whenever we want using either regular strings (`"hello"`), or `bytes` strings (`b"hello"`). When there is no more data to be transferred, we tell the client using `resp.eof()`, and we are done.

This asynchronous behavior of sending data at will does bring up an interesting question about the life cycle of the request. Since we are slightly getting ahead of ourselves, if you are interested to see how middleware behaves with streaming responses, jump ahead to *Chapter 6, Operating Outside the Response Handler*.

As I am sure you can probably imagine from our simple example, by having the `resp.send()` method available to us, we now have the freedom to execute asynchronous calls as desired. Of course, a silly example to illustrate our point would be to add a loop with some time delays:

```
@app.get("/")
async def handler(request: Request):
    resp = await request.respond()
    for _ in range(4):
        await resp.send(b"Now I'm free, free-falling")
        await asyncio.sleep(1)
    await resp.eof()
```

In the next section, we will see a more useful and complex example when we start sending SSE. But first, let's get back to our goal. We wanted to send the actual song. Not just metadata, not just the lyrics, but the actual music file so that we can listen to it through our web application.

File streaming

The simplest method by which to do this is with the `file_stream` convenience wrapper. This method takes care of all the work for us. It will asynchronously read the file contents, send the data in chunks to the client, and wrap up the response:

```
from sanic.response import import file_stream

@app.route("/herecomesthesun")
async def handler(request):
    return await file_stream("/path/to/herecomesthesun.mp4")
```

Now it is time to open the browser, turn up the volume, hit our web page, and enjoy.

Okay, so perhaps relying upon the browser to be our media player is not the best UI. *What if we want to embed the song content and have an actual player UI inside our frontend?* Of course, HTML and design are outside the scope of the book. But you can at least get started using the following:

```
<audio controls src="http://localhost:7777/herecomesthesun" />
```

Sanic will default to sending chunks of 4,096 bytes using this method. You might find it desirable to increase or decrease that number:

```
return await file_stream("/path/to/herecomesthesun.mp4", chunk_
size=8192)
```

Also, it is worth mentioning that Sanic does some work under the hood to attempt to figure out what kind of file you are sending. This is so that it can properly set up the content-type header. If it is unable to figure it out, then it will fall back to text/plain. Sanic will look at the file extension and try and match it against the operating system's MIME type definitions.

Server-sent events for push communication

Now that we know we can control the flow of information from the server, we are entering the territory of being able to build some great features for our web applications.

In the old days, when our application wanted to check the state of something, it would need to poll the web server by repeatedly sending the same request over and over again. We talked about building a music web application. We learned how we could display content, get information, and even stream some content to listen to music. Of course, the next step is to make the application social because we want to share our music with our friends. We want to add a feature that will list who is online and the name of the song they are listening to. Refreshing the page constantly would work but is a bad experience. Polling constantly by sending the same request over and over again also works, but this eats up resources and is also not a great experience.

What would be better is if our server simply notified the browser when someone comes online or when their music player changes. This is what SSE provides: a simple set of instructions for our server to send push notifications to the browser.

The basic unit of the SSE protocol is the *event*, which is a single line of text that contains a field and some body:

```
data: foo
```

In this case, the field is `data`, and the body is `foo`. A message can consist of one or more events separated by a single newline character: \n. Here is an example:

```
data: foo
data: bar
```

When a browser receives this message, it will be decoded as `foo\nbar`. A message should be terminated by the server by sending two newline characters: \n\n.

The SSE protocol has five basic fields:

Field	Description	Example
`<null>`	This should be treated as a comment.	`: This is a comment`
`event`	This is a description of the type of event being sent.	`event: songplaying`
`data`	The body of the message; often, this is either plain text or JSON.	`data: All Along the Watchtower`
`id`	This is a self-created event ID for tracking.	`id: 123456ABC`
`retry`	This is the reconnection time in milliseconds.	`Retry: 1000`

Table 5.2 – An overview of the allowed SSE fields

When creating an endpoint for SSE, I would suggest that you keep this table handy. In the following section, we will learn how we can craft messages to be sent by our server to comply with the SSE protocol utilizing these field types.

Starting with the basics

Before diving into how we can implement SSE from Sanic, we need a frontend application that can understand how to process these events. We are not too concerned about how to build the SSE client. There is a prebuilt frontend HTML client that you can find in the GitHub repository at `https://github.com/PacktPublishing/Web-Development-with-Sanic/blob/main/Chapter05/sse/index.html`. Just grab the code to follow along.

To deliver the client, we will store the prebuilt HTML client as `index.html` and use the existing tools we know to serve that file. To make sure that we cover the blank root path (`/`), we will also redirect it to our `index.html` file:

```
app.static("/index.html", "./index.html", name="index")
```

```
@app.route("/")
```

```
def home(request: Request):
    return redirect(request.app.url_for("index"))
```

With the preceding code in place, you should be able to serve your application and navigate to `http://localhost:7777/`. There, you should see the simple SSE frontend. Now that we have a client, let's build the SSE server endpoint to go along with it:

```
@app.get("/sse")
async def simple_sse(request: Request):
    headers = {"Cache-Control": "no-cache"}
    resp = await request.respond(
        headers=headers,
        content_type="text/event-stream"
    )

    await resp.send("data: hello\n\n")
    await asyncio.sleep(1)
    await resp.send("event: bye\ndata: goodbye\n\n")
    await resp.eof()
```

Most of this should look familiar. We already saw how we can control the sending of chunks to the client; here, we are just doing it in a more structured pattern. We start by creating a `response` object with our required headers using `request.respond`. Then, at periodic intervals, we will dispatch data to the frontend using `resp.send`.

Note that we are doing this in a very deliberate pattern using the SSE requirements for how data and events should be sent and making sure to add line breaks (\n), as required by the SSE specification. Of course, this simple proof-of-concept is far from being a feature-complete build that can be used for our music application. Let's examine how we can make it a bit better by creating a mini framework to make our job of formatting these SSE messages easier.

Building some SSE objects

To create our SSE framework, we will start by building some basic objects to help in creating our messages. SSE messages are indeed simple, so perhaps this is a bit overkill. On the other hand, making sure we use line breaks and field names appropriately sounds like a recipe for disaster. So, a few thoughtful steps upfront should go a long way for us.

The first thing we will build are some objects to create properly formatted fields:

```python
class BaseField(str):
    name: str

    def __str__(self) -> str:
        return f"{self.name}: {super().__str__()}\n"


class Event(BaseField):
    name = "event"


class Data(BaseField):
    name = "data"


class ID(BaseField):
    name = "id"


class Retry(BaseField):
    name = "retry"


class Heartbeat(BaseField):
    name = ""
```

Notice how we are starting our inheritance with `str`. This will make our objects operate as strings, with some auto-formatting involved:

```python
>>> print(Event("foo"))
event: foo
```

Moving on, we need a convenient way to compose fields together into a single message. It also would need to have proper string formatting, which means an additional \n at the end:

```python
def message(*fields: BaseField):
    return "".join(map(str, fields)) + "\n"
```

Now, looking at this, we should see a properly formatted SSE message:

```
>>> print(f"{message(Event('foo'), Data('thing'))}".encode())
b'event: foo\ndata: thing\n\n'
```

The next step is to try out our new building blocks and see if they send messages, as expected, to the frontend:

```
@app.get("/sse")
async def simple_sse(request: Request):
    headers = {"Cache-Control": "no-cache"}
    resp = await request.respond(headers=headers, content_
type="text/event-stream")

    await resp.send(message(Data("hello!")))
    for i in range(4):
        await resp.send(message(Data(f"{i=}")))
        await asyncio.sleep(1)
    await resp.send(message(Event("bye"), Data("goodbye!")))
    await resp.eof()
```

Let's pause for a moment and recap what it is we are trying to achieve. The goal is to send notifications to the browser when a certain event happens. So far, we have identified two events: another user logs into (or out of) the system, and a user starts (or stops) listening to a song. When one of these events is triggered, our stream should broadcast the notification back to the browser. To achieve this, we will build a pubsub.

A **pubsub** is a design paradigm where you have two actors: a *publisher* and a *subscriber*. It is the job of the publisher to send messages and the job of the subscriber to listen for messages. In our scenario, we want the stream to be the subscriber. It will listen for incoming messages, and when it receives one, it will know that it should dispatch the SSE.

Since we are still working out exactly how we want our notification system to work, we are going to keep it simple. The mechanism for our pubsub will be a simple `asyncio.Queue` implementation. Messages can come in, and messages can be consumed.

It should be noted that this design pattern will be limited. *Remember when we decided that we were going to run our development server with two workers?* The reason for doing that was to keep horizontal scaling in mind. What we are about to do will absolutely break this and will not work on a distributed system.

Therefore, to make this production worthy, we will need a new plan for how to distribute messages across a cluster. We will get there in *Chapter 11, A Complete Real-World Example*, which has our complete example.

1. First, we need to set up a single queue. We will do this with a listener:

```
@app.after_server_start
async def setup_notification_queue(app: Sanic, _):
    app.ctx.notification_queue = asyncio.Queue()
```

Now, when our application starts, anywhere we have access to the application instance, we can also access the notification queue.

> **Important Note**
>
> As we start building more complex applications, we are going to see more usage of `ctx` objects. These are convenient locations that Sanic provides for us—the developers—to do with as necessary. It is a storage location for *stuff*. Sanic almost never makes use of them directly. Therefore, we are free to set any properties we want on the object.

2. Next, we will create our subscriber. This instance will listen to the queue, and send messages when it finds a message on the queue that has not been dispatched:

```
class Notifier:
    def __init__(
        self,
        send: Callable[..., Coroutine[None, None, None]],
        queue: asyncio.Queue,
    ):
        self.send = send
        self.queue = queue

    async def run(self):
        await self.send(message(Heartbeat()))
        while True:
            fields = await self.queue.get()
            if fields:
                if not isinstance(fields, (list, tuple)):
                    fields = [fields]
                await self.send(message(*fields))
```

As you can see, our `run` operation consists of an infinite loop. Inside that loop, `Notifier` will pause and wait until there is something inside the queue. When there is, it removes the item from the queue and continues through the current iteration of the loop, which is to send that item. However, before we send our loop, we are going to send a single heartbeat message. This will flush out any startup events so that our client will clear out its queue. This is not necessary, but I think it is a helpful practice to get into.

3. To implement this in our endpoint, it should look like this:

```
@app.get("/sse")
async def simple_sse(request: Request):
    headers = {"Cache-Control": "no-cache"}
    resp = await request.respond(
        headers=headers,
        content_type="text/event-stream"
    )
    notifier = Notifier(resp.send, request.app.ctx.
notification_queue)
    await notifier.run()
    await resp.eof()
```

A `response` object is created using `request.respond`. Then, we create `Notifier` and let it run. At the very end, `eof` is called to close the connection.

> **Important Note**
>
> To be entirely upfront, the previous code sample is somewhat flawed. I intentionally left it simple for the sake of making a point. However, since there is no way to break out of the infinite loop, there is really no way the server would ever close the connection itself. This makes the inclusion of `eof` a bit of a moot point. It is nice to have it there as an example, but as written, this code will only ever be stopped client-side by navigating away from the endpoint.

4. The easy part should now be to push messages into the queue. We can do this on a separate endpoint, as follows:

```
@app.post("login")
async def login(request: Request):
    request.app.ctx.notification_queue.put_nowait(
```

```
        [Event("login"), Data("So-and-so just logged
in")]
    )

    return text("Logged in. Imagine we did something
here.")
```

5. We can now test this out! Open your browser and navigate to `http://localhost:7777/index.html`.

   You should see something like this:

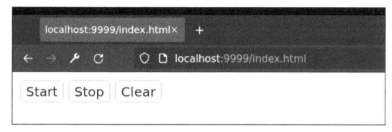

Figure 5.2 – The HTML to test out SSE

6. Once we **start** the stream by clicking on the start button, switch back to a terminal, and hit our fake login endpoint:

```
$ curl localhost:7777/login -X POST
Logged in. Imagine we did something here.
```

Did you see what just happened in the browser? Go ahead and do it again. Feel free to play around with this example to get a feel for how the different components are working.

Do you think you can build the fake "start playing music" endpoint? Just make sure that if you are going to use `Event`, your frontend application knows about it by using `eventSource.addEventListener`. Take a look at the `index.html` file and you will see the `login` event. I suggest you pause and take some time to dig into this code to see how the different components are working together to facilitate the exchange of data. *What kinds of amazing things could you build using this?*

In *Chapter 6, Operating Outside the Response Handler*, we will come back to this same example and explore another way we could achieve it through the use of signals. Also, I should probably point out that using `asyncio.Queue` in this instance has another disadvantage: it will really only work in a single browser. *Since our consumer (`Notifier`) drains down the queue, what happens when multiple browsers are running simultaneously?*

Well, only the first one gets the message. Again, this solution is far too simplistic for real-world usage, but hopefully, it has gotten the ideas flowing in terms of how you could build something more robust. To be entirely transparent, in situations like this, I really like to fall back to Redis. If you are familiar with Redis, you might know that it has a pubsub built into it. With the right Python library interface, you can easily solve both of the problems that the `asyncio.Queue` implementation gave us: it can be used to push messages to multiple subscribers at once, and it can be used in a distributed system where multiple publishers are pushing into it. Maybe before continuing to *Chapter 11, A Complete Real-World Example*, try to see whether you can make it work in our current example.

If nothing else, I hope you got excited when you saw the message pop up in your browser. For me, it is still really interesting and fun to see messages being pushed into a browser session. SSE is a super simple solution that can solve some potentially complex problems, which, ultimately, lead to a powerful feature set. Being able to push data into a web browser truly helps an application feel like it is transforming from a web page into a web application.

The downside of this implementation is that they are still only one-sided. To get two-way asynchronous communication, we need websockets.

Websockets for two-way communication

You have almost definitely experienced websockets on your favorite web applications before. They are a tool that helps to create a super-rich layer of user experience and can be used in a wide variety of contexts. While SSEs are, essentially, just an open stream that has not yet been terminated, websockets are something completely different.

Plain vanilla HTTP is just a specification (or protocol) for how messages can be formatted and transmitted over a TCP connection between machines. Websockets are a separate protocol complete with directions on how messages should be formatted, sent, received, and more. The specification for them is really quite involved, and we could probably devote an entire book to just discussing websockets. Instead, we will simply focus on their implementation within Sanic. The one technical detail about websockets that is worth mentioning is that they begin their life as a normal HTTP connection. The request comes in and asks the server for permission to upgrade its connection to a websocket. Then, the client and server do a bit of a dance together to iron out the details. When the negotiations are all done, we have an open socket where messages can be passed back and forth. Think of it as a two-lane highway where messages can pass by each other on their way to either end.

Perhaps the easiest way to conceptualize a websocket is to think of a chat application. We have a single endpoint on the backend server. Two separate web browsers connect to the server, and each is somehow connected so that when one pushes a message in, the server pushes that message out the other side. In this way, both clients are able to send and receive messages irrespective of what else is happening.

This is true asynchronous web behavior. Sanic uses `async`/`await` to leverage and optimize server efficiency for performance. However, the side benefit is that it also allows Sanic to offer an almost effortless mechanism to implement websockets. While we will not get into the details of how this works, you should be aware that Sanic makes use of the Python `websockets` package under the hood. It is a fantastic project, and it will be helpful to look at their documentation when building your own websocket endpoint in Sanic: `https://websockets.readthedocs.io`.

In the last section, we started to make our music player application social by sharing information about who was logged in. Let's turn up the social aspect by adding in a chat feature. Here, our goal is to have two different users access the web application through their web browser and to be able to communicate with one another. Since the applications we are developing right now are only available on the local network, we will simulate this by opening two browsers side by side. We should still be able to pass text messages back and forth between the two web browsers.

Just like with the SSE example, you can grab the frontend code from the GitHub repository, so we do not have to worry about those implementation details: `https://github.com/PacktPublishing/Python-Web-Development-with-Sanic/tree/main/Chapter05/websockets`. You should copy that code and place it into an `index.html` file just like we did with our last example. Once that has been done, make sure you have created both the `static` route and the bare-root (`/`) endpoint. Once this is complete, you can access the websocket frontend in the browser, and we can begin to create the backend for our websocket chatroom:

1. The first thing we are going to create is a `Client` class. When someone enters the application, the frontend will immediately open the websocket. The `Client` class will be a holding place for us to be able to keep track of who is in the application and how we can send them messages. Therefore, we need a unique identifier and a callable in which to send messages:

    ```
    class Client:
        def __init__(self, send) -> None:
            self.uid = uuid4()
            self.send = send
    ```

```
def __hash__(self) -> int:
    return self.uid.int
```

As you can see, we are going to keep track of the incoming session by assigning each client a UUID.

2. Instantiate this `Client` object inside the `websocket` handler:

```
@app.websocket("/chat")
async def feed(request, ws):
    client = Client(ws.send)
```

3. Next, we need to create our `ChatRoom` instance. This will be a global instance that exists during the lifetime of the application. Its role will be to keep track of all of the clients that have entered or exited. When someone tries to send a message, it will be responsible for publishing that message to the remaining clients.

Similar to our SSE example, the implementation I am about to show you is limited in that it cannot be run across a distributed cluster. It will function great in just a single instance. This is because we are registering the clients on a single instance in memory. To build a more scalable application to be used in a production environment, we should use something such as Redis or RabbitMQ to distribute the message across multiple Sanic instances. If you are interested in seeing what this distributed feed would look like, please take a look at a GitHub Gist that I have created (`https://gist.github.com/ahopkins/5b6d380560d8e9d49e25281ff964ed81`) as a demonstration. For now, we will stick with the simpler single-server implementation that looks like this:

```
class ChatRoom:
    def __init__(self) -> None:
        self.clients: Set[Client] = set()


    def enter(self, client: Client):
        self.clients.add(client)


    def exit(self, client: Client):
        self.clients.remove(client)


    async def push(self, message: str, sender: UUID):
        recipients = (client for client in self.clients
```

```
        if client.uid != sender)
            await asyncio.gather(*[client.send(message) for
    client in recipients])
```

This interface has a mechanism to add and remove clients, along with a method to push events to registered clients. One thing that is important to point out is that we do not want to send the message back to the person that sent it. That would be a little bit awkward and slightly annoying for the user to constantly have their own messages fed back to them. Therefore, we will filter out the sending client.

4. Remember that the ChatRoom instance is a single object that lives for the lifetime of the application instance. *So, where do you think it should be instantiated?* That's right, a listener:

```
@app.before_server_start
async def setup_chatroom(app, _):
    app.ctx.chatroom = ChatRoom()
```

5. Now, all we need to do is wire it up:

```
@app.websocket("/chat")
async def feed(request, ws):
    try:
        client = Client(ws.send)
        request.app.ctx.chatroom.enter(client)

        while True:
            message = await ws.recv()
            if not message:
                break
            await request.app.ctx.chatroom.push(message,
    client.uid)

    finally:
        request.app.ctx.chatroom.exit(client)
```

When the user enters, we add them to the chatroom. Then, the request enters an infinite loop and waits for a message to be received from the websocket. This is very similar, in concept, to the SSE implementation that we saw in the last section. When a message is received on the current websocket, it is passed to the ChatRoom object, which is responsible for sending it to all of the other registered clients. *It seems too easy, right?*

Now to test it out: open two web browsers. Navigate each of the web browsers to the frontend web application that you created earlier. They should connect to the websocket backend, and they should be all set up and ready to begin chatting. When you type a message in one browser, it should appear in the other. Take some time to send messages back and forth, and dissect how this is operating under the hood. Have fun chatting with yourself:

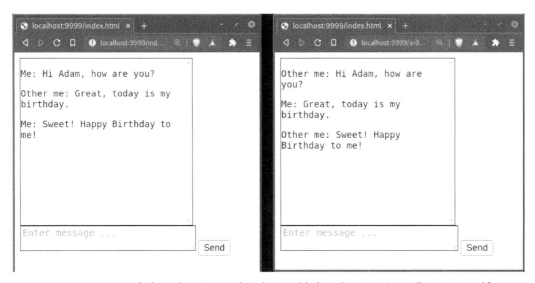

Figure 5.3 – Two side-by-side HTML websocket-enabled applications (me talking to myself)

The code that you need to run the preceding websocket HTML application (and the backend chatroom code that we just looked at) is available on GitHub at `https://github.com/PacktPublishing/Web-Development-with-Sanic/blob/main/Chapter05/websockets/index.html`.

In the next section we will discuss adding headers and cookies to responses.

Setting response headers and cookies

We have talked a lot about headers. They are super important when building web applications and are, generally, a fundamental part of application design. When building your application responses, you will likely find reasons to add handlers to your response objects. This could be for security purposes such as **Cross-Origin Resource Sharing** (**CORS**) headers, a Content-Security-Policy, or informational and tracking purposes. And, of course, there are cookies, which are their own special kind of headers that receive special treatment in Sanic.

Recall some of the earlier examples (such as the SSE example) where we actually set the headers. It is such an easy and intuitive process, so perhaps you did not even notice. Whenever we build a response object, all we need to do is pass a dictionary with key/value pairs:

```
text("some message", headers={
    "X-Foobar": "Hello"
})
```

That's really all there is to it! Bear in mind that you will not be required to set your own headers all of the time. Sanic takes care of some of them for you, including the following:

- `content-length`
- `content-type`
- `connection`
- `transfer-encoding`

Responding with a request ID

One pattern that is particularly helpful is to set an `x-request-id` header on every response. It then makes a habit of using `request.id` to log or trace a request through your application, so it becomes easier to track what is happening when you inevitability need to debug something.

We want to ensure our response includes the header:

```
@app.route("/")
async def handler(request):
    ...
        return text("...", headers={"x-request-id": request.id})
```

That is a simple example. As you have probably come to realize by now, that simple example might get tedious if we want to do it on all of our requests. *Do you want to try to come up with a solution for adding that to all responses?* Again, decorators and middleware are potential tools. We will come back to this, and you will see some implementations for setting this globally, in the full example, in *Chapter 11, A Complete Real-World Example.*

To truly make this useful, we should set up our logging to include the request ID. We have not discussed this much yet, but Sanic includes a default logger for you to use. It might be helpful for you to use that logger and override the default logging format to include the request ID. If you want to know more about how to set this up, jump ahead to *Chapter 6, Operating Outside the Response Handler*.

Setting response cookies

One of the most important types of headers you can set on an individual response would be the **cookie headers**. Since they are so prominent, and they could require a bit of complexity to set up, you can avoid having to use `Set-Cookie` directly.

Essentially, response cookies are a key/value pair that gets concatenated into a single string in the response but then is interpreted by the browser. It is yet another shared language in the conversation of the web. So, while a single cookie could be as simple as `flavor=chocolatechip`, that shared language allows us to set a whole bunch of metadata on top of the simple example.

Before we get to the metadata, let's look at the simple example:

```
@app.get("/ilikecookies")
async def cookie_setter(request):
    resp = text("Yum!")
    resp.cookies["flavor"] = "chocolatechip"
    return resp
```

It seems fairly straightforward. Let's see what it does to our response:

```
$ curl localhost:7777/ilikecookies -i
HTTP/1.1 200 OK
Set-Cookie: flavor=chocolatechip; Path=/
content-length: 4
connection: keep-alive
content-type: text/plain; charset=utf-8

Yum!
```

Our response headers now have this additional line that instructs the browser to create and store a cookie:

```
Set-Cookie: flavor=chocolatechip; Path=/
```

So, what's the deal with that `Path`*?* That is the cookie metadata at play. Cookies can have several different types of metadata attached to them, including `Path`, which is added by default. Here are some of the meta-values that we can add:

Meta value and type	Description
`expires: datetime`	The time at which a browser should discard the cookie.
`path: str`	The path in which the cookie will be applied.
`comment: str`	A comment.
`domain: str`	The domain in which the cookie will be applied.
`max-age: str`	The number of seconds before a browser should discard the cookie.
`secure: bool`	Whether it should only be sent over HTTPS.
`httponly: bool`	Whether it can be accessed by JavaScript.
`samesite: str`	Where it can be sent from; values can be lax, strict, or none.

Table 5.3 – Cookie meta-fields

When we set up our `flavor` cookie, it seemed like we were just adding a string value to a dictionary that looked like this:

```
{
    "flavor": "chocolatechip"
}
```

That is not really the case. The `response.cookies` object is, in fact, a `CookieJar` object, which is itself a special kind of `dict`. When we set up a new key/value on that `CookieJar` object, it is, in fact, creating a `Cookie` object. *Huh?*

Let's see what happens when we do the following:

```
resp.cookies["flavor"] = "chocolatechip"
```

This code looks like you are just adding a string value to a regular dictionary. It is more like you are creating `Cookie("flavor", "chocolatechip")` and then putting it into `CookieJar()`. To clean up some of the complexity involved with managing these instances, Sanic lets us just work with strings. We should keep this in mind when we go to set the metadata, which is what we will do next.

Let's imagine we have a cookie that should time out. After a while, we want the browser session to forget it existed. This might—for example—be useful with a session cookie. We set a value that identifies a browser session with a particular user. Storing it in a cookie means that, on subsequent requests, we can identify who the person is. However, by setting `Max-Age`, we can control the length of time that person can use the application before they need to log in again:

```
resp.cookies["session"] = "somesessiontoken"
resp.cookies["session"]["max-age"] = 3600
```

The same applies for all the other meta fields:

```
resp.cookies["session"]["secure"] = True
resp.cookies["session"]["httponly"] = True
resp.cookies["session"]["samesite"] = "Strict"
```

If we put this all together, our cookie headers will ultimately look like this:

```
Set-Cookie: flavor=chocolatechip; Path=/
Set-Cookie: session=somesessiontoken; Path=/; Max-Age=3600;
Secure; HttpOnly; SameSite=Strict
```

The last thing we should look at is how we should delete cookies. When you want to remove a cookie, it is tempting to just use `del`, as you might do with any other dictionary object. The problem is that this only works so far. Usually, what you want to do instead is tell the browser that it needs to remove the cookie so that the browser does not send it back in future requests. The easiest method to accomplish this is by setting the maximum age of the cookie to `0`:

```
resp.cookies["session"]["max-age"] = 0
```

Now you should feel comfortable adding and deleting cookies to and from responses. It might be a good opportunity to create a response handler and use your browser's cookie inspection tools to see how cookies can be set, manipulated, and deleted from your server.

Summary

Now that we have learned how to manipulate both the request and the response, we can build some really powerful applications. Whether we are building an HTML-based website, a JSON-powered web API, a streaming content application, or a combination of them all, Sanic provides us with the tools we need.

One of the first things we discussed is that Sanic tries hard to not obstruct the build of an application. We, as developers, have the freedom to build with different tools and layer them together to build a truly unique platform. This is very much prevalent when we realize the freedom given to the developer regarding the response object. *Do you need to write bytes directly?* Sure. *Do you want to use a specific templating engine?* Not a problem!

Now that we have a basic understanding of how to handle the life cycle of an HTTP connection from a request through to a response, we can start to see what else we have at our disposal. In the next chapter, we will take a deeper dive into some of the concepts that have already been introduced, such as middleware, background tasks, and signals. Combining these basic building blocks will help us build not only a powerful application, but also one that is easy to maintain, update, and expand.

6
Operating Outside the Response Handler

The basic building block of application development within Sanic is the response handler, which is sometimes known as a **route handler**. Those terms can be used interchangeably and mean the same thing. It is the function that Sanic runs when a request has been routed to your application to be handled and responded to. This is where business logic and **HyperText Transfer Protocol** (**HTTP**) logic combine to allow the developer to dictate how responses should be delivered back to the client. It is the obvious place to start when learning how to build with Sanic.

However, response handlers alone do not provide enough power to create a polished application experience. To build out an application that is polished and professional, we must break outside the handler to see which other tools Sanic has to offer. It is time to think about the HTTP request/response cycle as not being confined to a single function. We will broaden our scope so that responding to a request is not the responsibility of just the handler, but the entire application. We already got a taste of this when we caught a glimpse of middleware.

In this chapter, we are going to cover the following topics:

- Making use of `ctx`
- Altering requests and responses with middleware
- Leveraging signals for intra-worker communication
- Mastering HTTP connections
- Implementing proper exception handling
- Background task processing

Of course, not all projects will need features such as these, but when used in the right place, they can be extremely powerful. Have you ever worked on a **Do It Yourself** (**DIY**) project around your home and not quite had the right tools for the job? It can be super frustrating and inefficient when you need a Phillips head screwdriver, but all you have are flat head screwdrivers. Not having the right tool for the job can make your task harder, but it also sometimes decreases the quality of the work that you can perform.

Think of the features that we explore in this chapter as tools. There is a common saying you may have heard: "*If you are holding a hammer, then every problem looks like a nail.*" Luckily for us, we have a bunch of tools, and our job now is to learn how to use them. We are about to go explore the Sanic tool belt and see what kinds of problems we can solve.

Technical requirements

In this chapter, you should have at your disposal the same tools available as in the previous chapters in order to be able to follow along with the examples (**integrated development environment** (**IDE**), modern Python, and curl).

You can access the source code for this chapter on GitHub at `https://github.com/PacktPublishing/Python-Web-Development-with-Sanic/tree/main/Chapter06`.

Making use of ctx

Before we begin with the tool belt, there is one more concept that we must become familiar with. It is fairly ubiquitous in Sanic, and you will see it in a lot of places. I am talking about: `ctx`. What is it?

It stands for *context*. These ctx objects can be found in several places, and it is impractical to build a professional-grade Sanic web application without making good use of them. What they enable is the passing of state from one location in your application to another. They exist for your own usage as a developer, and you should feel free to use them however you wish. That is to say that the ctx objects are yours to add information to without worrying about name collisions or otherwise impacting the operation of Sanic.

The most common example that comes to mind is your database connection object. You create it once, but you want to have access to it in many places. How does this work? Have a look at the following code snippet:

```
@app.before_server_start
async def setup_db(app, loop):
    app.ctx.db = await setup_my_db()
```

Now, anywhere you can access the application instance, you can access the db instance. For example, you can access it inside a function somewhere, like this:

```
from sanic import Sanic


async def some_function_somewhere():
    app = Sanic.get_app()

    await app.ctx.db.execute(...)
```

Or, perhaps you need it in your route handler, as illustrated here:

```
bp = Blueprint("auth")


@bp.post("/login")
async def login(request: Request):
    session_id = await request.app.ctx.db.execute(...)
    ...
```

Here is a list of all locations that have a `ctx` object:

Object	Description	Example
Sanic	Available during the entire lifetime of your worker instance. It is worker-specific, meaning that if you run multiple workers, it will not keep them synchronized. Best used for connection management or other things that need to be made available throughout the lifetime of the application instance.	`app.ctx`
Blueprint	Available on a `Blueprint` instance as long as the blueprint exists. This might be helpful if you have some specific data that needs to be available for the entire worker lifetime but want to control its access to anything attached to that particular blueprint.	`bp.ctx`
Request	Available for the duration of a single HTTP request. Helpful for adding details in middleware, and then making it available in the handler or other middleware. Common uses include session identifiers (IDs) and user instances.	`request.ctx`
ConnInfo	Available for the duration of an entire HTTP connection (potentially multiple requests). Be careful with this one, particularly if you use a proxy. It usually should not be used for sensitive information.	`request.conn_info.ctx`
Route	Available on Route and Signal instances. This is the one exception where Sanic actually does store some details on the `ctx` object.	`request.route.ctx`

Table 6.1 – Sanic features with a ctx object

We will continue to come back to `ctx` objects often. They are a very important concept in Sanic to allow the passing of arbitrary data and objects. Not all of them are created equal, and you will likely find yourself using `app.ctx` and `request.ctx` much more often than any of the others.

Now that we have this basic building block behind us, we will see what it actually looks like to pass these objects around. In the next section regarding middleware, we will see how the `Request` object—and therefore also `request.ctx`—can be accessed in multiple places from your application.

Altering requests and responses with middleware

If you have been following along with the book up until now, the concept of middleware should be familiar. This is the first tool in the tool belt that you should become familiar with.

Middleware is snippets of code that can be run before and after route handlers. Middleware comes in two varieties: request and response.

Request middleware

The request middleware executes in the order in which it was declared, before the route handler, as shown here:

```
@app.on_request
async def one(request):
    print("one")


@app.on_request
async def two(request):
    print("two")


@app.get("/")
async def handler(request):
    print("three")
    return text("done")
```

When we try to reach this endpoint, we should see the following in the Terminal:

```
one
two
three
(sanic.access)[INFO][127.0.0.1:47194]: GET http://
localhost:7777/  200 4
```

But this only tells a part of the story. Sometimes, we may need to add some additional logic to only *some parts* of our application. Let's pretend we are working on building an e-commerce application. As with other online stores, we will need to build a shopping cart that holds products that are going to be purchased. For the sake of our example, we will imagine that when the user logs in, we create a cart in our database and store a reference to it in a cookie. We discussed how to add a cookie to a Sanic response object in *Chapter 5*, *Building Response Handlers*, in the *Setting response headers and cookies* section. To achieve the goal of setting the cookie on login, it could look something like this:

```
@app.post("/login")
async def login(request):
    user = await do_some_fancy_login_stuff(request)
    cart = await generate_shopping_cart(request)
    response = text(f"Hello {user.name}")
    response.cookies["cart"] = cart.uid
    return responses
```

Don't get too tied up in the details here. The point is that on every subsequent request, there will be a cookie called `cart` that we can use to fetch data from our database.

Now, suppose that we want all endpoints on our `/cart` path to have access to the shopping cart. We might have endpoints for adding items, removing items, changing quantities, and so on. However, we will always need access to the cart. Rather than repeating the logic in every handler, we can do it once on the blueprint. Adding middleware to all the routes on a single blueprint looks and functions similarly to application-wide middleware, as the following code snippet illustrates:

```
bp = Blueprint("ShoppingCart", url_prefix="/cart")


@bp.on_request
async def fetch_cart(request):
    cart_id = request.cookies.get("cart")
    request.ctx.cart = await fetch_shopping_cart(cart_id)


@bp.get("/")
async def get_cart(request):
    print(request.ctx.cart)
    ...
```

As we would expect, every endpoint that is attached to the `ShoppingCart` blueprint will fetch the cart before it runs the handler and stores it in the local request context. I am sure you can see the value in this sort of pattern. Where you can identify a group of routes that need similar functionality, sometimes it is best to pull that out into middleware. Doing this will make solving bugs or adding new features much easier, as you only have a single function to update and not all of the individual route handlers. This is a good time to also point out that this works also with blueprint groups. We could change the middleware to the following and have the same impact:

```python
group = Blueprint.group(bp)


@group.on_request
async def fetch_cart(request):
    cart_id = request.cookies.get("cart")
    request.ctx.cart = await fetch_shopping_cart(cart_id)
```

Just as we would expect, endpoints that are within that blueprint group will now have the shopping cart accessible to them.

Knowing that we can execute middleware that is both application-wide and blueprint-specific leads to an interesting question: in what order is it applied? No matter the order in which it is declared, all application-wide middleware will *always* run before blueprint-specific middleware. To illustrate this point, we will use an example here that mixes the two types:

```python
bp = Blueprint("Six", url_prefix="/six")


@app.on_request
async def one(request):
    request.ctx.numbers = []
    request.ctx.numbers.append(1)


@bp.on_request
async def two(request):
    request.ctx.numbers.append(2)


@app.on_request
async def three(request):
    request.ctx.numbers.append(3)
```

```
@bp.on_request
async def four(request):
    request.ctx.numbers.append(4)


@app.on_request
async def five(request):
    request.ctx.numbers.append(5)


@bp.on_request
async def six(request):
    request.ctx.numbers.append(6)


@app.get("/")
async def app_handler(request):
    return json(request.ctx.numbers)


@bp.get("/")
async def bp_handler(request):
    return json(request.ctx.numbers)


app.blueprint(bp)
```

As you can see in this example, we interspersed declaring application and blueprint middleware by alternating between them: first, application middleware, then blueprint middleware, and so on. While the code lists the functions in sequential order (1, 2, 3, 4, 5, 6), our output will not be in sequence. You should be able to anticipate how our endpoints will respond, with the application numbers appended before the blueprint numbers. Sure enough, that is the case, as we can see here:

```
$ curl localhost:7777
[1,3,5]

$ curl localhost:7777/six
[1,3,5,2,4,6]
```

It is also really helpful to point out that since middleware is just passing along the Request object, subsequent middleware has access to whatever changes earlier middleware performed. In this example, we created a list of numbers in one function, which was then available to all of the middleware.

Response middleware

On the other side of the HTTP life cycle, we have response middleware. The same rules for request middleware apply, as outlined here:

- It is executed based upon the order of declaration, although it is reverse order!
- Response middleware can be both application-wide or blueprint-specific.
- All application-wide middleware will run before any blueprint-specific middleware.

In the last section, we counted from 1 through 6 using middleware. We will take the exact same code (order is important!), but change from `request` to `response`, as follows:

```python
bp = Blueprint("Six", url_prefix="/six")


@app.on_response
async def one(request, response):
    request.ctx.numbers = []
    request.ctx.numbers.append(1)


@bp.on_response
async def two(request, response):
    request.ctx.numbers.append(2)


@app.on_response
async def three(request, response):
    request.ctx.numbers.append(3)


@bp.on_response
async def four(request, response):
    request.ctx.numbers.append(4)


@app.on_response
async def five(request, response):
    request.ctx.numbers.append(5)


@bp.on_response
async def six(request, response):
    request.ctx.numbers.append(6)
```

```
@app.get("/")
async def app_handler(request):
    return json(request.ctx.numbers)


@bp.get("/")
async def bp_handler(request):
    return json(request.ctx.numbers)
```

Now, when we hit our endpoint, we will see a different order, as illustrated here:

```
$ curl localhost:7777
500 — Internal Server Error
============================
'types.SimpleNamespace' object has no attribute 'numbers'

AttributeError: 'types.SimpleNamespace' object has no attribute
'numbers' while handling path /
Traceback of __main__ (most recent call last):

  AttributeError: 'types.SimpleNamespace' object has no
attribute 'numbers'
    File /path/to/sanic/app.py, line 777, in handle_request
    response = await response


    File /path/to/server.py, line 48, in app_handler
    return json(request.ctx.numbers)
```

Uh oh—what happened? Well, since we did not define our ctx.numbers container until the response middleware, it was not available inside the handlers. Let's make a quick change. We will create that object inside of a request middleware. For the sake of our example, we will create our response from our last middleware and ignore the response from the handler. In the following example, the last middleware to respond will be the first blueprint response middleware declared:

```
@bp.on_response
async def complete(request, response):
    return json(request.ctx.numbers)
```

```python
@app.on_request
async def zero(request):
request.ctx.numbers = []


@app.on_response
async def one(request, response):
    request.ctx.numbers.append(1)


@bp.on_response
async def two(request, response):
    request.ctx.numbers.append(2)


@app.on_response
async def three(request, response):
    request.ctx.numbers.append(3)


@bp.on_response
async def four(request, response):
    request.ctx.numbers.append(4)


@app.on_response
async def five(request, response):
    request.ctx.numbers.append(5)


@bp.on_response
async def six(request, response):
    request.ctx.numbers.append(6)


@bp.get("/")
async def bp_handler(request):
    request.ctx.numbers = []
    return json("blah blah blah")
```

Take a close look at the preceding code. We still have a mixture of application and blueprint middleware. We create a numbers container inside of the handler. Also, it is important to note that we are using the exact same ordering that we used for the request middleware that yielded 1, 3, 5, 2, 4, 6. The changes here merely show us how the response middleware reverses its order. Can you guess what order our numbers will be in? Let's check here:

```
$ curl localhost:7777/six
[5,3,1,6,4,2]
```

First, all of the application-wide response middleware runs (in reverse order of declaration). Second, all of the blueprint-specific middleware runs (in reverse order of declaration). Keep this distinction in mind when you are creating your response middleware if it is connected with blueprint-specific middleware.

Whereas a common use case for request middleware is to add some data to the request object for further processing, this is not so practical for response middleware. Our preceding example is a bit odd and impractical. What, then, is response middleware good for? Probably the most common use case is setting headers and cookies.

Here is a simple (and very common) use case:

```
@app.on_response
async def add_correlation_id(request: Request, response:
HTTPResponse):
    header_name = request.app.config.REQUEST_ID_HEADER
    response.headers[header_name] = request.id
```

Why would you want to do this? Many web **application programming interfaces** (**APIs**) use what is known as a *correlation ID* to help identify individual requests. This is helpful for logging purposes, for tracking a request as it trickles through various systems in your stack, and also for clients that are consuming your API to keep track of what is happening. Sanic latches onto this principle and will set the `request.id` value automatically for you. This value will either be the incoming correlation ID from the incoming request headers or a unique value generated per request. By default, Sanic will generate a **universally unique ID** (**UUID**) for this value. You usually need not worry about this unless you want to use something other than a UUID for correlating web requests. If you are interested in how you can override Sanic's logic for generating these, check out *Chapter 11, A Complete Real-World Example*.

Coming back to our aforementioned example, we see that we are simply grabbing that value and appending it to our response headers. We can now see it in action here:

```
$ curl localhost:7777 -i
HTTP/1.1 200 OK
X-Request-ID: 1e3f9c46-1b92-4d33-80ce-cca532e2b93c
content-length: 9
connection: keep-alive
content-type: text/plain; charset=utf-8

Hello, world.
```

This small snippet is something I would highly encourage you to add to all of your applications. It is extremely beneficial when you pair it with request ID logging. This is also something we will add to our application in *Chapter 11, A Complete Real-World Example.*

Responding early (or late) with middleware

When we explored the response middleware-ordering example from the last section, did you notice something peculiar happening with our responses? You may have seen this:

```
@bp.on_response
async def complete(request, response):
    return json(request.ctx.numbers)


...


@bp.get("/")
async def bp_handler(request):
    request.ctx.numbers = []
    return json("blah blah blah")
```

We had a nonsensical response from the handler, but it was not returned. That is because in our middleware we returned an HTTPResponse object. Whenever you return a value from middleware—whether request or response—Sanic will assume that you are trying to end the HTTP life cycle and return immediately. Therefore, you should *never* return anything from middleware that meets the following criteria:

- Is not an HTTPResponse object

- Is not intended to interrupt the HTTP life cycle

This rule, however, does not apply to None values. You can still return None if you simply want to halt the execution of the middleware, as follows:

```
@app.on_request
async def check_for_politeness(request: Request):
    if "please" in request.headers:
        return None
    return text("You must say please")
```

> **Tip**
>
> A good thing to know about HTTP headers is that they are case-insensitive. Even though we did a check for the please header, we could just as well have received the same result using the following code:
>
> ```
> if "Please" in request.headers:
> ```

Let's see how this middleware plays out now when we access the endpoint, as follows:

```
$ curl localhost:7777/show-me-the-money
You must say please


$ curl localhost:7777/show-me-the-money -H
"Please: With a cherry on top"
```

The second request, it was allowed to proceed because it had the correct header. Therefore, we can see that returning None is also acceptable from middleware. If you are familiar with using continue inside of a Python loop, it has roughly the same impact: halt the execution and move onto the next step.

> **Important Note**
>
> Even though we were looking for the please value in the request headers, we were able to pass Please and for it to still work since headers are always case-insensitive.

Middleware and streaming responses

There is one more *gotcha* that you should know about middleware. Remember how we simply said that the middleware basically wraps before and after the route handler? This is not entirely true.

In truth, the middleware wraps the generation of the response. Since this *usually* happens in the return statement of a handler, that is why we take the simplistic approach.

This point can be easily seen if we revisit the *Chapter 5, Building Response Handlers* example with our streaming handler. Here is where we started:

```
@app.get("/")
async def handler(request: Request):
    resp = await request.respond()
    for _ in range(4):
        await resp.send(b"Now I'm free, free-falling")
        await asyncio.sleep(1)
    await resp.eof()
```

Let's add some print statements and some middleware so that we can examine the order of execution, as follows:

```
@app.get("/")
async def handler(request: Request):
    print("before respond()")
    resp = await request.respond()
    print("after respond()")

    for _ in range(4):
        print("sending")
        await resp.send(b"Now I'm free, free-falling")
        await asyncio.sleep(1)

    print("cleanup")
    await resp.eof()
    print("done")

@app.on_request
async def req_middleware(request):
    print("request middleware")

@app.on_response
async def resp_middleware(request, response):
    print("response middleware")
```

Now, we will hit the endpoint and look at our Terminal logs, as follows:

```
request middleware
before respond()
response middleware
after respond()
sending
(sanic.access)[INFO][127.0.0.1:49480]: GET http://
localhost:7777/  200 26
sending
sending
sending
cleanup
done
```

As we would expect, the request middleware runs first, and then we begin the route handler. But the response middleware runs immediately after we call `request.` `respond()`. For most use cases of response middleware (such as adding headers), this should not matter. It will, however, pose a problem if you absolutely must execute some bit of code *after* the route handler is complete. If this is the case, then your solution is to use signals, which we will explore in the next section. Specifically, we will see in the *Using built-in signals* section that the `http.lifecycle.response` signal will help us to execute the code after the handler in this situation.

Signals are sometimes a great replacement for middleware. While middleware is essentially a tool that allows us to extend business logic outside the confines of the route handler and to share it among different endpoints, we will learn that signals are more like breakpoints that allow us to inject code into the Sanic life cycle.

Leveraging signals for intra-worker communication

In general, Sanic tries to make it possible for developers to extend its capabilities to create custom solutions. This is the reason that when interfacing with Sanic, there are several options to inject custom classes to overtake, change, or otherwise extend its functionality. For example, did you know that you could swap out its HTTP protocol to essentially turn Sanic into a **File Transfer Protocol** (**FTP**) server (or any other **Transmission Control Protocol** (**TCP**)-based protocol)? Or, maybe you want to extend the router capabilities?

These sorts of customizations are rather advanced. We will not cover them in this book since for most use cases, it is the equivalent of hanging a picture nail on your wall with a sledgehammer.

The Sanic team introduced signals as a method to extend the functionality of the platform in a more user-friendly format. Very intentionally, setting up a signal handler looks and feels like a route handler, as illustrated in the following code snippet:

```
@app.signal("http.lifecycle.begin")
async def connection_begin(conn_info):
    print("Hello from http.lifecycle.begin")
```

You may be asking: *What exactly is this, and how can I use it?* In this example, we learn that `http.lifecycle.begin` is an event name. When Sanic opens an HTTP connection to a client, it dispatches this signal. Sanic will then look to see if any handlers are waiting for it and run them. Therefore, all we did was set up a handler to attach to that event. We will dig a little more into pre-defined events in this chapter, but first, let's have a closer examination of the structure and operation of signals.

Signal definitions

All signals are defined by their event name, which is composed of three segments. We just saw a signal event called `http.lifecycle.begin`. Obviously, the three segments are `http`, `lifecycle`, and `begin`. An event will *only* ever have three segments.

This is important to know because even though Sanic ships with a bunch of signals out of the box, it also allows us to create our own signals along the way. Therefore, we will need to follow the pattern. It is helpful to think of the first segment as a namespace, the middle as a reference, and the last as an action, sort of like this:

```
namespace.reference.action
```

Thinking in these terms helps me conceptualize them. I like to think of them as routes. In fact, they actually are! Under the hood, Sanic deals with signal handlers the same way as it does with route handlers because they inherit from the same base class.

If a signal is essentially a route, does that mean it can look for dynamic path parameters too? Yes! Check this out:

```
@app.signal("http.lifecycle.<foo>")
async def handler(**kwargs):
    print("Hello!!!")
```

Go hit any route in your application now, and we should see the following in our Terminal:

```
[DEBUG] Dispatching signal: http.lifecycle.begin
Hello!!!
[DEBUG] Dispatching signal: http.lifecycle.read_head
Hello!!!
[DEBUG] Dispatching signal: http.lifecycle.request
Hello!!!
[DEBUG] Dispatching signal: http.lifecycle.handle
Hello!!!
Request middleware
response middleware
[DEBUG] Dispatching signal: http.lifecycle.response
Hello!!!
[INFO] [127.0.0.1:39580]: GET http://localhost:7777/  200 20
[DEBUG] Dispatching signal: http.lifecycle.send
Hello!!!
[DEBUG] Dispatching signal: http.lifecycle.complete
Hello!!!
```

Before continuing on to see what kinds of signals are available, there is one more thing we need to be aware of: the condition. The `app.signal()` method accepts a keyword argument called `condition` that can help in limiting events that match on it. Only an event that is dispatched with the same condition will be executed.

We will look at a concrete example here:

1. Start by adding some request middleware, like this:

    ```
    @app.on_request
    async def req_middleware(request):
        print("request middleware")
    ```

2. Then, add a signal to attach to our middleware (this is a built-in, as we will see later), as follows:

    ```
    @app.signal("http.middleware.before")
    async def handler(**kwargs):
        print("Hello!!!")
    ```

3. Now, let's go take a look at our Terminal after we hit an endpoint, as follows:

   [DEBUG] Dispatching signal: http.middleware.before
   request middleware

   Hmmm—we see that the signal was dispatched and that our middleware ran, but our signal handlers did not. Why? `http.middleware.*` events are special in that they will only run when a specific **condition** is met. Therefore, we need to amend our signal definition to include the required condition.

4. Change your signal to add the condition, like this:

   ```
   @app.signal("http.middleware.before", condition={"attach_
   to": "request"})
   async def handler(**kwargs):
       print("Hello!!!")
   ```

5. Hit the endpoint again. We should now see the text as anticipated, as illustrated here:

   [DEBUG] Dispatching signal: http.middleware.before
   Hello!!!
   request middleware

Conditions are something that you can also add to your custom signal dispatches (keep reading ahead to the *Custom signals* section to learn more). It would look like this:

```
app.dispatch("custom.signal.event", condition={"foo": "bar"})
```

Most signal use cases will not need this approach. However, if you find the need for additional control on signal dispatching, it might just be the right tool for the job. Let's turn our attention back to Sanic's built-in signals and see what other kinds of events we can attach signals to.

Using built-in signals

There are many built-in signals that we can use. Take a look at the following tables and dog-ear this page in the book. I highly encourage you to come back to these tables often and look at your options when trying to solve a problem. While the implementations and usages we come up with in this book may be small, it is your job to learn the process so that you can more effectively solve your own application needs.

First are the signals related to routing that will execute on every request. You can see these here:

Event name	Arguments	Description
http.routing.before	request	When Sanic is ready to resolve the incoming path to a route
http.routing.after	request, route, kwargs, handler	Immediately after a route has been found

Table 6.2 – Available built-in routing signals

Second, we have the signals that are specifically related to the request/response life cycle, listed here:

Event name	Arguments	Description
http.lifecycle.begin	conn_info	When an HTTP connection is established
http.lifecycle.read_head	head	After an HTTP head is read, but before it is parsed
http.lifecycle.request	request	Immediately upon the creation of a Request object
http.lifecycle.handle	request	Before Sanic begins to handle a request
http.lifecycle.read_body	body	Every time bytes are read from a request body
http.lifecycle.exception	request, exception	When an exception is raised in a route handler or middleware
http.lifecycle.response	request, response	Just before a response is sent
http.lifecycle.send	data	Every time data is sent to an HTTP transport
http.lifecycle.complete	conn_info	When an HTTP connection is closed

Table 6.3 – Available built-in request/response life cycle signals

Third, we have the events that wrap around each middleware handler. These are not likely signals that you will use often. Instead, they primarily exist for the benefit of Sanic plugin developers. You can see them listed here:

Event name	Arguments	Conditions	Description

`http.middleware.before`	`request,`
	`response`
		`{"attach_to":`
		`"request"} or`
		`{"attach_to":`
		`"response"}`
			Before each middleware runs

`http.middleware.after`	`request,`
	`response`
		`{"attach_to":`
		`"request"} or`
		`{"attach_to":`
		`"response"}`
			After each middleware runs

Table 6.4 – Available built-in middleware signals

Finally, we have the server events. These signals are a one-to-one match with the listener events. Although you can call them as any other signal, there is a convenient decorator for each of them, as indicated in the descriptions in the following table:

Event name	Arguments	Description
`server.init.before`	`app, loop`	Before a server starts up (equivalent to `app.before_server_start`)
`server.init.after`	`app, loop`	After a server starts up (equivalent to `app.after_server_start`)
`server.shutdown.before`	`app, loop`	Before a server shuts down (equivalent to `app.before_server_stop`)
`server.shutdown.after`	`app, loop`	After a server shuts down (equivalent to `app.after_server_stop`)

Table 6.5 – Available built-in server life cycle signals

I want to share an anecdote that exemplifies the power of signals. I do a lot of support for Sanic users. If you have spent any time looking over the community resources (either the forums or the Discord server), you likely have seen me helping developers solve their problems. I really do enjoy this aspect of being involved in **open source software** (**OSS**).

On one occasion, I was contacted by someone who was having trouble with middleware. The goal was to use response middleware to log out helpful information about responses as they were being delivered from the server. The problem is that when an exception is raised in the middleware, it will halt the rest of the middleware from running. Therefore, this individual was not able to log every response. The requests that raised an exception in other response middleware never made it to the logger. The solution—as you have probably guessed—was to use signals. In particular, the `http.lifecycle.response` event worked perfectly for this use case.

To illustrate the point, here is some code:

1. Set up two middleware, one for logging and one for causing an exception. Remember—they need to be in reverse order from how you want them to run. Here's how to do this:

    ```
    @app.on_response
    async def log_response(request, response):
        logger.info("some information for your logs")



    @app.on_response
    async def something_bad_happens_here(request, response):
        raise InvalidUsage("Uh oh")
    ```

2. When we hit any endpoint, `log_response` will *never* be run.

3. To solve this, change `log_response` from middleware into a signal (which is as easy as changing the decorator), as follows:

    ```
    @app.signal("http.lifecycle.response")
    async def log_response(request, response):
        logger.info("some information for your logs")
    ```

4. Now, when we access the endpoint and experience the exception, we still get our logs as expected, as illustrated here:

    ```
    [ERROR] Exception occurred in one of response middleware
    handlers
    Traceback (most recent call last):
      File "/home/adam/Projects/Sanic/sanic/sanic/request.
    py", line 183, in respond
        response = await self.app._run_response_middleware(
      File "_run_response_middleware", line 22, in _run_
    ```

```
response_middleware
    from ssl import Purpose, SSLContext, create_default_
context
  File "/tmp/p.py", line 23, in something_bad_happens_
here
    raise InvalidUsage("Uh oh")
sanic.exceptions.InvalidUsage: Uh oh
[DEBUG] Dispatching signal: http.lifecycle.response
[INFO] some information for your logs
[INFO] [127.0.0.1:40466]: GET http://localhost:7777/  200 3
```

We can also use this exact same signal to solve one of our earlier problems. Remember when we were examining response middleware and had somewhat surprising results with a streaming handler? Earlier in the chapter, in the *Middleware and streaming responses* section, we noticed that the response middleware was actually called when the response object was created, not after the handler completed. We could use `http.lifecycle.response` to wrap up after our lyrics are done streaming, as follows:

```
@app.signal("http.lifecycle.response")
async def http_lifecycle_response(request, response):
    print("Finally... the route handler is over")
```

This might be another good time for you to put the book down and do some exploration. Go back to that earlier example with the streaming handler and play around with some of these signals. Take a look at the arguments they receive and think about how you might make use of them. It is also, of course, important to understand the order in which they are dispatched.

After you complete that, we will take a look at creating custom signals and events.

Custom signals

So far, we have been looking specifically at built-in signals, but they are sort of a narrow implementation of what Sanic signals have to offer. While it is helpful to think of them as breakpoints that allow us to insert functionality into Sanic itself, in truth, there is a more general concept at play.

Signals allow for intra-application communication. Because they can be dispatched asynchronously as background tasks, it can become a convenient method for one part of your application to inform another that something has happened. This introduces another important concept of signals: they can be dispatched as inline or as tasks.

So far, every single example we have seen with built-in signals is inline—that is to say that Sanic will halt the processing of a request until the signals are complete. This is how we can add functionality into the lifecycle while maintaining a consistent flow.

This might not always be desirable. In fact, often, when you want to implement your own solution with custom signals, having them run as a background task gives the application the ability to continue responding to the request while it goes and does something else.

Let's take logging, for example. Imagine that we are back in our example where we are building an e-commerce application. We want to augment our access logs to include information about the authenticated use (if any) and the number of items they have in their shopping cart. Let's take our earlier middleware example and convert it to signals, as follows:

1. We need to create a signal to pull the user and shopping cart information onto our request object. Again, we just need to change the first line so that the code looks like this:

    ```
    @app.signal("http.lifecycle.handle")
    async def fetch_user_and_cart(request):
        cart_id = request.cookies.get("cart")
        session_id = request.cookies.get("session")
        request.ctx.cart = await fetch_shopping_cart(cart_id)
        request.ctx.user = await fetch_user(session_id)
    ```

2. For the sake of our example, we want to throw together some quick models and fake getters, like this:

    ```
    @dataclass
    class Cart:
        items: List[str]


    @dataclass
    class User:
        name: str


    async def fetch_shopping_cart(cart_id):
        return Cart(["chocolate bar", "gummy bears"])


    async def fetch_user(session_id):
        return User("Adam")
    ```

3. This will be enough to get our example operational, but we want to be able to see it. For now, we will add a route handler that just outputs our `request.ctx` object, as follows:

```
@app.get("/")
async def route_handler(request: Request):
    return json(request.ctx.__dict__)
```

4. We should now see that our fake user and cart are available, as expected. The following snippet confirms this is the case:

```
$ curl localhost:7777 -H 'Cookie: cart=123&session_
id=456'
{
  "cart": {
    "items": [
      "chocolate bar",
      "gummy bears"
    ]
  },
  "user": {
    "name": "Adam"
  }
}
```

5. Since we want to use our own access logs, we should turn off Sanic's access logs. Back in *Chapter 2*, *Organizing a Project*, we decided we were going to run all of our examples like this:

```
$ sanic server:app -p 7777 --debug --workers=2
```

We are going to change that now. Add `--no-access-logs`, as follows:

```
$ sanic server:app -p 7777 --debug --workers=2
--no-access-logs
```

6. Now, we are going to add our own request logger. But to illustrate the point we are trying to make, we will manually make our signal take a while to respond, as shown in the following code snippet:

```
@app.signal("http.lifecycle.handle")
async def access_log(request):
    await asyncio.sleep(3)
```

```
    name = request.ctx.user.name
    count = len(request.ctx.cart.items)
    logger.info(f"Request from {name}, who has a cart
with {count} items")
```

7. When you access the endpoint, you will see the following output in your logs. You should also experience a delay before the logging appears and before your response is delivered:

    [DEBUG] Dispatching signal: http.lifecycle.request

    [DEBUG] Dispatching signal: http.lifecycle.handle

    [INFO] Request from Adam, who has a cart with 2 items

8. To fix this, we will create a custom signal for our logger and dispatch the event from `fetch_user_and_cart`. Let's make the following changes:

```
@app.signal("http.lifecycle.request")
async def fetch_user_and_cart(request):
    cart_id = request.cookies.get("cart")
    session_id = request.cookies.get("session")
    request.ctx.cart = await fetch_shopping_cart(cart_id)
    request.ctx.user = await fetch_user(session_id)
    await request.app.dispatch(
        "olives.request.incoming",
        context={"request": request},
        inline=True,
    )


@app.signal("olives.request.incoming")
async def access_log(request):
    await asyncio.sleep(3)
    name = request.ctx.user.name
    count = len(request.ctx.cart.items)
    logger.info(f"Request from {name}, who has a cart
with {count} items")
```

9. This time, when we go and access the endpoint, there are two things you need to pay attention to. First, your response should return almost immediately. The delayed response we experienced earlier should be gone. Second, the delay in the access log should remain.

What we have effectively done here is take any **input/output (I/O)** wait time in the logging away from the request cycle. To do this, we created a custom signal. That signal was called `olives.request.incoming`. There is nothing special about this—it is entirely arbitrary. The only requirement, as we discussed, is that it has three parts.

To execute the signal, we just need to call `app.dispatch` with the same name, as follows:

```
await app.dispatch("olives.request.incoming")
```

Because we wanted to have access to the `Request` object in `access_log`, we used the optional argument context to pass the object.

So, why did the `http.lifecycle.handle` signal delay the response but `olives.request.incoming` did not? Because the former was executed *inline* and the latter as a background task. Under the hood, Sanic calls dispatch with `inline=True`. Go ahead and add that to the custom dispatch to see how that impacts the response. Once again, both the logging and the response are now delayed. You should use this when you want your application to pause on the dispatch until all signals attached to it are done running. If that order is not important, you will achieve more performance if you leave it out.

There are a few more arguments that `dispatch` takes that might be helpful for you. Here is the function signature:

```
def dispatch(
    event: str,
    *,
    condition: Optional[Dict[str, str]] = None,
    context: Optional[Dict[str, Any]] = None,
    fail_not_found: bool = True,
    inline: bool = False,
    reverse: bool = False,
):
```

The arguments that this function accepts are outlined here:

- `condition`: Used as seen with the middleware signals to control additional matching (we saw this used by the `http.middleware.*` signals).

- `context`: Arguments that should be passed to the signal.

- `fail_not_found`: What if you dispatch an event that does not exist? Should it raise an exception or fail silently?

- `inline`: Run in a task or not, as discussed already.

- `reverse`: When there are multiple signals on an event, what order should they run in?

Signals are not the only way that you can take action on an event in Sanic. There are also tools that will allow you to wait for an event in arbitrary locations in your code. In the next section, we will look at how this can be accomplished.

Waiting on events

The last helpful thing about dispatching a signal event is that it can also be used like `asyncio` events to block until it is dispatched. The use case for this is different than with dispatching. When you dispatch a signal, you are causing some other operation to occur, usually in a background task. You should wait on a signal event when you want to pause an existing task until that event happens. This means that it will block the currently existing task, whether that is a background task or the actual request that is being handled. If this is used inside of a request/response life cycle—for example, if it were inside of a route handler or middleware—then the entire request would be blocked until the event is resolved. This may or may not be your desired behavior, so you should understand its impact.

The easiest way to show this is with a super simple loop that runs constantly in your application. Follow these next steps:

1. Set up your loop as shown in the following code snippet. Notice that we are using `app.event` with our event name. For simplicity, we are using a built-in signal event, but it could also be a custom one. For this to work, we would just need an `app.signal` method to be registered with the same name:

```
async def wait_for_event(app: Sanic):
    while True:
        print("> waiting")
        await app.event("http.lifecycle.request")
        print("> event found")


@app.after_server_start
async def after_server_start(app, loop):
    app.add_task(wait_for_event(app))
```

2.   Now, when we hit our endpoint, we should see this in the logs:

```
> waiting
[INFO] Starting worker [165193]
[DEBUG] Dispatching signal: http.lifecycle.request
> event found
> waiting
```

This might be a helpful tool especially if your application uses WebSockets. You might, for example, want to keep track of the number of open sockets. Feel free to turn back to the WebSockets example and see if you can integrate some events and signals into your implementation.

One more helpful use case is where you have a number of things that need to happen in your endpoint before you respond. You want to push off some work to a signal, but ultimately, it does need to be complete before responding.

We could do something like this. Set up the following handlers and signals:

```
@app.signal("registration.email.send")
async def send_registration_email(email, request):
    await asyncio.sleep(3)
    await request.app.dispatch("registration.email.done")


@app.post("/register")
async def handle_registration(request):
    await do_registration()

    await request.app.dispatch(
        "registration.email.send",
        context={
            "email": "alice@bob.co",
            "request": request,
        },
    )

    await do_something_else_while_email_is_sent()

    print("Waiting for email send to complete")
```

```
await request.app.event("registration.email.done")
print("Done.")


return text("Registration email sent")
```

Now, when we look at the Terminal, we should see this:

```
do_registration
Sending email
do_something_else_while_email_is_sent
Waiting for email send to complete
Done.
```

Since we know that sending the email will be an expensive operation, we send that off to the background while continuing with processing the request. By using app.event, we were able to wait for the registration.email.done event to be dispatched before responding that the email had in fact been sent.

One thing that you should make note of is that in this example, there is not actually a signal attached to registration.email.done. Out of the box, Sanic will complain and raise an exception. If you would like to use this pattern, you have three options, as outlined here:

1. Register a signal, like this:

```
@app.signal("registration.email.done")
async def noop():
    ...
```

2. Since we do not need to actually execute anything, we do not need a handler, so we can execute the following code:

```
app.add_signal(None, "registration.email.done")
```

3. Tell Sanic to automatically create all events when there is a dispatch, regardless of whether there is a registered signal. Here's how to do this:

```
app.config.EVENT_AUTOREGISTER = True
```

Now that we know there are several ways to control the execution of business logic within an HTTP life cycle, we will next explore some other things we can do to exploit our newfound tools.

Mastering HTTP connections

Earlier, in *Chapter 4*, *Ingesting HTTP Data*, we discussed how the HTTP life cycle represents a conversation between a client and a server. The client requests information, and the server responds. In particular, we likened it to a video chat with bi-directional communication. Let's dig into this analogy a little deeper to expand our understanding of HTTP and Sanic.

Rather than thinking about an HTTP request as the video chat, it is better to think of it as an individual conversation or—better yet—a single question and answer. It could go something like this:

> **Client**: Hi, my session ID is 123456, and my shopping cart ID is 987654. Can you tell me what other items I can buy?

> **Server**: Hi, Adam—you have pure olive oil and extra virgin olive oil in your cart already. You can add balsamic vinegar or red wine vinegar.

Sanic is a "performant" web framework because it is capable of having these conversations with multiple clients at the same time. While it is fetching the results for one client, it can begin conversations with other clients, like this:

> **Client 1**: What products do you sell?

> **Client 2**: How much does a barrel of olive oil cost?

> **Client 3**: What is the meaning of life?

By being capable of corresponding within multiple video chat sessions simultaneously, the server has become more efficient at responding. But what happens when one client has multiple questions? Starting and stopping the video chat for each *conversation* would be time-consuming and costly, as illustrated here:

> *Start video chat*

> **Client**: Here are my credentials—can I log in?

> **Server**: Hi, Adam—nice to see you again. Here is a session ID: 123456. Goodbye.

> *Stop video chat*

> *Start video chat*

> **Client**: Hi, my session ID is 123456. Can I update my profile information?

> **Server**: Oops, bad request. Looks like you did not send me the right data. Goodbye.

> *Stop video chat*

Every time that the video chat starts and stops, we are wasting time and resources. HTTP/1.1 sought to solve this problem by introducing persistent connections. This is accomplished with the `Keep-Alive` header. We do not need to worry specifically about how this header works from the client or server, as Sanic will take care of responding appropriately.

What we do need to understand is that it exists and that it includes a timeout. This means that Sanic will not close the connection to the client if another request comes within some timeout period. Here's an illustration of this:

Start video chat

Client: Here are my credentials—can I log in?

Server: Hi, Adam—nice to see you again. Here is a session ID: 123456.

Server: *waiting...*

Server: *waiting...*

Server: *waiting...*

Server: Goodbye.

Stop video chat

We have now created efficiency within a single video chat to allow for multiple conversations.

There are two practical concerns we need to think about here, as follows:

* How long should the server wait?

* Can we make the connection more efficient?

Keep-Alive within Sanic

Sanic will keep HTTP connections alive by default. This makes operations more performant, as we saw earlier. There may, however, be instances where this is undesirable. Perhaps you *never* want to keep these connections open. If you know that your application will never handle more than one request per client, then perhaps it is wasteful to use precious memory to keep open a connection that will never be reused. To turn it off, just set a configuration value on your application instance, like this:

```
app.config.KEEP_ALIVE = False
```

As you can probably guess, even the most basic web applications will never fall into this category. Therefore, even though we have the ability to turn off `KEEP_ALIVE`, you probably should not.

What you are more likely going to want to change is the timeout. By default, Sanic will keep connections open for 5 seconds. This may not seem long, but it should be long enough for most use cases without being wasteful. This is, however, Sanic just making a complete guess. You are more likely to know and understand the needs of your application, and you should feel free to tune this number to your needs. How? Again, with a simple configuration value, as illustrated here:

```
app.config.KEEP_ALIVE_TIMEOUT = 60
```

To give you some context, here is a snippet from the Sanic user guide that provides some insight into how other systems operate:

Apache httpd server default keepalive timeout = 5 seconds

Nginx server default keepalive timeout = 75 seconds

Nginx performance tuning guidelines uses keepalive = 15 seconds

IE (5-9) client hard keepalive limit = 60 seconds

Firefox client hard keepalive limit = 115 seconds

Opera 11 client hard keepalive limit = 120 seconds

Chrome 13+ client keepalive limit > 300+ seconds

Source: `https://sanic.dev/en/guide/deployment/configuration.html#keep-alive-timeout`

How do you know if you should increase the timeout? If you are building a **single-page application** (**SPA**) where your API is meant to power a JavaScript frontend, there is a high likelihood that your browser will make a lot of requests. This is generally the nature of how these frontend applications work. This would be especially true if you expect users to click a button, browse through some content, and click some more. The first thing that comes to my mind would be a web portal-type application where a single user might need to make dozens of calls within a minute, but they might be spaced out by some interval of browsing time. In this case, increasing the timeout to reflect the expected usage might make sense.

This does not mean that you should increase it too far. First, as we have seen previously, browsers generally have a limit on the maximum amount of time they will hold a connection open. Second, going too far with connection length can be wasteful and harmful to your memory performance. It is a balance that you are after. There is no one good answer, so you may need to experiment to see what works.

Caching data per connection

If you are thinking about ways you might exploit some of these tools for your application's needs, you might have noticed a potential efficiency you can create. Back at the beginning of this chapter, there is a table that lists all of the context (`ctx`) objects that are available to you in Sanic. One of them is connection-specific.

This means that not only are you able to create stateful requests, but you can also add state into a single connection. Our simple example will be a counter. Follow these next steps:

1. Start by creating a counter when the connection is established. We will use a signal for this, as follows:

    ```
    from itertools import count


    @app.signal("http.lifecycle.begin")
    async def setup_counter(conn_info):
        conn_info.ctx._counter = count()
    ```

2. Next, we will increment the counter on every request using middleware, like this:

    ```
    @app.on_request
    async def increment(request):
        request.conn_info.ctx.count = next(
            request.conn_info.ctx._counter
        )
    ```

3. Then, we will output that in our request body so that we can see what this looks like. Here's the code to do this:

    ```
    @app.get("/")
    async def handler(request):
        return json({"request_number": request.conn_info.ctx.
    count})
    ```

4. Now, we will issue multiple requests using `curl`. To do that, we just give it the **Uniform Resource Locator** (**URL**) multiple times, like this:

    ```
    $ curl localhost:7777 localhost:7777
    {"request_number":0}
    {"request_number":1}
    ```

This is, of course, a trivial example, and we could get that information from Sanic easily enough by executing the following code:

```
@app.get("/")
async def handler(request):
    return json(
        {
            "request_number": request.conn_info.ctx.count,
            "sanic_count": request.protocol.state["requests_
count"],
        },
    )
```

This could be extremely useful if you have some data that might be expensive to obtain but want it available for all requests. Coming back to our earlier roleplay model, it would be as if your server fetched some details when the video chat started. Now, every time the client asks a question, the server already has the details on hand in the cache.

> **Important Note**
>
> This does come with a warning. If your application is exposed through a proxy, it could be connection pooling. That is to say that the proxy could be taking requests from differing clients and bundling them together in one connection. Think of this as if your video chat session were not in someone's private home, but instead in the foyer of a large university dormitory. Anyone could walk up to the single video chat session and ask a question. You might not be guaranteed to have the same person all the time. Therefore, before you expose any sort of sensitive details on this object, you must know that it will be safe. A best practice might just be to keep the sensitive details on `request.ctx`.

Handling exceptions like a pro

In an ideal world, our applications would never fail, and users would never submit bad information. All endpoints would return a `200 OK` response all the time. This is, of course, pure fantasy, and no web application would be complete if it did not address the possibility of failures. In real life, our code will have bugs, there will be edge cases not addressed, and users will send us bad data and misuse the application. In short: our application will fail. Therefore, we must think about this constantly.

Sanic does, of course, provide some default handling for us. It includes a few different styles of exception handlers (**HyperText Markup Language** (**HTML**), **JavaScript Object Notation** (**JSON**), and text), and can be used both in production and development. It is of course unopinionated, and therefore likely inadequate for a decently sized application. We will talk more about fallback error handling in the *Fallback handling* section later. As we just learned, handling exceptions in an application is critical to the quality (and ultimately security) of a web application. We will now learn more about how to do that in Sanic.

Implementing proper exception handling

Before we look at how to handle exceptions with Sanic, it is important to consider that a failure to properly address this could become a security problem. The obvious way would be through inadvertent disclosure of sensitive information, which is known as *leaking*. This occurs when an exception is raised (by mistake or on purpose by the user) and your application reports back, exposing details about how the application is built or the data stored.

In a real-world worst-case scenario, I once had an old forgotten endpoint that no longer worked in one of my web applications. No one used it anymore, and I simply forgot that it existed or was even still live. The problem was that the endpoint did not have proper exception handling and errors were directly reported as they occurred. That means even *Failure to connect to database XYZ using username ABC and password EFG* messages were flowing right to anyone that accessed the endpoint. Oops!

Therefore, even though we do not discuss security concerns in general until *Chapter 7, Dealing with Security Concerns*, it does extend into the current exploration of exception handling. There are two main concerns here: providing exception messages with tracebacks or other implementation details, and incorrectly using 400 series responses.

Bad exception messages

While developing, it is super helpful to have as much information about your request as possible. This is why it would be desirable to have exception messages and tracebacks in your responses. When you are building your applications in debug mode, you will get all of these details, but make sure you turn it off in production! Just as I wish my applications only served a `200 OK` response all the time, I wish I never stumbled onto a website that accidentally leaked debug information to me. It happens out there in the wild, so be careful not to fall into that mistake.

What is perhaps more common is failing to properly consider the content of errors when responding. When writing messages that will reach the end user, keep in mind that you do not want to accidentally disclose implementation details.

Misusing statuses

Closely related to bad exceptions are exceptions that leak information about your application. Imagine that your bank website has an endpoint of `/accounts/id/123456789`. They do their due diligence and properly protect the endpoint so that only you can access it. That is not a problem. But what happens to someone that cannot access it? What happens when I try to access your bank account? Obviously, I would get a `401 Unauthorized` error because it is not my account. However, as soon as you do that, the bank is now acknowledging that 123456789 is a legitimate account number. Therefore, I *highly* encourage you to use the following information and commit it to memory:

Status	Description	Sanic exception	When to use
400	Bad Request	`InvalidUsage`	When any user submits data in an unexpected form, or they otherwise did something your application does not intend to handle
401	Unauthorized	`Unauthorized`	When an unknown user has not been authenticated—in other words, you do not know who the user is
403	Forbidden	`Forbidden`	When a known user does not have permissions to do something on a KNOWN resource
404	Not Found	`NotFound`	When any user attempts access on a hidden resource

Table 6.6 – Sanic exceptions for common 400 series HTTP responses

Perhaps the biggest failure here is when people inadvertently expose the existence of a hidden resource with a `401` or `403` error code. Your bank should have instead sent me a `404` error code and directed me to a `page not found` response. This is not to say that you should always favor a `404` error code, but it is to your benefit from a security perspective to think about who could be accessing the information, and what they should or should not know about it. Then, you can decide which error response is appropriate.

Responses through raising an exception

One of the most convenient things about exception handling in Sanic is that it is relatively trivial to get started. Remember—we are just coding a Python script here, and you should treat it like you might anything else. What should you do when something goes wrong? Raise an exception! Here is an example:

1. Make a simple handler—we will ignore the return value here since we do not need it to prove our point. Use your imagination for what could be beyond the . . . shown here:

    ```
    @app.post("/cart)
    async def add_to_cart(request):
        if "name" not in request.json:
            raise InvalidUsage("You forgot to send a product
    name")
        . . .
    ```

2. Next, we will submit some JSON to the endpoint, leaving out the name property. Make sure to use -i so that we can inspect the response headers, as illustrated in the following code snippet:

    ```
    $ curl localhost:7777/cart -X POST -d '{}' -i
    HTTP/1.1 400 Bad Request
    content-length: 83
    connection: keep-alive
    content-type: text/plain; charset=utf-8

    400 — Bad Request
    ==================
    You forgot to send a product name
    ```

Take note of how we received a 400 response but did not actually return a response from the handler. This is because if you raise any exception from sanic. exceptions, it could be used to return an appropriate status code. Furthermore, you will find that many of the exceptions in that module (such as InvalidUsage) have a default status_code value. This is why when you raise InvalidUsage, Sanic will respond with a 400 error code. You could, of course, override the status code by passing a different value. Let's see how that would work.

3. Set up this endpoint and change status_code to something other than 400, as illustrated here:

```
@app.post("/coffee")
async def teapot(request):
    raise InvalidUsage("Hmm...", status_code=418)
```

4. Now, let's access it as follows:

```
$ curl localhost:777/coffee -X POST -i
HTTP/1.1 418 I'm a teapot
content-length: 58
connection: keep-alive
content-type: text/plain; charset=utf-8

418 — I'm a teapot
==================
Hmm...
```

As you can see, we passed the 418 status code to the exception. Sanic took that code and properly converted it to the appropriate HTTP response: 418 I'm a teapot. If you did not catch the HTTP humor when we discussed it earlier, you can look it up in **Request for Comments (RFC)** *7168*, § *2.3.3* (https://datatracker.ietf.org/doc/html/rfc7168#section-2.3.3).

Here is a reference of all of the built-in exceptions and their associated response codes:

Exception	Status
HeaderNotFound	400 Bad Request
InvalidUsage	400 Bad Request
Unauthorized	401 Unauthorized
Forbidden	403 Forbidden
FileNotFound	404 Not Found
NotFound	404 Not Found
MethodNotSupported	405 Method Not Allowed
RequestTimeout	408 Request Timeout
PayloadTooLarge	413 Request Entity Too Large
ContentRangeError	416 Request Range Not Satisfiable

Exception	Status
`InvalidRangeType`	416 Request Range Not Satisfiable
`HeaderExpectationFailed`	417 Expectation Failed
`ServerError`	500 Internal Server Error
`URLBuildError`	500 Internal Server Error
`ServiceUnavailable`	503 Service Unavailable

Table 6.7 – Sanic exceptions with built-in HTTP responses

It is, therefore, a really good practice to make usage of these status codes. An obvious example might be when you are looking up something in your database that does not exist, as illustrated in the following code snippet:

```
@app.get("/product/<product_id:uuid>")
async def product_details(request, product_id):
    try:
        product = await Product.query(product_id=product_
id)
    except DoesNotExist:
        raise NotFound("No product found")
```

Using Sanic exceptions is perhaps one of the easiest solutions to getting appropriate responses back to the users.

We could, of course, go one step further. We can make our own custom exceptions that subclass from the Sanic exceptions to leverage the same capability.

5. Create an exception that subclasses one of the existing Sanic exceptions, as follows:

```
from sanic.exceptions import InvalidUsage


class MinQuantityError(InvalidUsage):
    ...
```

6. Raise it when appropriate, like this:

```
@app.post("/cart")
async def add_to_cart(request):
    if request.json["qty"] < 5:
        raise MinQuantityError(
```

```
                      "Sorry, you must purchase at least 5 of this
       item"
                )
```

7. Here, see the error when we have a bad request (fewer than 5 items):

```
$ curl localhost:777/cart -X POST -d '{"qty": 1}' -i
HTTP/1.1 400 Bad Request
content-length: 98
connection: keep-alive
content-type: text/plain; charset=utf-8


400 — Bad Request
==================
Sorry, you must purchase at least 5 of this item
```

Using and reusing exceptions that inherit from SanicException is highly encouraged. It not only is a good practice because it provides a consistent and clean mechanism for organizing your code, but it also makes it easy to provide the appropriate HTTP responses.

So far throughout this book, when we have hit an exception with our client (such as in the last example), we have received a nice textual representation of that error. In the next section, we will learn about the other types of exception output and how we can control this.

Fallback handling

Let's face it: formatting exceptions is mundane. There is little doubt that using the skills we have learned so far, we could build our own set of exception handlers. We know how to use templates, catch exceptions, and return HTTP responses with an error status. But creating those takes time and a lot of boilerplate code.

This is why it is nice that Sanic offers three different exception handlers: HTML, JSON, and plain text. For the most part, the examples in this book have used the plain text handlers only because this has been a more suitable form for presenting information in a book. Let's go back to our example where we raised a NotFound error and see what it might look like with each of the three types of handlers.

HTML

1. Let's set up our endpoint to raise an exception, as follows:

```
@app.get("/product/<product_name:slug>")
async def product_details(request, product_name):
    raise NotFound("No product found")
```

2. Tell Sanic to use HTML formatting. We will look into configurations in more detail in *Chapter 8*, *Running a Sanic Server*. For now, we will just set the value right after our Sanic instance, like this:

```
app = Sanic(__name__)
app.config.FALLBACK_ERROR_FORMAT = "html"
```

3. Open up a web browser and go to our endpoint. You should see something like this:

Figure 6.1 – Example 404 HTML page in Sanic

JSON

1. Use the same setup as before, as shown here, but change the fallback format to json:

```
app.config.FALLBACK_ERROR_FORMAT = "json"
```

2. This time, we will access the endpoint with curl, as follows:

```
$ curl localhost:7777/product/missing-product
{
```

```
        "description": "Not Found",
        "status": 404,
        "message": "No product found"
    }
```

Instead of the nicely formatted HTML that we saw with the previous example, our exception has been formatted into JSON. This is more appropriate if your endpoint will—for example—be used by a JavaScript browser application.

Text

1. Again using the same setup, we will change the fallback format to `text`, as follows:

    ```
    app.config.FALLBACK_ERROR_FORMAT = "text"
    ```

2. We will again use `curl` to access the endpoint, like this:

    ```
    $ curl localhost:7777/product/missing-product
    404 — Not Found
    ===============
    No product found
    ```

As you can see, there are three convenient formatters for our exceptions that may be appropriate in different circumstances.

Auto

The previous three examples used `FALLBACK_ERROR_FORMAT` to show that there are three types of built-in error formats. There is a fourth option for setting `FALLBACK_ERROR_FORMAT`: `auto`. It would look like this:

```
app.config.FALLBACK_ERROR_FORMAT = "auto"
```

When the format is set to `auto`, Sanic will look at the routing handler and the incoming request to determine what is likely to be the most appropriate handler to use. For example, if a route handler always uses the `text()` response object, then Sanic will assume that you want the exceptions to also be formatted in text format. The same applies to `html()` and `json()` responses.

Sanic will even go one step further than that when in `auto` mode. It will analyze the incoming request to look at the headers to make sure that what it *thinks* is correct matches with what the client said that it wants to receive.

Manual override per route

The last option we have is to set the error format on an individual route inside of the route definition. This would allow us to be specific and deviate from the fallback option if needed. Follow these steps:

1. Consider the example where we set the fallback to `html`, as shown here:

    ```
    app.config.FALLBACK_ERROR_FORMAT = "html"
    ```

2. Let's now change our route definition from the beginning of this section to look like the following with a specific defined `error_format` value:

    ```
    @app.get("/product/<product_name:slug>", error_
    format="text")
    async def product_details(request, product_name):
        raise NotFound("No product found")
    ```

3. As you might already be able to guess, we will *not* see a formatted HTML page, but instead will see the plain text from earlier, as illustrated here:

    ```
    $ curl localhost:7777/product/missing-product
    404 — Not Found
    ================
    No product found
    ```

If you are using Sanic to develop an API to power a browser-based or mobile **user interface** (**UI**), then you likely will not need to have route-level overrides. In this instance, you would usually want to see the `FALLBACK_ERROR_FORMAT` value for the entire application. This pattern, however, could be helpful if you have some endpoints that will be returning HTML content. In the next section, we will take exceptions one step further to see how we can intercept them to provide appropriate responses to our end users.

Catching exceptions

Although Sanic conveniently handles a lot of exceptions for us, it goes without saying that it cannot anticipate every error that could be raised in an application. We thus need to think about how we want to handle exceptions that come from outside of Sanic or, rather, how to handle exceptions that are not manually raised by our application using one of the Sanic exceptions that conveniently adds a response code.

Returning to our e-commerce example, let's imagine that we are using a third-party vendor for handling our credit card transactions. They have conveniently provided us with a module that we can use to process credit cards. When something goes wrong, their module will raise a `CreditCardError` response. Our job now is to make sure that our application is ready to handle this error.

Before we do that, however, let's see why this is important, as follows:

1. Imagine that this is our endpoint:

```
@app.post("/cart/complete")
async def complete_transaction(request):

    ...

    await submit_payment(...)

    ...
```

2. Now, we access the endpoint, and if there is an error, we get this response:

```
$ curl localhost:7777/cart/complete -X POST
500 — Internal Server Error
==============================
The server encountered an internal error and cannot
complete your request.
```

That is not a very helpful message. If we look at our logs, however, we might see this:

```
[ERROR] Exception occurred while handling uri: 'http://
localhost:7777/cart/complete'
Traceback (most recent call last):
  File "handle_request", line 83, in handle_request
    """
  File "/path/to/server.py", line 19, in complete_
transaction
    await submit_payment(...)
  File "/path/to/server.py", line 13, in submit_payment
    raise CreditCardError("Expiration date must be in
format: MMYY")
CreditCardError: Expiration date must be in format: MMYY
[INFO] [127.0.0.1:58334]: POST http://localhost:7777/cart/
complete  500 144
```

That error looks potentially far more helpful to our users. It specifically has information that might be pertinent to return to the user.

One solution could, of course, just be to catch the exception and return the response that we want, like this:

```
@app.post("/cart/complete")
async def complete_transaction(request):
    ...
    try:
        await submit_payment(...)
    except CreditCardError as e:
        return text(str(e), status=400)
    ...
```

This pattern is not ideal, however. It would require a lot of extra code when we need to catch every potential exception in various locations in the application to cast them to responses. This also would turn our code into a giant mess of try/except blocks and make things harder to read and, ultimately, maintain. In short, it would go against some of the development principles we established early on in this book.

A better solution would be to add an application-wide exception handler. This tells Sanic that anytime this exception bubbles up, it should catch it and respond in a certain way. It looks very much like a route handler, as we can see here:

```
@app.exception(CreditCardError)
async def handle_credit_card_errors(request, exception):
    return text(str(exception), status=400)
```

Sanic has now registered this as an exception handler and will use it anytime that a CreditCardError response is raised. Of course, this handler is super simplistic, but you might imagine that it could be used for the following: extra logging, providing request context, sending out an emergency alert notification to your **development-operations (DevOps)** team at 3 a.m., and so on.

> **Tip**
> Error handlers are not limited to your application instance. Just as with other regular route handlers, they can be registered on your blueprint instances to be able to customize error handling for a specific subset of your application.

Exception handling is an incredibly important part of application development. It is an immediate differentiator between amateur applications and professional applications. We now know how we can use exceptions to provide not only helpful messages to our users but also to provide proper HTTP response codes. We now move on to another topic (background processing) that can really help to take your applications to the next level.

Background task processing

There comes a time in the development of most applications where the developers or users start to notice the application is feeling a bit slow. Some operations seem to take a long time and it is harming the usability of the rest of the application. It could be computationally expensive, or it could be because of a network operation reaching out to another system.

Let's imagine that you are in this scenario. You have built a great application and an endpoint that allows users to generate a **Portable Document Format** (**PDF**) report with the click of a button, showing all kinds of fancy data and graphs. The problem is that to retrieve all the data and then crunch the numbers seems to take 20 seconds. That's an eternity for an HTTP request! After spending time squeezing as much performance out of the report generator as you can, you are finally at the conclusion that it runs as fast as it can. What can you do?

Push it to the background.

When we say *background processing*, what we really mean is a solution that allows the current request to complete without having finalized whatever needs to be done. In this example, it would mean completing the request that *starts* the report generation before it is actually finished. Whenever and wherever you can, I recommend pushing work to the background. Earlier, in the *Waiting on events* section of this chapter, we saw a use case for sending out registration emails in the background. Indeed, the usage of signals (as described earlier) is a form of background processing. It is, however, not the only tool Sanic provides.

Adding tasks to the loop

As you may already know, one of the cornerstones of the `asyncio` library is tasks. They are essentially the unit of processing that is responsible for running asynchronous work on the loop. If the concept of a task or task loop is still foreign to you, it might be a good time to do a little research on the internet before continuing.

In a typical scenario, you can generate a task by getting access to the event loop and then calling `create_task`, as seen here:

```
import asyncio

async def something():
    ...
async def main():
loop = asyncio.get_running_loop()
loop.create_task(something())
```

This is probably not new to you, but what this does is start running `something` in a task outside of the current one.

Sanic adds a simple interface for creating tasks, as shown here:

```
async def something():
    ...
app.add_task(something)
```

This is probably the simplest form of background processing and is a pattern that you should get comfortable using. Why use this over `create_task`? For these three reasons:

- It is easier since you do not need to fetch the loop.
- It can be used in the global scope before the loop has started.
- It can be called or not called, and also with or without the application instance as an argument.

To illustrate the flexibility, contrast the previous example with this:

```
from sanic import Sanic
from my_app import something

app = Sanic("MyAwesomeApp")
app.add_task(something(app))
```

> **Tip**
> If the task is not called, as in the first example, Sanic will introspect the function to see if it expects the app instance as an argument and inject it.

`asyncio` tasks are very helpful, but sometimes you need a more robust solution. Let's see what our other options are.

Integrating with an outside service

If there is work to be done by your application, but it is outside of the scope of your API for whatever reason, you might want to turn to an off-the-shelf solution. This comes in the form of another service that is running somewhere else. The job of your web API now is to feed work into that service.

In the Python world, the classic framework for this kind of work is `Celery`. It is of course not the only option, but since this book is not about deciding what to use, we will show `Celery` as an example because it is widely used and known. In short, `Celery` is a platform with workers that read messages from a queue. Some client is responsible for pushing work to the queue, and when a worker receives the message, it executes the work.

For `Celery` to operate, it runs a process on a machine somewhere. It has a set of known operations that it can perform (that are also called *tasks*). To initiate a task, an outside client needs to connect to it through a broker and send instructions to run the task. A basic implementation might look like this:

1. We set up a client to be able to communicate with the process. A common place to put this is on the `application.ctx` object to make it usable anywhere in the application, as illustrated in the following code snippet:

    ```
    from celery import Celery


    @app.before_server_start
    def setup_celery(app, _):
        app.ctx.celery = Celery(...)
    ```

2. To use it, we simply call the client from the route handler to push some work to `Celery`, like this:

    ```
    @app.post("/start_task")
    async def start_task(request):
        task = request.app.ctx.celery.send_task(
            "execute_slow_stuff",
            kwargs=request.json
        )
        return text(f"Started task with {task.id=}",
    status=202)
    ```

An important thing to point out here is that we are using a `202 Accepted` status to tell whoever requested it that the operation has been accepted for processing. No guarantee is being made that it is done or will be done.

After examining `Celery`, you may be thinking that it is overkill for your needs, but `app.add_task` does not seem to be enough. Next, we look at how you could develop your own in-process queue system.

Designing an in-process task queue

Sometimes, the *obvious* goldilocks solution for your needs is to build something entirely confined to Sanic. It will be easier to manage if you have only one service to worry about instead of multiples. You may still want to keep the idea of *workers* and a *task queue* without the overhead required in implementing a service such as `Celery`. So, let's build something that you can hopefully use as a launching point for something even more amazing in your applications.

Before we go any further, let's change the name from *task queue* to *job queue*. We do not want to confuse ourselves with `asyncio` tasks, for example. For the rest of this section, the word *task* will relate to an `asyncio` task.

To begin, we will develop a set of needs for our job queue, as follows:

- There should be one or more *workers* that are capable of executing jobs outside of the request/response cycle.

- They should execute jobs in a **first-in, first-out** (**FIFO**) order.

- The completion order of jobs is not important (for example, job A starts before job B, but it does not matter which one finishes first).

- We should be able to check on the state of a job.

Our strategy to achieve this will be to build out a framework where we have a *worker* that is itself a background task. Its job will be to look for jobs inside of a common queue and execute them. The concept is very similar to `Celery`, except we are handling it all within our Sanic application with `asyncio` tasks. We are going to walk through the source code to accomplish this, but not all of it. Implementation details not relevant to this discussion will be skipped here. For full details, please refer to the source code in the GitHub repository at `https://github.com/PacktPublishing/Web-Development-with-Sanic/tree/main/Chapter06/inprocess-queue`.

Follow these next steps:

1. To begin, let's set up a very simple application with a single blueprint, as follows:

    ```
    from sanic import Sanic
    from job.blueprint import bp


    app = Sanic(__name__)
    app.config.NUM_TASK_WORKERS = 3
    app.blueprint(bp)
    ```

2. That blueprint will be the location where we will attach some listeners and our endpoints, as illustrated here:

    ```
    from sanic import Blueprint
    from job.startup import (
        setup_task_executor,
        setup_job_fetch,
        register_operations,
    )
    from job.view import JobListView, JobDetailView

    bp = Blueprint("JobQueue", url_prefix="/job")

    bp.after_server_start(setup_job_fetch)
    bp.after_server_start(setup_task_executor)
    bp.after_server_start(register_operations)

    bp.add_route(JobListView.as_view(), "")
    bp.add_route(JobDetailView.as_view(), "/<uid:uuid>")
    ```

As you can see, we have three listeners that we need to run: `setup_job fetch`, `setup_task_executor`, and `register_operations`. We also have two views: one is a list view and the other a detail view. Let's take each of these items in turn to see what they are, as follows:

1. Since we want to store the state of our tasks, we need some sort of a datastore. To keep things really simple, I created a file-based database called `FileBackend`, as illustrated in the following code snippet:

    ```
    async def setup_job_fetch(app, _):
        app.ctx.jobs = FileBackend("./db")
    ```

2. The functionality of this job management system will be driven from our job queue, which will be implemented with `asyncio.Queue`. So, we next need to set up our queue and workers. Here's the code we need to accomplish this:

```
async def setup_task_executor(app, _):
    app.ctx.queue = asyncio.Queue(maxsize=64)
    for x in range(app.config.NUM_TASK_WORKERS):
        name = f"Worker-{x}"
        print(f"Starting up executor: {name}")


    app.add_task(worker(name, app.ctx.queue, app.ctx.jobs))
```

After creating our queue, we create one or more background tasks. As you can see, we are simply using Sanic's `add_task` method to create a task from the `worker` function. We will see that function in just a moment.

3. The last listener we need will set up an object that will be used to hold all of our potential operations, as illustrated in the following code snippet:

```
async def register_operations(app, _):
    app.ctx.registry = OperationRegistry(Hello)
```

To remind you, an `Operation` is something that we want to run in the background. In this example, we have one operation: `Hello`. Before looking at the operation, let's look at the two views.

4. The list view will have a `POST` call that is responsible for pushing a new job into the queue. You can also imagine that this would be an appropriate place to make an endpoint that lists all of the existing jobs (paginated, of course). First, it will need to get some data from the request, with the following code:

```
class JobListView(HTTPMethodView):
    async def post(self, request):
        operation = request.json.get("operation")
        kwargs = request.json.get("kwargs", {})
        if not operation:
            raise InvalidUsage("Missing operation")
```

Here, we perform some very simple data validation. In a real-world scenario, you might want to do some more to make sure that the request JSON conforms to what you are expecting.

5. After validating the data, we can push information about the job to the queue, like this:

```
uid = uuid.uuid4()
await request.app.ctx.queue.put(
    {
        "operation": operation,
        "uid": uid,
        "kwargs": kwargs,
    }
)
return json({"uid": str(uid)}, status=202)
```

We created a UUID. This unique ID will be used both in storing the job in our database and retrieving information about it later. Also, it is important to point out that we are using the `202 Accepted` response since it is the most appropriate form.

6. The detail view is very simple. Using the unique ID, we simply look it up in the database and return it, like this:

```
class JobDetailView(HTTPMethodView):
    async def get(self, request, uid: uuid.UUID):
        data = await request.app.ctx.jobs.fetch(uid)
        return json(data)
```

7. Coming back to our `Hello` operation, we will build it now, like this:

```
import asyncio
from .base import Operation

class Hello(Operation):
    async def run(self, name="world"):
        message = f"Hello, {name}"
        print(message)
        await asyncio.sleep(10)
        print("Done.")
        return message
```

As you can see, it is a simple object that has a `run` method. That method will be called by the worker when running a job.

8. The worker is really nothing more than an `async` function. Its job will be to run a never-ending loop. Inside that loop, it will wait until there is a job in the queue, as illustrated in the following code snippet:

```
async def worker(name, queue, backend):
    while True:
        job = await queue.get()
        if not job:
            break


        size = queue.qsize()
        print(f"[{name}] Running {job}. {size} in queue.")
```

9. Once it has the information about how to run a job, it needs to create a job instance and execute it, as follows:

```
job_instance = await Job.create(job, backend)


async with job_instance as operation:
    await job_instance.execute(operation)
```

Once this is complete, we can finally start to interact with the API. Now is your chance to play with the server that we just created. Here are a few helpful `curl` commands you can try on your own:

```
$ curl localhost:7777/job -X POST -d '{"operation": "hello"}'
$ curl localhost:7777/job -X POST -d '{"operation": "hello",
"kwargs": {"name": "Adam"}}'
$ curl localhost:7777/job/<UID FROM PREVIOUS COMMANDS>
```

A couple of final things to say about this solution: one of its biggest faults is that it has no recovery. If your application crashes or restarts, there is no way to continue processing a job that had already begun. In a true task management process, this is usually an important feature. Therefore, in the GitHub repository, in addition to the source used to build this solution, you will find source code for a *subprocess* task queue. I will not walk you through the steps to build it since it is largely a similar exercise, with a lot of the same code. However, it differs from this solution in two important ways: it does have the ability to recover and restart an unfinished job, and instead of running in `asyncio` tasks, it leverages Sanic's process management listeners to create a subprocess using multiprocessing techniques. Please take some time to look through the source code there as you continue to learn and work your way through this book.

Summary

In my opinion, one of the biggest leaps that you can make as an application developer is devising strategies to abstract a solution to a problem and reusing that solution in multiple places. If you have ever heard of the **Don't Repeat Yourself** (**DRY**) principle, this is what I mean. Applications are seldom ever *complete*. We develop them, maintain them, and change them. If we have too much repetitive code or code that is too tightly coupled to a single use case, then it becomes more difficult to change it or adapt it to different use cases. Learning to generalize our solutions mitigates this problem.

In Sanic, this means taking logic out of the route handlers. It is best if we can minimize the amount of code in individual handlers, and instead place that code in other locations where it can be reused by other endpoints. Did you notice how the route handlers in the final example in the *Designing an in-process task queue* section had no more than a dozen lines? While the exact length is not important, it is helpful to keep these clean and short and place your logic somewhere else.

Perhaps one of the biggest takeaways from this chapter should be that there is usually not a single way to do something. Often, we can use a mixture of methodologies to achieve our goal. It is then the job of the application developer to look at the tool belt and decide which tool is best for any given situation.

For this reason, as a Sanic developer, you should learn how to devise strategies to respond to web requests outside of the route handler. In this chapter, we learned about some tools to help you accomplish this using middleware, built-in and custom signals, connection management, exception handling, and background processing. Again, think of these as your core tools in your tool belt. Got a screw that needs tightening? Pull out your middleware. Need to drill a hole in some wood? Time to grab for signals. The more familiar you become with basic building blocks such as these in Sanic, the greater your understanding will be of how to piece together a professional-grade application.

It is your job now to play with these and internalize them on your way to becoming a better developer.

We have scratched the surface of security-related issues. In the next chapter, we will take a closer look at how we can protect our Sanic applications.

7
Dealing with Security Concerns

When you're building a web application, it may be very tempting to sit down, plan out your functionality, build it, test it, and only then come back to think about security. For example, when you're building a **single-page application** (**SPA**), you may not even consider CORS until the first time you see this message in the browser while testing:

Cross-Origin Request Blocked: The Same Origin Policy disallows reading the remote resource at $somesite.

To a large extent, this is how we have been building in this book. We see a feature and build it. Anytime we have come across a potential security issue in this book, we have pushed it to a later date. Finally, we are at the point where we will learn how to deal with security issues in Sanic. The topic of web security is, of course, extremely broad, and it is beyond the scope of this book to provide an exhaustive study.

Instead, in this chapter, we will cover the following topics:

- Setting up an effective CORS policy
- Protecting applications from CSRF
- Protecting your Sanic app with authentication

In particular, we want to gain a basic understanding of the security issues so that we can build Sanic solutions to solve them. The bigger takeaway from this chapter will be to make you feel comfortable enough with these topics that they do not become afterthoughts. When these issues are broken down, we will see that building them into application design from the beginning will make them more effective and less burdensome to implement.

Technical requirements

The requirements for this chapter will, once again, build upon what we used in the previous chapters. Since web security often includes interaction between frontend JavaScript applications and backend Python applications, we will look at some examples that use JavaScript that are widely available in major web browsers. You can find all of the source code for this chapter at `https://github.com/PacktPublishing/Python-Web-Development-with-Sanic/tree/main/Chapter07`.

In addition, we are going to use three common (and battle-tested) security libraries: `cryptography`, `bcrypt`, and `pyjwt`. If you do not already have them installed in your virtual environment, you can add them now by running the following code:

```
$ pip install cryptography bcrypt pyjwt
```

Let's begin with setting up a CORS policy.

Setting up an effective CORS policy

If you are building a web application where the server exclusively responds to requests on a single computer, and that computer is physically disconnected from the internet, perhaps this section is not as relevant to you. For anyone else, pay attention! To be clear, you are part of "anyone else." This is important stuff.

In simple terms, **cross-origin resource sharing** (**CORS**) is a fancy way of saying *accessing one domain from another domain with a browser*. Without an effective strategy to handle this, your application could open up a security risk for your users.

What is the security issue with ineffective CORS?

The modern web uses a lot of JavaScript in web browsers. This enables all kinds of interactive and quality user experiences. One of those capabilities is to issue requests for data on behalf of the user without them knowing about it. This feature is one of the biggest differentiators between web applications today and web applications from the late 1990s. Requesting data while the user is on a website is what makes web pages feel like applications; that is, it makes them interactive and engaging.

So, imagine that you have a hypothetical application that appears to the user as `https://superawesomecatvideos.com`. It is a super successful website, and lots of people like to come to visit it to see their favorite cat videos. If it starts requesting information in the background (because of a hacker attack, or otherwise) from `https://mybank.com`, well, we do not want to allow that to succeed. There is no reason that the Super Awesome Cat Videos website should be able to access anything from My Bank, especially if I have an authenticated web session on My Bank's website.

For this reason, web browsers will not allow this by default because of the **same-origin policy**. This means that web applications may only interact with resources that are of the same origin. An origin is comprised of the following components:

- The HTTP scheme
- The domain
- The port

Let's look at some examples of URLs that are and are not considered to be same-origin:

URL A	URL B	Same-Origin?
`http://sacv.com`	`http://sacv.com`	Yes
`http://sacv.com`	`http://sacv.com/about`	Yes, the path does not matter
`http://sacv.com`	`https://sacv.com`	No, different HTTP schemes
`http://sacv.com`	`http://sacv.com:8080`	No, different ports
`http://sacv.com`	`http://api.sacv.com`	No, different domains

Table 7.1 – Comparison of URLs and their same-origin status

We'll assume that our Super Awesome Cat Video website also has `sacv.com` as its domain. For example, if `https://superawesomecatvideos.com` wants to load `https://superawesomecatvideos.com/catvid1234.mp4`, then that is fine. When the only difference is the path or resource being loaded, the URLs are considered same-origin. In our example, both URLs contain the same HTTP scheme, domain, and port designation. However, when the same website, `https://superawesomecatvideos.com`, tries to fetch data from `https://api.superawesomecatvideos.com/videos`, uh oh—error time. These are the sorts of potential attack vectors that the same-origin policy is meant to protect you from. So, the question becomes: how can we allow legitimate cross-origin requests without allowing *all* cross-origin requests? The answer is that we essentially need to create a whitelist and let the browser know which origins our server will accept requests from.

Let's build a super simple example that will show us the problem. We are going to build two web servers here. One will be a stand-in for the frontend application, while the other will be the backend, which is meant to feed data to the frontend:

1. We will begin by building and running a simple API endpoint that looks no different from anything we have seen previously. Stand up the application using the same method we have already used. Here is what your endpoint should look like:

```
@app.get("/<name>")
async def handler(request, name):
    return text(f"Hi {name}")
```

You should now have a Sanic server running on port 7777 using what we have already learned. You can test it out by accessing http://localhost:7777/Adam.

2. Create a directory somewhere and add a file called index.html to it. For my example, it will be /path/to/directory:

```
<!DOCTYPE html>
<html lang="en">

    <head>
        <meta charset="utf-8" />
        <meta name="viewport" content="width=device-
width, initial-scale=1">
        <title>CORS issue</title>

    <body>
        <h1>Loading...</h1>
        <script>
            const element = document.querySelector("h1")
            fetch("http://localhost:7777/Adam")
            .then(async response => {
                const text = await response.text()
                element.innerHTML = text
            })
        </script>
    </body>
```

As you can see, this application will run a background request to our application that is running on `http://localhost:7777`. After it gets the necessary content, it will dump it on the screen in place of the `Loading ...` text.

3. To run this application, we are going to use a neat little trick that Sanic includes called *Sanic Simple Server*. Instead of building a Sanic application, we will point the Sanic CLI at a directory and it will serve that for us as a website:

```
$ sanic -s /path/to/directory
```

> **Tip**
>
> This is a super helpful tool to keep in your back pocket, even when you're not building a Sanic application. While developing, I often find a need to quickly stand up a web application to view static content in a browser. This could be useful when you're building an application that only uses static content, or when you're building a JavaScript application and you need a development server.

4. Open a web browser and go to this application, which should be running at `http://localhost:8000`. You should see something like this:

Figure 7.1 – The web application with a CORS issue

Uh oh—something went wrong. Our application is throwing an error:

Cross-Origin Request Blocked: The Same Origin Policy disallows reading the remote resource at http://localhost:7777/Adam. (Reason: CORS header 'Access-Control-Allow-Origin' missing).

For those of you getting into web development, this experience will be your first with CORS. What on Earth does this mean? What is a *Cross-Origin Request*, and why is it blocked? What is a CORS header? And, most importantly, how do I make this go away?! This last question is the one that bothers me. We are not going to *make it go away*—we are going to understand what this means, why the browser decided to throw up a roadblock, and then move on to creating a solution.

The naive web developer, upon seeing this error, will immediately go online to search for how to deal with this, find a mess of partial or way-too-in-depth information, and then move on without ever understanding the actual issue. Making it go away will get you back to developing since the error is no longer blocking your progress, but it will not solve the problem. By doing this, you have just created a new one. To become a better developer, you are not going to just implement an off-the-shelf solution without understanding it. Instead, you will pause to learn what is happening and why. Maybe you have found this issue yourself; if not, you surely will at some point. Whether you have or have not "solved" this problem in the past, we are going to take some time to learn the rationale behind this error before coming up with an appropriate—or rather, *obvious*—solution. Once you peel back the layers of CORS, you will see that it starts to make a lot of sense and can become simple to master.

I was one such naive person that searched for this error, clicked the first link, copied and pasted a solution that made the error go away, and then moved on with life, not giving it another thought. The browser no longer complained: problem solved. At least that is what I thought. I did not think about the consequences of my action and the security hole I had introduced. What was that security hole masquerading as a fix? The solution I found was to add a simple header, and I gave it no further thought: `Access-Control-Allow-Origin: *`. *DO NOT DO THIS!* I did not know any better and I moved on, never thinking twice about CORS, except that it was the pesky thing in browsers that seemed to cause me problems.

The issue here is that the frontend application is trying to access details from another origin: hence, *cross-origin*. When I added that header, I was effectively disabling the same-origin protection that the browser was creating. The * symbol means to *allow this application to request any cross-origin information it wants*.

My browser had created a castle for protection. Rather than learn about how to effectively handle CORS, I decided to drop the drawbridge, open all of the gates, and send the guards home to their beds.

What should I have done? Let's find out.

Developing a strategy for effectively dealing with CORS

My strategy to completely disable the browser's defenses was not the best approach. It was the easy way out, the lazy way out, and the irresponsible way out. What I should have done is go to a resource such as the one that Mozilla provides and read up on the issue (`https://developer.mozilla.org/en-US/docs/Web/HTTP/CORS`). If I did, then this would have caught my attention:

Who should read this article?

Everyone, really.

Oh, *everyone* should read it? If you have not read it, you have the opportunity to take a different path than me and read it now. I'm not kidding. Please do yourself a favor: put a bookmark in this book and go read that web page. Then, come back here. I promise we will wait for you. It is fairly easy to understand and is an authoritative resource to keep in your back pocket.

According to the official HTTP specification, the `OPTIONS` method *allows a client to determine the options and/or requirements associated with a resource, or the capabilities of a server, without implying a resource action* (`https://datatracker.ietf.org/doc/html/rfc7231#section-4.3.7`). In other words, it gives an HTTP client the ability to check what an endpoint may require from it before sending an actual request. If you have ever built a browser-based web application, or if you intend to, this method is extremely important. So, as we dive into what CORS headers are, we will also revisit and heavily make use of our `OPTIONS` handler from *Chapter 3, Routing and Intaking HTTP Requests*. Go back to that section to reacquaint yourself with how we will automatically attach `OPTIONS` handlers to all of our routes.

Understanding CORS headers

Solving these cross-origin access issues can be accomplished by applying response headers. So, we will need to learn what some of these headers are and when they should be applied and used. Our job in this section will be to build HTTP responses with some basic CORS headers that we can use in our applications. We could take the easy way out and install one of the third-party packages on PyPI that will automatically add the headers for us.

However, I only suggest that you do this for a production application. CORS issues can be complex, and implementing a *trusted* solution should bring some level of comfort and peace of mind. However, relying upon one of these packages without knowing the basics is only slightly better than my first solution of disabling the same-origin policy completely.

Here are some of the common CORS response headers we should know about:

- **Access-Control-Allow-Origin**: This is used by the server to tell the client which origins it will and will not accept cross-origin requests from.

- **Access-Control-Expose-Headers**: This is used by the server to tell the browser which HTTP headers it can allow JavaScript to access safely (meaning they do not contain sensitive data).

- **Access-Control-Max-Age**: This is used by the server to tell the client how long it can cache the results of a **preflight request** (see the next section to learn what a preflight request is).

- **Access-Control-Allow-Credentials**: This is used by the server to tell the client whether it can or cannot include credentials when sending in requests.

- **Access-Control-Allow-Methods**: This is used by the server in preflight requests to tell the client what HTTP methods it will accept on a given endpoint.

- **Access-Control-Allow-Headers**: This is used by the server in preflight requests to tell the client which HTTP headers it will allow it to add.

Understanding preflight requests

In certain scenarios, before a browser tries to dispatch a cross-origin request, it will issue what is known as a **preflight request**. This is a request to the same domain and endpoint as the intended resource that happens before the actual call, except with an OPTIONS HTTP method. The goal of this request is to get access to the CORS headers to learn what the server will and will not allow. If the browser determines that the response is not "safe," it will not allow the cross-origin request.

When will a browser decide to issue a preflight request? Mozilla provides a great overview on their CORS page (https://developer.mozilla.org/en-US/docs/Web/HTTP/CORS#simple_requests). In summary, a preflight request will *not* be issued by the browser when the following is true:

- It is a GET, HEAD, or POST method.

- It does not contain any manually set headers, except for Accept, Accept-Language, Content-Language, or Content-Type.

- The request headers include Content-Type, and it is set to one of application/x-www-form-urlencoded, multipart/form-data, or text/plain.

- There are no JavaScript event listeners on the request.

- The response is not going to be streamed by the client.

These requests are generally meant to cover the scenarios that are encountered by *normal* web traffic: navigating to a page, submitting an HTML form, and basic AJAX requests. As soon as your application starts adding functionality that is typical of most single-page web applications, you will begin to notice your browser issuing preflight requests. In this context, the two most common types of requests that trigger preflight requests are as follows:

- JavaScript applications that inject custom headers (`Authorization`, `X-XSRF-Token`, `Foobar`, and so on)

- JavaScript applications that submit JSON data using `Content-Type: application/json`

You may be wondering: why does this matter? It is important to understand this so that we know when we need to respond with each of the six CORS response headers we saw in the previous section.

Solving CORS with Sanic

So far, we have entirely avoided using any third-party plugins; that is, we have steered away from any implementation that would require us to `pip install` a solution. This has been a conscious decision so that we learn the principles that are needed to build our web applications before we just outsource the solution to someone else. While this is still valid here and is the reason we are about to handle CORS requests by hand, it is also important to point out that this is a problem that has been solved already. The officially supported `sanic-ext` package and the community-supported `sanic-cors` package are both reputable options for implementing CORS protection.

With that said, let's think about each of the six response headers and *when* and *how* we will need to implement them. We have some headers we want to add regardless of the type of request, and some that will `only` be added to preflight requests. We will need a standard and repeatable method for adding response headers in these two scenarios. What is our go-to strategy for that? Middleware.

Let's start with the following basic piece of middleware and add some code to it:

```python
def is_preflight(request: Request) -> bool:
    return (
        request.method == "OPTIONS"
        and "access-control-request-method" in request.headers
    )
```

```
@app.on_response
async def add_cors_headers(request: Request, response:
HTTPResponse) -> None:
    # Add headers here on all requests


    if is_preflight(request):
        # Add headers here for preflight requests
        ...
```

We are doing two things to determine that a request is indeed a preflight request:

1. First, we know that the browser will always issue it as an OPTIONS request.

2. Second, the browser will always attach a request header called Access-Control-Request-Method with the value of the type of HTTP request that it is about to send.

To simulate a preflight request, we will use the following curl request, which adds the two headers we need to trigger the preflight request response (the Origin header and the Access-Control-Request-Method header):

```
$ curl localhost:7777 -X OPTIONS -H "Origin: http://mysite.com"
-H "Access-Control-Request-Method: GET" -i
```

The last thing we need is some ability to add OPTIONS as a viable HTTP method for every existing route in our application. This is something that sanic-ext adds, and we will learn about an easy way to accomplish this in *Chapter 11*, *A Complete Real-World Example*, using that package. But first, you may recall that this is something that we built way back in *Chapter 3*, *Routing and Intaking HTTP Requests*. We will reuse the code that looped through all of our defined routes and added an OPTIONS endpoint. You can find it in that chapter in the *Blanket support for OPTIONS and HEAD* section.

With this established, we will look at each response header to understand them more completely.

Access-Control-Allow-Origin

This header alone is arguably the most important one to add. It is also the one that is most tempting to be the nuclear option that just disables CORS protection completely, as discussed earlier. Unless you have a specific reason to accept requests from any browser origin, you should avoid *.

Instead, the value should be the address that you anticipate requests to be coming from. You should *NOT* just recycle the incoming request's `Origin` header and apply that. This is effectively the same as `*`. Instead, it is a good practice to have a predefined list of allowed origins and cross-reference the incoming `Origin` with those. If there is not a match, simply do not add any CORS headers.

Here is the first snippet we will add to our middleware to do that:

```
origin = request.headers.get("origin")
if not origin or origin not in request.app.config.ALLOWED_
ORIGINS:
    return


response.headers["access-control-allow-origin"] = origin
```

Make sure that you set the configuration `ALLOWED_ORIGINS` value as well. You can do this wherever you create your app instance:

```
app = Sanic(__name__)
app.config.ALLOWED_ORIGINS = ["http://mysite.com"]
```

As you can see, we will add this to all the responses that are coming from the browser. How do we know it is a browser request? Because we can expect that browsers will add the `Origin` header.

Access-Control-Expose-Headers

The `Access-Control-Expose-Headers` header lets the server control which headers are exposed to JavaScript access. It is a security measure that provides whitelist control over what information is available to the in-browser application.

Let's start by adding some tests to the browser. For these examples, we will use a basic HTML structure that's similar to the one we used previously:

1. We will start by setting up our HTML. The goal here is to read the `foobar` header in JavaScript and output it on the screen. Save the following HTML as `test.html`:

    ```
    <body>
        <h1>CORS Testing</h1>
        <h2 id="foobar">Loading...</h2>
        <script>
            const element = document.querySelector("#foobar")
            fetch("http://localhost:7777/").then(async
    ```

```
response => {
        const text = await response.text()
        element.innerHTML = 'foobar='${response.
headers.get("foobar")}''
    })
    </script>
</body>
```

2. You can now start this as a simple server using the following command in the
    directory where the test.html file is located:

    ```
    $ sanic -s ./
    ```

    You should now be able to open your web browser to http://
    localhost:8000/test.html. It is fine if it appears to not be working yet.

3. We need to set an endpoint on a web application to add the header. Set up a new
    server with this endpoint:

    ```
    @app.get("/")
    async def handler(request: Request):
        response = text("Hi")
        response.headers["foobar"] = "hello, 123"
        return response
    ```

4. Next, we should add a simple configuration to allow CORS:

    ```
    app.config.ALLOWED_ORIGINS = ["http://localhost:8000"]


    @app.on_response
    async def add_cors_headers(request: Request, response:
    HTTPResponse) -> None:
        # Add headers here on all requests

        origin = request.headers.get("origin")
        if not origin or origin not in request.app.config.
    ALLOWED_ORIGINS:
            return

        response.headers["access-control-allow-origin"] =
    origin
    ```

You should now be able to start this server as normal.

5. To verify our curiosity, we will double-check the response with `curl` to make sure the header is being sent:

```
$ curl localhost:7777 -i
HTTP/1.1 200 OK
foobar: hello, 123
content-length: 2
connection: keep-alive
content-type: text/plain; charset=utf-8

Hi
```

6. Now, open your browser to `http://127.0.0.1:8000/test.html` again. You should see the following output:

```
CORS Testing
foobar='null'
```

Here, the browser was able to make a CORS request but was blocked from accessing the header. If we want to allow it, then we need to be explicit.

7. So, let's head back to the `add_cors_headers` middleware that we were building and add the following snippet:

```
response.headers["access-control-expose-headers"] =
"foobar"
```

Do not forget that since we are testing this on a browser, we need to set the `ALLOWED_ORIGINS` configuration value appropriately:

```
app.config.ALLOWED_ORIGINS = ["http://127.0.0.1:7777"]
```

This time, when you access the browser, you should see that the JavaScript was able to reach in and get the value from the `foobar` header:

```
CORS Testing
foobar='hello, 123'
```

So, if you intend to use any sort of metadata on the client side of your application, you will need to use `access-control-expose-headers`. The code for this example can be found at `https://github.com/PacktPublishing/Python-Web-Development-with-Sanic/tree/main/Chapter07/corsresponse/access-control-expose-headers`.

Access-Control-Max-Age

When a browser *does* issue a preflight request, it can cache that response so that the next time it makes the same request, it does not need to hit the server. This performance improvement can be controlled (to some extent) by the server using `Access-Control-Max-Age`, which specifies the length of time (in seconds) that the preflight request can be cached.

Typically, web browsers will set a maximum value for this. If you try to set it to some absurdly large number, they will drop it down to their predefined maximum value. For this reason, I usually recommend going with a value that is around 10 minutes. Some browsers will allow you to go up to 24 hours, but that is probably about the maximum allowed.

We will see this now in our middleware:

```
response.headers["access-control-max-age"] = 60 * 10
```

Access-Control-Allow-Credentials

This header is for preflight requests only. So, the snippet we will add here needs to go inside our `is_preflight(request)` block.

When a JavaScript application makes a request, it must *explicitly* make a call that allows credentials to be sent. If not, then the browser will not include them in the request. The server can then play its part and tell the browser that the request that includes credentials is—or is not—safe to expose to the JavaScript application.

To allow it, we can set the header like this:

```
response.headers["access-control-allow-credentials"] = "true"
```

Access-Control-Allow-Methods

So far, there hasn't been a need for any plugins. Adding these CORS headers has been fairly straightforward. The next part, however, is something that could become a little more tricky.

The `Access-Control-Allow-Methods` header is meant to be a warning to the browser during the preflight request about what HTTP methods the browser is allowed to send to the endpoint cross-origin. A lot of applications disable this protection by allowing everything:

```
response.headers[
    "access-control-allow-methods"
] = "get,post,delete,head,patch,put,options"
```

This is a simple solution. It is less harmful than that first CORS solution I came across that allowed any origin. But we can still do better.

To have dynamic methods that match the *actual* endpoint possibilities, we are going to change some things around in our code:

1. Remember how we are defining a request as a preflight one? Let's do that upfront in the request middleware:

    ```
    @app.on_request
    async def check_preflight(request: Request) -> None:
        request.ctx.preflight = is_preflight(request)
    ```

2. Next, when we generate the handlers for our OPTIONS requests, we will inject a list of all of the allowed methods, like this:

    ```
    from functools import partial


    @app.before_server_start
    def add_info_handlers(app: Sanic, _):
        app.router.reset()
        for group in app.router.groups.values():
            if "OPTIONS" not in group.methods:
                app.add_route(
                    handler=partial(
                        options_handler,
                        methods=group.methods
                    ),
                    uri=group.uri,
                    methods=["OPTIONS"],
                    strict_slashes=group.strict,
                    name="options_handler",
                )
        app.router.finalize()
    ```

3. Now that we have access to the preflight check in our options handler, we can do
    our check and add the headers there. We can also take the list of methods that were
    passed in and concatenate them into a comma-delimited list. This should now
    provide an automated set of OPTIONS endpoints with the HTTP methods that
    will be used:

```
async def options_handler(request, methods):
    resp = response.empty()
    if request.ctx.preflight:
        resp.headers["access-control-allow-credentials"]
= "true"
        resp.headers["access-control-allow-methods"] =
",".join(methods)
    resp.headers["vary"] = "origin"
    origin = request.headers.get("origin")
    if not origin or origin not in request.app.config.
ALLOWED_ORIGINS:
        return

    resp.headers["access-control-allow-origin"] = origin
    return resp
```

4. We will look at the preflight response using curl to see all of our headers:

```
$ curl localhost:7777 -X OPTIONS -H "Origin: http://
mysite.com" -H "Access-Control-Request-Method: GET" -i
HTTP/1.1 204 No Content
access-control-allow-credentials: true
access-control-allow-methods: GET
vary: origin
access-control-allow-origin: http://mysite.com
connection: keep-alive
```

You can access a working version of this code by going to this book's GitHub repository
at https://github.com/PacktPublishing/Python-Web-Development-
with-Sanic/tree/main/Chapter07/corsresponse/access-control-
allow-methods.

Access-Control-Request-Headers

The final header we are concerned with here is `Access-Control-Request-Headers`, and it is also one that should be sent in preflight responses. It tells the browser which non-standard headers can be sent in the cross-origin request.

If the JavaScript wanted to send a header called `counting`, then it would do this:

```
fetch("http://localhost:7777/", {
    headers: {counting: "1 2 3"}
})
```

However, because this would trigger a preflight request, the browser will fail with a CORS error because the server has not explicitly allowed counting as an acceptable header.

To do this, we can enable it in our preflight block:

```
resp.headers["access-control-allow-headers"] = "counting"
```

Our review of CORS headers has added a lot of code. To see the completed version, please check out this book's GitHub repository: `https://github.com/PacktPublishing/Python-Web-Development-with-Sanic/tree/main/Chapter07/corsresponse`. Now that we have completed our CORS review, we will continue by looking at a related topic: CSRF.

Protecting applications from CSRF

The next step in our journey is handling **cross-site request forgery** (**CSRF**). It should also be noted that this often also carries the acronym **XSRF**. If you see these two on the web, they refer to the same issue. So, what is the issue?

Do you remember that suspiciously awkward email that you received that says **Click here to claim your $500 prize**? That link likely brings you to a malicious website that's controlled by someone that is trying to hack you. They may have placed some links or caused you to do something on their site that sends off a background request to a legitimate website to do something bad. If your application is not protected from CSRF attacks like this, it could be possible for that bad actor to make your users change their passwords without them even knowing!

Thwarting these attacks can be done on both sides. Your users could, of course, take better care not to open the email in their spam box. But you, as a responsible web application developer, also have a responsibility to protect your users.

Solutions that do not work

Cookies. You may be surprised if you skipped ahead to peak at the solution that I have offered, as you will see that it does include cookies. Indeed, cookies can play a part in solving the problem. However, they are a flawed security measure and *cannot* be the answer to the CSRF problem by themselves.

How would this even work? Imagine that you set a session ID in a cookie. It is a decently good mixture of random characters, so it would be impractical for someone to guess it correctly. The problem is that cookies are sent with every request and based not on where the request is initiated, but where it is headed. So, if your browser sees that it has a cookie in storage for `yourapplication.com`, then even if the request was initiated at `h4ck3rsp4r4d1se.com`, the browser would send the cookies.

It should also be noted that introducing TLS and reading the `Origin` header are not sufficient solutions either. Of course, these are useful and valid things your application should do, but they alone do not add protection from CSRF. The `Origin` header, for example, could easily be spoofed.

Solutions that do work

Now that we know what will not protect us from CSRF attacks, we can look into a few solutions that will work and help protect our web applications. These are not mutually exclusive, and I suggest that you consider implementing them all in one form or another. Your decision will be contextual, of course, but here are some good practices to keep in mind while protecting your application from CSRF attacks.

Do not change state on GET

This is incredibly important. We discussed this issue back in *Chapter 3*, *Routing and Intaking HTTP Requests*, but GET requests should not be state-changing. This means that the application should not take any direction from a GET request to go do something. These should be for information purposes only. By removing GET from the hackers' arsenal, we are forcing them into using JavaScript exploits on their malicious websites.

The reason that we want to allow this is because the browser has some built-in security measures we know about and can use to our advantage. First, from within the browser, the `Origin` header cannot be spoofed.

Let's say our bad website had the following code in it:

```
fetch("http://localhost:7777/", {
    headers: {origin: "http://localhost:7777"}
})
```

If you went to `somebadwebsite.com`, the origin would still be `http://`
`somebadwebsite.com`. That is why CORS protection works. By disallowing stateful
changes from `GET` requests, we make it so that a hack such as this will not work:

```
<img src="http://yourapplication.com/
changepassword?password=kittens123">
```

Forcing the hacker into JavaScript—especially JavaScript requests that are forced into
issuing preflight requests—gives us some more control, as we are about to see.

Cookies

The next helpful solution involves cookies.

Wait, what? Cookies were in the *do not work* solution category; what gives?

We just said that we want to force malicious attackers to use JavaScript in their exploits.
This is because we also know that browser cookies have a feature that we can control:
`HttpOnly`. When a server creates a cookie, it can decide whether or not JavaScript
should be able to access that cookie. This means that the cookie will continue to be sent
on every web request when enabled, but it will be inaccessible to any JavaScript code. This
makes it an ideal location for storing secure credentials such as session tokens. Besides
this, cookies are subject to what is known as **cross-site scripting** (**XSS**) attacks. This is an
attack where a hacker can extract secure details from a frontend browser using JavaScript.

Important Note

If your browser application can access some piece of information with
JavaScript, so can a hacker.

We also mentioned that there was a problem that cookies for `yourapplication.com`
can still be sent unknowingly from `h4ck3rsp4r4d1se.com`. Since JavaScript, when it
is allowed to access cookies, can only do so on the current domain, we have another tool
in our belt we can use while building our solution.

When a user logs in, if we set two cookies (one for the session and one for the CSRF
protection) we can set the `HttpOnly` value based on the intended usage. The session
cookie remains inaccessible, and the cookie that is set aside for CSRF protection could
be JavaScript accessible. We could then require that the JavaScript uses that cookie's
value when sending in a request. This will work because the JavaScript that is running
on `h4ck3rsp4r4d1se.com` will not be able to access cookies that are marked for
another domain.

What should the value of this cookie be? Well, anything that could be impossible to guess. It is best to keep that value user-specific so that you can verify its contents and be assured that the token is authentic. Also, the value should change and not be static. This will help make it more difficult for any would-be attackers. This dual cookie method is not 100% fault-proof. However, it should be reasonably secure for most applications' needs. The problem is when your users start accidentally downloading malware that is capable of circumventing the browser's protection. We'll leave that issue aside as it is an issue that we cannot control, and it requires a much more in-depth conversation that's beyond the scope of this book.

It should be noted that we do not necessarily care that the CSRF token could be compromised and used by a bad actor. That's fine—even if they could access it, they have no way to then send it with both the correct origin and the correct session token.

Form fields

There is another form of CSRF protection that other frameworks use. For example, Django made the idea of injecting some hidden HTML into the page popular:

```
<input type="hidden" name="csrftoken" value="SOMETOKEN" />
```

This value would then be included in form responses or read into the request in some expected way. This is essentially the same idea that I am proposing here. The only difference is that instead of injecting the value into a hidden—although JavaScript-accessible location—input, we are storing it in a cookie. Both solutions will ultimately depend on what happens when that value is sent back to the server. We will look at this in the next section.

Putting a solution into practice

Now that we have a general idea of our approach, let's recap to be clear. We want to allow stateful changes to be made in our application for authenticated users. To achieve confidence that the changes are coming from our users and not hackers, we will allow these changes when the following is true:

- The HTTP method is POST, PATCH, PUT, or DELETE.
- The origin of the incoming request matches what we would expect.
- The incoming request has a cookie that was stored with HttpOnly.
- The incoming request has a valid CSRF token.

To accomplish our goal, we need to decide where we will put the code that is going to accomplish this goal. Here, we come back to the debate we have discussed a few times already: decorators or middleware. There is not a correct choice, and the answer will, of course, depend on what you are building.

For our example, we will build it as a decorator. When we come to authentication in the next section, it will become more clear why we are using the decorator pattern here. If you think middleware works for you, go ahead and try to rebuild this as middleware. Both options are legitimate patterns and may serve your needs in different circumstances. To be honest, however, I usually find the decorator pattern to be more easily adoptable and that it has a broader set of use cases. Here are the steps:

1. To start, we will make a barebones decorator. To make this job easier, you can grab a decorator template from the Sanic User Guide at `https://sanic.dev/en/guide/best-practices/decorators.html#templates`:

   ```python
   def csrf_protected(func):
       def decorator(f):
           @wraps(f)
           async def decorated_function(request, *args,
   **kwargs):

               response = f(request, *args, **kwargs)
               if isawaitable(response):
                   response = await response

               return response

           return decorated_function

       return decorator(func)
   ```

   When there is a CSRF failure, the correct response should be 403 Forbidden. We will make a custom exception that we can raise whenever this happens:

   ```python
   from sanic.exceptions import Forbidden

   class CSRFFailure(Forbidden):
       message = "CSRF Failure. Missing or invalid CSRF
   token."
   ```

2. Thinking about our goals and our requirements, we want a way to determine that the request is coming from a browser. This is because a browser request will be subject to CSRF protection. There is no need to implement it on direct access API requests. I like to do this by adding an HttpOnly cookie to every request if it does not exist. The value is completely irrelevant. The only thing we care about is that the value was sent. The same goes for the origin header. If an Origin header was sent, we will assume it is a browser request and subject it to the stiffer requirements we will impose next. This is a belt and suspenders approach since it is a bit duplicative. However, it does give you an idea of the types of strategies you should be thinking about when you're designing your solutions:

```
@app.on_request
async def check_request(request: Request):
    request.ctx.from_browser = (
        "origin" in request.headers or
        "browser_check" in request.cookies
    )


@app.on_response
async def mark_browser(_, response: HTTPResponse):
    response.cookies["browser_check"] = "1"
    response.cookies["browser_check"]["domain"] =
"mydomain.com"
    response.cookies["browser_check"]["httponly"] = True
```

> Tip
>
> Marking the browser_check cookie on every request is overkill. I generally recommend doing this on a landing page or somehow catching the case when there is an Origin and no cookie to set it. I will leave this to your discretion to determine an appropriate place and method for setting this cookie. If you control the frontend application, you may even consider setting it there. The point of this cookie is just to give us an additional indication that this is not a direct access API request.

3. Looking at our list of requirements again, let's add some code to `decorated_`
    `function` of our decorator to ensure that the origin matches. This is necessary
    because since we already know that when the request is coming from the browser's
    JavaScript, this value cannot be spoofed:

    ```
    origin = request.headers.get("origin")
    if request.ctx.from_browser and origin not in app.config.
    ALLOWED_ORIGINS:
        raise CSRFFailure
    ```

4. The next requirement that we have is to make sure that an `HttpOnly` token is
    present. For now, we will just use our `browser_check` cookie. This could also be
    satisfied with a session cookie if you have one:

    ```
    origin = request.headers.get("origin")
    if request.ctx.from_browser and (
        origin not in app.config.ALLOWED_ORIGINS
        or "browser_check" not in request.cookies
    ):
        raise CSRFFailure
    ```

5. Lastly, we need to verify our CSRF token. I know we have not discussed what one is
    or how to generate one, so we haven't gotten to the verification bit yet. We will get
    there soon. Until then, let's simply add a function to round out our decorator:

    ```
    origin = request.headers.get("origin")
    if request.ctx.from_browser and (
        origin not in app.config.ALLOWED_ORIGINS
        or "browser_check" not in request.cookies
        or not csrf_check(request)
    ):
        raise CSRFFailure
    ```

Finally, let's turn to CSRF tokens. For our implementation, we are going to use a Fernet token. This is a method of encrypting some bit of text with a secret key so that it cannot be changed or read without that key. We are going to set this token in a cookie that will explicitly *not* be HttpOnly. We want the frontend JavaScript application to read this value and send it back to the application via the headers. When the potentially harmful state-changing request comes in, we will verify that the header and the cookie match. We will also extract the payload of the Fernet token and validate its contents. The actual value of that token will be stored in a second cookie that will be HttpOnly. The purpose of this dual cookie and dual submit verification is to protect our applications from various types of attacks that may otherwise compromise our strategy. The solution may sound much more complicated than it is, so let's look at some code to start piecing this together:

1. We will begin by setting up some configuration values that we will need:

    ```
    app.config.CSRF_REF_PADDING = 12
    app.config.CSRF_REF_LENGTH = 18
    app.config.CSRF_SECRET =
    "DZsM9KOs6YAGluhGrEo9oWw4JKTjdiOot9Z4gZ0dGqg="
    ```

    > **Important Note**
    >
    > It should come as no surprise that you should never, *never*, *NEVER* hardcode a secret like this in your applications. This is for example purposes only. Instead, you should be injecting secret values via environment variables or some other, more secure, method than this.

2. We need a function that will generate our CSRF reference value and token. To accomplish this, we will use the cryptography library that we mentioned at the beginning of this chapter. It is battle-tested and reliable. It should be the first place that you turn to for all your cryptographic needs in Python. Here's the code:

    ```
    from base64 import b64encode
    from cryptography.fernet import Fernet


    def generate_csrf(secret, ref_length, padding) ->
    Tuple[str, str]:
        cipher = Fernet(secret)
        ref = os.urandom(ref_length)
        pad = os.urandom(padding)
    ```

```
        pretoken = cipher.encrypt(ref)


    return ref.hex(), b64encode(pad + pretoken).
decode("utf-8")
```

As you can see, this is fairly simple. We create the cipher object using our secret. Then, as per the recommendation from the cryptography library, we use the operating system's random generator logic with os.urandom to make our reference value, plus some extra fluff. The reference is encrypted, and our token is then padded and returned, along with the reference value.

3. Creating the reverse to verify our token is a matter of performing these steps in reverse, and then comparing the encrypted value to the passed referenced value:

```
def verify_csrf(secret, padding, ref, token):
    if not ref or not token:
        raise InvalidToken("Token is incorrect")

    cipher = Fernet(secret)
    raw = b64decode(token.encode("utf-8"))
    pretoken = raw[padding:]
    encoded_ref = cipher.decrypt(pretoken)

    if ref != encoded_ref.hex():
        raise InvalidToken("Token is incorrect")
```

4. We will need a way to make sure these values exist as cookies. So, we will generate them in middleware for this example. However, it may be logical to perform this function on a login endpoint instead:

```
@app.on_response
async def inject_csrf_token(request: Request, response:
HTTPResponse):
    if (
        "csrf_token" not in request.cookies
         or "ref_token" not in request.cookies
    ):
        ref, token = generate_csrf(
            request.app.config.CSRF_SECRET,
            request.app.config.CSRF_REF_LENGTH,
```

```
                        request.app.config.CSRF_REF_PADDING,
        )

            response.cookies["ref_token"] = ref
            response.cookies["ref_token"]["domain"] =
    "localhost"
            response.cookies["ref_token"]["httponly"] = True

            response.cookies["csrf_token"] = token
            response.cookies["csrf_token"]["domain"] =
    "localhost"
```

Remember, the plan is for `csrf_token` to be JavaScript-accessible. We want the incoming request to not only include this in a cookie value but also have this value injected into the HTTP headers. This can only be done by JavaScript running on our applications because of the same-origin policy. CORS to the rescue! So, do not forget to whitelabel the request header, which we will see next: `X-XSRF-Token`.

Remember that, back in our `@csrf_protected` decorator, one of the checks was `csrf_check(request)`. Let's finally uncover what that function is:

```python
def csrf_check(request: Request):
    csrf_header = request.headers.get("x-xsrf-token")
    csrf_cookie = request.cookies.get("csrf_token")
    ref_token = request.cookies.get("ref_token")

    if csrf_header != csrf_cookie:
        raise CSRFFailure

    try:
        verify_csrf(
            request.app.config.CSRF_SECRET,
            request.app.config.CSRF_REF_PADDING,
            ref_token,
            csrf_cookie,
        )
    except InvalidToken as e:
```

```
     raise CSRFFailure from e

  return True
```

There should be three values that we care about: the two cookies we just set, and the incoming X-XSRF-Token header. This header, as we already know, will be generated on the client side by extracting the cookie and injecting the value into the header. It should now be simply a matter of verifying the following:

- That the cookie and the header match
- That the protected HttpOnly reference value is the same as the encrypted value

If that all checks out, we can be confident that the request is genuine.

> **Tip**
>
> You may be wondering why I chose XSRF here instead of X-CSRF-Token, or even just CSRF-Token for the header name. The reason is that some frontend frameworks automatically add this header injection for your client side. Since it is not important what the header is called from our perspective, we may as well play nice with some other tooling that likes it named this way.

Samesite cookies

You may be familiar with a newer concept in CSRF protection known as **samesite** cookies. This is a value that can be appended to the cookie that provides extra directions to the browser about how to treat that cookie. In short, by setting this value on the cookies on the server, we allow the application to dictate to the browser when it is and is not acceptable to send the cookie. This alone *nearly* mitigates the issues with CSRF, but it should *NOT* be used by itself as the solution.

The **Open Web Application Security Project (OWASP)**—a nonprofit foundation that promotes the enhancement of security practices online—specifically states that the samesite attribute "*should not replace having a CSRF token. Instead, it should co-exist with that token to protect the user more robustly.*" (https://cheatsheetseries.owasp.org/cheatsheets/Cross-Site_Request_Forgery_Prevention_Cheat_Sheet.html#samesite-cookie-attribute)

We will now learn about samesite cookie protection and how to integrate it into our solution. Three values are allowed: None, Lax, and Strict.

Samesite=None

Cookies that use Samesite=None should only be considered for non-security-related cookies. This is because they will be sent with every request, no matter what site they are originating from. So, if you are on the hacker's website, that hacker will be able to submit requests on your behalf to other sites that you have visited and make use of the cookies you have on your computer. Not cool.

But for the right kind of cookies, this is not an issue. So long as the value has nothing to do with security or sessions, this is acceptable. It should be noted, however, that for this to work, it will also only be allowed when the cookie is marked as Secure; that is, it is only allowed to be passed across https requests. In your production-level code, you should be doing this regardless. You are using TLS encryption, right? If not, we will see a super simple solution to this in *Chapter 8, Running a Sanic Server*, and *Chapter 10, Implementing Common Use Cases with Sanic*.

Setting Samesite=None is as simple as doing the following:

```
response.cookies["myfavorite"] = "chocolatechip"
response.cookies["myfavorite"]["domain"] = "mydomain.com"
response.cookies["myfavorite"]["samesite"] = None
response.cookies["myfavorite"]["secure"] = True
```

This will result in the following cookie:

```
Set-Cookie: myfavorite=chocolatechip; Path=/; Domain=mydomain.
com; SameSite=None; Secure
```

Samesite=Lax

This is the default in most modern web browsers now. You should not, however, rely upon that fact, and it is certainly still a best practice to do so explicitly.

What does this value mean? It means that the cross-site POST requests we have been worried about will not include the cookies (which is a big part of CSRF protection). However, it will allow them in some contexts. To be sent in a cross-site request, the request must provide top-level navigation (think of this as the address bar in the browser), and the HTTP method must be GET or HEAD.

This boils down to protection from AJAX requests but allowing the cookie to be sent when someone navigates to the site from a third-party link. This makes a lot of sense and is probably what you want to use for a lot of your cookies.

For example, if your session cookies were *not* set to Lax (and instead were Strict) when someone clicked a link from another website that brought them to your site, they would not appear as logged in. However, once they started clicking around, their session would suddenly appear. This might be an awkward experience for the user. So, it is suggested that session management and authentication cookies should be Lax for most typical applications. If you are building a secured banking application, you probably have no use for someone to link to a secured banking page, so maybe Lax is not the right answer. However, it is generally acceptable to use Lax for authentication.

As we mentioned previously, you do not need to explicitly state the samesite attribute anymore, but being explicit is better than being implicit:

```
response.cookies["session_token"] = session_token
response.cookies["session_token"]["domain"] = "localhost"
response.cookies["session_token"]["httponly"] = True
response.cookies["session_token"]["samesite"] = "lax"
response.cookies["session_token"]["secure"] = True
```

This will generate a cookie that looks like this:

```
Set-Cookie: session_token=<TOKEN>; Path=/; Domain=localhost;
HttpOnly; SameSite=lax; Secure
```

Samesite=Strict

As we alluded to in the previous section, a Strict cookie will only be sent when the request originated from the correct site. This means that the user must be on your application, and *then* submit the request. In my opinion, this sounds like the type of request that is state-changing. Do you see where I am heading with this?

In my opinion (and you will undoubtedly come across different opinions), CSRF protection cookies should be Samesite=Strict. There is no legitimate use case (at least not in my applications) where I can think that I would not want my user to be on my application first before initiating the types of requests that I am trying to protect. You may have different needs, so this may not work for you. If Lax makes sense, then go with it. I'll stick to this:

```
response.cookies["ref_token"] = ref
response.cookies["ref_token"]["domain"] = "localhost"
response.cookies["ref_token"]["httponly"] = True
response.cookies["ref_token"]["samesite"] = "strict"
```

```
response.cookies["ref_token"]["secure"] = True

response.cookies["csrf_token"] = token
response.cookies["csrf_token"]["domain"] = "localhost"
response.cookies["csrf_token"]["samesite"] = "strict"
response.cookies["csrf_token"]["secure"] = True
```

As you can probably guess, our cookies now look like this:

```
Set-Cookie: ref_token=<TOKEN>; Path=/; Domain=localhost;
HttpOnly; SameSite=strict; Secure
Set-Cookie: csrf_token="<TOKEN>"; Path=/; Domain=localhost;
SameSite=strict; Secure
```

> **Important Note**
>
> As we mentioned previously, support for samesite cookies is not universal. You should check a website such as CanIUse to see if any browsers you are targeting do not implement it: `https://caniuse.com/same-site-cookie-attribute`. Also, a "same" site in this context does include subdomains. There is a public list of addresses that are considered "top-level" for this context, which does not completely line up with .com, .org, .io, and so on. For example, two websites on github.io are not considered samesites. For the full list, go to `https://publicsuffix.org`.

In our review of CSRF, we mentioned session tokens and authentication a lot, but we have not looked at them yet. Although this is an incredibly deep topic in itself, we will explore how you can implement authentication in your applications using Sanic.

Protecting your Sanic app with authentication

When many people think about a web application, what comes to mind is some type of platform on the web where they log in to do... *something*. The activity is not what we care about here. When you are done reading this book, you are going to go off and build some amazing applications. What we care about are the journey and the process. The part of the process that we care about right now is logging in.

To be more specific and correct, what we are about to look at is **authentication**, not **authorization**. While these two ideas are very closely related, they are not the same and are not interchangeable. Authorization usually presumes that authentication has already happened.

So, what's the difference?

- **Authentication**: This answers the question: who are you?
- **Authorization**: This answers the question: what are you allowed to do?

To confuse matters even more, when **authentication** fails, a `401 Unauthorized` response occurs. This is super unfortunate naming from the early days of the internet. Am **authorization** failure returns a `403 Forbidden` response.

In 2020, I spoke at EuroPython about access control issues. The slides and a link to the YouTube presentation are on my GitHub page: `https://github.com/ahopkins/europython2020-overcoming-access-control`. If you have about 30 minutes to watch a riveting presentation about this thrilling topic, it's a "don't miss" opportunity.

The presentation covers this authentication/authorization topic, but also largely tries to answer the question, *what are the different methods for protecting my API?*

It answers this by comparing session-based authentication with non-session-based authentication (that is, stateless authentication). We will review both of these strategies here, but we will also specify how to implement API keys (which is not covered in the aforementioned presentation).

To do this, a set of questions need to be answered. Before we dive into how to implement some of the most common strategies with Sanic, we will review some of the questions that you should ask yourself before deciding on a strategy:

1. *Who will consume the API?*

    You should think about whether the API is going to be used by other applications or scripts, or by actual people. Will it be used by programmers who are integrating it into their applications? Or will it be used to power a mobile application? Does a frontend JavaScript application need to access it?

    The reason you should care is that you must have an understanding of the technical abilities, but also the weaknesses, of your intended use case. If your API will only ever be consumed by other backend scripts and applications, then you will have an easier time securing it. Most of that stuff we talked about cookies is highly irrelevant, and CORS is a non-issue.

    On the other hand, if you intend to power a browser-based SPA, then you will likely need a more robust authentication strategy than simple API keys.

2. *Do you have control over the client?*

   The core of this question is whether you (or your organization) will be the consumer of the API. Contrast this with an API that is meant to be consumed by integrations and other applications, and you should see that this can have a difference on how you control access. For example, if you are building a microservice that is not exposed to the internet, but only exists within a highly controlled network, you have a different set of security concerns than the API that powers your bank's website.

3. *Will this power a web browser frontend application?*

   This is a bit of a subset of the first question, but it is important enough to think about on its own. The reason that this is so much of an issue is that the browser is flawed. When the internet was first created and web browsers were first being released, no one could quite predict the direction and level of importance that the internet would have. The security concerns—and the solutions to mitigate them—were born out of years of hackers attempting to exploit a system that was never really designed with a security-first mindset.

   For example, the fact that non-encrypted `http://` websites even exist in today's world is mind-boggling. This chapter has devoted a lot of energy to how to deal with certain security concerns that only exist because the web browser is broken. Therefore, knowing that there is even a possibility of frontend usage for your application should trigger warning bells early on that you *must* dedicate time and attention to this topic.

With those three questions in our mind, we will now look at three potential schemes for authenticating users. But first, I will provide another reminder that just because I do something some way here does not mean you should as well. Use your skills to take what is provided to build the solutions you need for your application. We are talking security, so maybe you should be careful before you stray too far. If you ever have a question about a strategy, feel free to bring the question to the community on Discord or in the forums.

Next, we will explore *some* of the strategies you may find.

Using API keys

By far, API keys are the simplest authentication scheme. They are easy to set up and easy for the end user to implement. This also means that they offer less security. However, this does not mean they should be overlooked. In the right context, API keys can be the right tool to get the job done, provided you take measures to mitigate any security concerns.

API keys go by many names, but they boil down to a simple concept: your application provides a secure persistent token. When the request is accompanied by that token, it works. If not, it fails. It is as simple as that. One of the main benefits—besides simplicity—is that the keys are easy to invalidate. Since you are storing the keys *somewhere*, all you need to do is change the stored value or remove it and that key will no longer work.

The reason that API keys are more susceptible to attacks is that they are single, persistent values. This means that it is, in theory, easier for the value to be brute-force attacked. A hacker could set up a machine and try every single combination until one works. Therefore, the first step in making sure that your API scheme is secure is to *use strong keys*. This means a high amount of entropy.

Once a sufficiently complex API key has been generated, it should be hashed before storing. *Do not encrypt* your keys. What is the difference between hashing and encrypting? When you "encrypt" data, it can be reversed. As we saw with Fernet encryption, we were able to reverse the process and decrypt the original value. This is a no-no for API keys. Hashing, on the other hand, is a one-way street. Once an API key has been hashed, there is no way to recover the original value. Therefore, to validate a value against it, you need to hash the incoming value using the same strategy and compare the result to the stored hash.

This may sound like password management, right? That is because you should treat an API key exactly as you would a password. This brings up the second potential security pitfall when using API keys: storage. *Never* store them in plain text, *never* store them in a format where the original value can be retrieved, and *never* store them so that the hashed value can easily be predicted.

Once you have the value of a newly generated key, you will add a "salt" before storing it. A password salt is a random bit of text that is added to a password so that when the password is hashed, it is done so in an unpredictable format. If you do not salt the password, then the hashed value can be cracked by comparing it to known hashes for common passwords.

Hackers keep databases of the hashed values of common passwords for this reason. Even though they may not be able to decrypt a hashed value, if you fail
to salt it, then it is super simple for them to backward engineer the value by simply looking at known values. Luckily, the `bcrypt` module makes this easy. Let's dive into some code:

1. We will begin by creating a function to generate an API key. To do this, we will use the `secrets` module that comes from the Python standard library. In our example, we will use `secrets.token_urlsafe` to generate the value. You could also use `secrets.token_hex`, but it will produce a slightly longer string to represent the same value. The reason I suggest using this library with its default settings is that the maintainers of Python will change the amount of entropy that's needed based on the current best practices. At the time of writing, the default is 32 bytes. If you feel more is required, feel free to increase that value:

    ```
    from secrets import token_urlsafe
    from bcrypt import hashpw, gensalt


    def generate_token():
        api_key = token_urlsafe()
        hashed_key = hashpw(api_key.encode("utf-8"),
    gensalt())
        return api_key, hashed_key
    ```

    We also used the `bcrypt` module to generate a salt. What this does is add random text, create a hash, and then repeat the cycle several times. By folding the hashed value with multiple rounds of salting, it becomes more difficult to compare it against a known value (it also becomes computationally more expensive, so setting the value too high may be super time-consuming). We will use `gensalt` with the default value of 12 rounds here.

2. You will need an endpoint that generates and stores these values. A typical implementation will have a frontend UI where the user clicks a button to generate the API key. The value is returned on screen just long enough for them to copy it. Once they navigate away, that value is gone and cannot be recovered. In the backend, this means that we need an endpoint that uses `generate_token`, sends the API key to the user, and stores the hashed key in the database:

    ```
    @app.post("/apikey")
    async def gen_handler(request: Request):
        api_key, hashed_key = generate_token()
    ```

```
user = await get_user_from_request(request)
await store_hashed_key(user, hashed_key)
return json({"api_key": api_key})
```

As a reminder, you can look back to *Chapter 4, Ingesting HTTP Data*, for strategies on how to extract data from the request to get the user. For example, in the preceding code, `get_user_from_request` is a stand-in to show that you would be pulling the user information based on the incoming request. Similarly, since we have not looked at how to interact with databases yet, `store_hashed_key` is just a stand-in to show that you would need to use the user and the hashed key to *somehow* store the value. If you would like to look at a functioning version of this endpoint that gets around this issue for demonstration purposes, check out the GitHub repository at `https://github.com/PacktPublishing/Python-Web-Development-with-Sanic/tree/main/Chapter07/apitoken`.

3. Now, we will create a new decorator to protect the endpoints with our API key. In this decorator, we will extract the user from the request and compare the hashed key to whatever the user has sent:

```
from bcrypt import checkpw
from sanic.exceptions import Unauthorized


def api_key_required(
    maybe_func=None,
    *,
    exception=Unauthorized,
    message="Invalid or unknown API key"
):
    def decorator(f):
        @wraps(f)
        async def decorated_function(request: Request,
*args, **kwargs):
            try:
                user = await get_user_from_
request(request)
                is_valid = checkpw(request.token.
encode("utf-8"), user.hashed_key)

                if not is_valid:
```

```
                    raise ValueError("Bad token")
            except ValueError as e
                raise exception(message) from e

        response = f(request, *args, **kwargs)
        if isawaitable(response):
            response = await response

        return response

    return decorated_function

    return decorator(maybe_func) if maybe_func else
decorator
```

One thing that is helpful to point out here is that Sanic will extract a token from the `Authorization` header for us. It is a very common scheme to send tokens in headers as so-called **bearer tokens**. They look like this:

```
Authorization: Bearer <token_here>
```

They can also look like this:

```
Authorization: Token <token_here>
```

So, to get access to that token value, all you need to use is `request.token` and Sanic will find it from either location.

4. Now, to implement this, all we need to do is wrap our endpoints:

```
@app.get("/protected")
@api_key_required
async def protected_handler(request):
    return text("hi")
```

Another thing to point out is the inherent security leak in failing to use the correct status codes and exception messages when something goes wrong. We mentioned this back in *Chapter 6, Operating Outside the Response Handler*, and it is worth knowing how to address that concern here. You may have noticed that we are allowing our decorator to pass in an `Exception` class and message. This is so that we have control over what information is sent to the end user. Providing this level of customization in the decorator can become a powerful tool when you're building out your application.

Now that we have seen how easy it is to implement *proper* API keys, the only remaining question is, when is it appropriate to use them?

Never use an API key to secure a browser-based UI.

The security that's afforded by the API key is not sufficient to handle all of the issues that the browser raises by storing credentials. This is only appropriate for integrations that are coming from outside scripts or applications. I like to refer to this as *direct access* since the API is being used directly by the client and not a web browser.

Because of this, I like to use the `check_request` middleware we created earlier in this chapter in conjunction with my authorization decorator. Since `@api_key_required` should never be valid for a request from the browser, I like to change the following code:

```
if not is_valid:
    raise exception(message)
```

I replace the preceding code with the following:

```
if request.ctx.from_browser or not is_valid:
    raise exception(message)
```

Now that we know how and when to use API keys, let's look at approaches to handling authentication in a scenario that is appropriate for web applications.

Understanding session-based versus non-session-based authentication

User sessions are perhaps the most common approach to handling authentication in web applications. A more recent strategy employs tokens known as **JSON Web Tokens (JWTs)**. In most other contexts, these are referred to as *stateful* versus *stateless*. User sessions are *stateful*, while JWTs are *stateless*. This is all true, but I like to refer to them as **session-based** and **non-session-based**. Call me a rebel, but I think that this more clearly describes what we are trying to achieve.

First, what is a session? If a user logs into your application, and you record that login in your database so that it can be invalidated at will, then you are creating a session. This means that so long as that record exists in your database, there is an active session that can be authenticated against that particular user.

Session-based authentication is very simple to implement on both the frontend and the backend. And, because it offers a high degree of security, it is the reason that it has become a default approach for many web applications. One of its huge benefits is that any active session can be inactivated at any time. Have you ever been on a web application (perhaps your email provider) that lists everywhere you are logged in? With a click of a button, you can log out of the other locations. This is helpful in case a session is compromised or hacked.

On the other hand, non-session-based authentication provides a great deal more flexibility. The typical example of a non-session-based token is a JWT. So, even though I am talking specifically about JWTs, they are not the only way to handle non-session-based authentication. The most critical component that is offered by this strategy is that the token itself is self-authentication. This means that a server only needs to look at the token to determine if it is genuine and if it has been tampered with.

Because of this, authenticating a JWT becomes highly portable. You can have one microservice that handles authentication and generating tokens, and then other services can verify them without having to involve the authentication service at all! This allows for very scalable architectures. This also highlights another benefit. Every time a session token is received, to authenticate it, you *must* make a round-trip to your storage engine. This means that every single API call includes at least one more network call to the database. This is completely avoided with self-authenticating tokens and can lead to overall performance benefits.

JWTs also have the benefit that they can be embedded with non-secret payloads. This often means that you can include a list of permissions or meta-information about a user that the frontend application can make use of.

This all sounds great, but the downside to JWTs is that, once issued, they cannot be invalidated. When they are created, they are given an expiration time. The token will remain valid until that time expires. This is the reason why these expiration times are usually quite short and usually measured in minutes (not hours or days, as may be typical of sessions). If a token expires every 10 minutes, it would be super inconvenient for a web application user to need to log in again so frequently. Therefore, JWTs are often accompanied by a refresh token. This token is a value that allows a user to exchange an expired JWT for a fresh new one.

Furthermore, session-based tokens are generally easier to protect from XSS attacks using the `HttpOnly` cookies that we saw earlier. Since JWTs are usually sent as bearer tokens, such as API keys, implementing them also means revisiting how we are going to protect them inside the browser. If your head is starting to spin thinking about all of the concerns that exist with trying to implement JWTs as both a secure and user-friendly approach, then you are not alone. Adding JWTs to an application is certainly much more involved than sessions. So, you must think about your specific application needs when deciding which strategy to use.

"*Hold up!*", you may be saying to yourself. "*If JWTs have so many benefits, why not just treat them like session tokens and store them as cookies? Also, we can get around invalidating tokens by comparing them against a black list! Then, we can make them longer and just add them to the blacklist when we want to log out or invalidate them. Both problems solved.*"

Yes, that is true. Let's look at both of those proposals in turn.

First, storing JWTs as cookie-like session tokens does work. However, you lose out on one of its big benefits: the authenticated payload. Remember that one of their benefits is that they can carry meta details that your frontend application could use. If they are stuck inside of an `HttpOnly` cookie, then that information is not available. (We will look at a way to address this when we look at JWT implementations.)

Second, if you are maintaining a blacklist of tokens to allow a token to be *revoked* or *invalidated*, then you are no longer using non-session-based authentication. Instead, you are using JWTs in a session-based scheme. This is acceptable and people do it. However, it makes your tokens less portable since they require a centralized store to be validated, and also include additional network calls. Implement this at your own risk.

We now turn to implementation strategies within Sanic. Because we have not looked at database implementation, we will still use some stand-in functions to get and store information when needed. Try to look over those details for now since we are focusing more on how to handle authentication, not persisting data. If you look at these examples on this book's GitHub repository, you will see some dummy versions of these functions, just to make the examples operational. Try not to get hung up on those details right now.

Using sessions

After reading the *Understanding session-based versus non-session-based authentication* section, you have decided that stateful sessions are the right option for your application. Super—you already know just about everything you need.

We have already learned how to handle passwords (the same as API keys). Therefore, implementing a login route should be simple.

We already know that the session token *doesn't* need to be accessible from JavaScript to combat XSS. Therefore, we will use `HttpOnly` cookies.

We also know that using an `HttpOnly` cookie by itself leaves an application vulnerable to CSRF attacks. Therefore, we will couple our implementation with the CSRF protection scheme we came up with earlier.

What's left? Not much. We need endpoints for the following purposes:

- Registering a user (who will be responsible for storing the password securely)
- Logging in (which takes a username and password and verifies it, just like in the API key example, then creates a session key, stores it, and sets it as a cookie)
- Logging out (which deletes the session from the database)

This is a great opportunity for you to take these requirements and try and build a solution. Put this book down and build these three endpoints. If you get stuck, don't worry—there is an example solution in this book's GitHub repository.

To protect your endpoints, you can use a similar approach but with a decorator. Do you remember the `@csrf_protected` decorator we built earlier? If you are using session-based authentication, then I suggest combining that decorator with the one we are building here. They compliment each other nicely and make it easier for you to protect your endpoints.

Here is how we will rebuild it:

1. First, we must add a block to our decorator that's similar to the API key decorator. This will raise an exception if session verification fails:

```
def session_protected(
    maybe_func=None,
    *,
    exception=Unauthorized,
    message="Invalid or unknown API key"
):
```

```
    def decorator(f):
        @wraps(f)
        async def decorated_function(request, *args,
**kwargs):

            origin = request.headers.get("origin")
            if request.ctx.from_browser and (
                origin not in app.config.ALLOWED_ORIGINS
                or "browser_check" not in request.cookies
                or not csrf_check(request)
            ):
                raise CSRFFailure

            session_token = request.cookies.get("session_
token")
            if not session_token or not await verify_
session(session_token):
                raise exception(message)

            response = f(request, *args, **kwargs)
            if isawaitable(response):
                response = await response

            return response

        return decorated_function

    return decorator(maybe_func) if maybe_func else
    decorator
```

2. The session's verification does depend on your database implementation. But, in general, it should look something like this:

```
async def verify_session(session_token):
    try:
        await get_session_from_database(session_token):
    except NotFound:
```

```
            return False
        return True
```

If the session token exists, then we can proceed. If it does not, then return `False`.

As you can see, sessions tend to be easy to implement once you have the basic functionality for storing and retrieving data from a database. We will now turn to the more complicated alternative.

JSON Web Tokens (JWTs)

So, you have read the *Understanding session-based versus non-session-based authentication* section and decided to implement JWTs. Now what? The problem that we need to solve is that using them to their full capacity within a frontend application poses two problems:

- How to store and send them to not compromise on functionality or security
- How to maintain a reasonable user experience without sacrificing security

We will address these issues in turn, and then develop a solution that gives us satisfaction in terms of both.

To cookie, or not to cookie?

There are two competing interests when you're deciding how to send the access token (please note that from here on out, the term *access token* is synonymous with *JWT*): usability and security. If we send the token via the headers, it would look like this:

```
Authentication: Bearer <JWT>
```

To accomplish this, we need some client-side JavaScript to read the value and inject it into our request:

```javascript
const accessToken = document.cookie
    .split('; ')
    .find(row => row.startsWith('access_token='))
    .split('=')[1]


fetch(url, {headers: {Authorization: 'Bearer ${accessToken}'}})
```

You should (by now) already be suspecting the problem with this: an XSS vulnerability! If our frontend application can access the token from JavaScript, then that means that any bad script can as well. Bummer.

Important Note

You may be thinking to yourself, why is the JWT being stored on the client side in a cookie and not in web storage (either `localStorage` or `sessionStorage`)? The reason is that both of those solutions are great for handling non-sensitive details. They are subject to the XSS attacks we are trying to prevent. You may see a lot of advice online suggesting that you use these for JWTs. *Don't do it!* The solution that is offered here will be much more secure and still not sacrifice usability. All it takes is a little extra work on the server side, so please be patient and do not rush off to this sub-standard alternative.

To fix this problem, we can use `HttpOnly` and let our application just send the cookie back by itself. In this situation, we will rely on the server to write and read the cookie as needed. But, in doing this, we cannot access the JWT payload. There is also the problem of CSRF that we have seen a few times already, but by now, you should already understand how to solve that problem. If not, please go back to the *Protecting applications from CSRF* section of this chapter.

One option may be to return the payload of the access token when you first log in. You could store these details in web storage safely and use them whenever you want. This may look something like this on the server:

```
@app.post("/login")
async def login(request):
    user = await authenticate_login_credentials(
        request.json["username"],
        request.json["password"],
    )
    access_token = generate_access_token(user)
    response = json({"payload": access_token.payload})
    response.cookies["access_token"] = access_token
    response.cookies["access_token"]["domain"] = "localhost"
    response.cookies["access_token"]["httponly"] = True
    response.cookies["access_token"]["samesite"] = "lax"
    response.cookies["access_token"]["secure"] = True
    return response
```

I support this approach, and it will certainly work. You can gain access to the payload, and you have a secure way to transport and store the access token.

A second option would be to use split cookies. More on that in just a bit. Feel free to skip ahead, or go back and reference that EuroPython talk I mentioned at the beginning of this chapter, where I discussed this approach.

"Your session expired after 10 minutes, please log in again"

Have you ever been on a website that does this? Usually, it is banking or financial applications because they are concerned about a user standing up from their computer and walking away to leave a logged-in session. Maybe this is your need, so great! You can rest comfortably with JWTs as a solution and expire your tokens often with no concern.

For most applications, however, this would lead to a terrible user experience.

Remember, the reason we are expiring our access tokens at such a short interval is to reduce the potential attack surface. If a token were to fall into the wrong hands, it can only be used for a very small window. The shorter the expiration, the more secure the token.

The solution to this problem requires a little bit of frontend complexity. But, I think it is worth the protection it affords. There are two solutions that you can choose from:

- Use JavaScript's `setInterval` to periodically send a request to refresh the token in the background that's unknown to the user.
- Wrap your JavaScript fetch call with a proper exception handler. It will catch the scenario where an expired token was submitted, send a request to refresh the token, and then retry the original request with the new token.

Feel free to choose the approach that works for you. This book's GitHub repository contains some sample JavaScript for implementing each strategy: `https://github.com/PacktPublishing/Python-Web-Development-with-Sanic/tree/main/Chapter07/accesstoken`.

To implement a refresh token, we will borrow some of the concepts we used earlier for making the API token. When a user logs in, we will continue to generate the access token, but we will also generate and store a refresh token by reusing the API token logic:

1. Create a login endpoint that also generates and stores a refresh token:

```
@app.post("/login")
async def login(request):
    user = await authenticate_login_credentials(
        request.json["username"],
```

```
        request.json["password"],
    )

    access_token = generate_access_token(user)
    refresh_token, hashed_key = generate_token()
    await store_refresh_token(user, hashed_key)

    response = json({"payload": access_token.payload})
    response.cookies["access_token"] = access_token
    response.cookies["access_token"]["domain"] =
"localhost"
    response.cookies["access_token"]["httponly"] = True
    response.cookies["access_token"]["samesite"] = "lax"
    response.cookies["access_token"]["secure"] = True
    response.cookies["refresh_token"] = refresh_token
    response.cookies["refresh_token"]["domain"] =
"localhost"
    response.cookies["refresh_token"]["httponly"] = True
    response.cookies["refresh_token"]["samesite"] =
"strict"
    response.cookies["refresh_token"]["secure"] = True

    return response
```

Go back to the *Using API keys* section to see the `generate_token` function.

2. To issue a new access token, we need to create a new endpoint that will validate the refresh token (like we did for the API token). As an added level of security (since a single point of authentication from the browser is not a good idea), we also will require a previously issued access token, even if it has already expired:

```
from bcrypt import checkpw
from sanic.exceptions import Forbidden
from sanic.response import empty

@app.post("/refresh")
async def refresh_access_token(request: Request) ->
HTTPResponse:
    user = await get_user_from_request(request)
```

```
access_token = request.cookies["access_token"]
refresh_token = request.cookies["refresh_token"]

if not user.refresh_hash:
    raise Forbidden("Invalid request")

is_valid_refresh = checkpw(
    refresh_token.encode("utf-8"),
    user.refresh_hash
)
is_valid_access = check_access_token(access_token,
allow_expired=True)

if not is_valid_refresh or not is_valid_access:
    raise Forbidden("Invalid request")

generated_access_token = generate_access_token(user)

response = empty()
response.cookies["access_token"] = generated_access_
token
response.cookies["access_token"]["domain"] =
"localhost"
response.cookies["access_token"]["httponly"] = True
response.cookies["access_token"]["samesite"] = "lax"
response.cookies["access_token"]["secure"] = True

return response
```

We have not learned how to validate JWT yet, so do not worry if you are not sure how to implement check_access_token. We will do that next.

Solving JWTs in browser-based applications

At this point, we should have an understanding of what we want to achieve. What we need to look at now is the following:

- How to generate the access token

- How to verify the access token (both with and without expiration)

- How to "split" the token to make it usable and secure

To generate the token, we will use `pyjwt`. The first thing we will need to do is create an application with a secret. Just like I did previously, I will hardcode it in my example, but you will get the value from an environment variable or an other secure method:

1. Set the secret and some other configuration values that we will need:

    ```
    from datetime import timedelta


    app.config.JWT_SECRET = "somesecret"
    app.config.JWT_EXPIRATION = timedelta(minutes=10)
    app.config.REFRESH_EXPIRATION = timedelta(hours=24)
    app.config.COOKIE_DOMAIN = "127.0.0.1"
    ```

2. Create a model that will hold our JWT details:

    ```
    from dataclasses import dataclass


    @dataclass
    class AccessToken:
        payload: Dict[str, Any]
        token: str

        def __str__(self) -> str:
            return self.token

        @property
        def header_payload(self):
            return self._parts[0]

        @property
        def signature(self):
            return self._parts[0]

        @property
        def _parts(self):
            return self.token.rsplit(".", maxsplit=1)
    ```

3. Generate the token with some payload. In JWT speak, a payload is essentially just a dictionary of values. It can contain a "claim," which is a special key-value pair that can be used to authenticate a token. If you get into JWT, I suggest that you dig deeper into some of the standard claims. In our example, the only one we are using is the expiration claim; that is, exp. Other than that, feel free to add whatever you want to the payload:

```python
import jwt


def generate_access_token(user: User, secret: str, exp:
int) -> AccessToken:
    payload = {
        "whatever": "youwant",
        "exp": exp,
    }
    raw_token = jwt.encode(payload, secret,
algorithm="HS256")
    access_token = AccessToken(payload, raw_token)
    return access_token
```

To verify the token, we can do the reverse. We have a use case for when we will accept an expired token (when using the refresh token). Therefore, we need a flag to allow us to skip the check for the exp claim:

```python
def check_access_token(
    access_token: str, secret: str, allow_expired: bool =
False
) -> bool:
    try:
        jwt.decode(
            access_token,
            secret,
            algorithms=["HS256"],
            require=["exp"],
            verify_exp=(not allow_expired),
        )
    except jwt.exceptions.InvalidTokenError as e:
        error_logger.exception(e)
        return False
```

4. Once you have generated the AccessToken object, it will be super easy to split it into two cookies. One of them will be JavaScript accessible, while the other will be HttpOnly. We also want the refresh token to be HttpOnly. Your login handler will look something like this:

```
access_token_exp = datetime.now() + request.app.config.
JWT_EXPIRATION
refresh_token_exp = datetime.now() + request.app.config.
REFRESH_EXPIRATION
access_token = generate_access_token(
    user,
    request.app.config.JWT_SECRET,
    int(access_token_exp.timestamp()),
)
refresh_token, hashed_key = generate_token()
await store_refresh_token(user, hashed_key)
response = json({"payload": access_token.payload})
```

5. Then, we must set all of our cookies with a convenience function. Pay attention to how these cookies are set with respect to httponly and samesite:

```
set_cookie(
    response,
    "access_token",
    access_token.header_payload,
    httponly=False,
    domain=request.app.config.COOKIE_DOMAIN,
    exp=access_token_exp,
)
set_cookie(
    response,
    "access_token",
    access_token.signature,
    httponly=True,
    domain=request.app.config.COOKIE_DOMAIN,
    exp=access_token_exp,
)
set_cookie(
    response,
```

```
        "refresh_token",
        refresh_token,
        httponly=True,
        samesite="strict",
        domain=request.app.config.COOKIE_DOMAIN,
        exp=refresh_token_exp,
    )
```

We now have all the building blocks needed to build out our endpoints and our decorator. It is time for you to put your skills to the test and try and piece together the endpoints from the knowledge provided in this chapter. Don't worry—there is a full solution in this book's GitHub repository, including the `set_cookie` convenience function that we used in the aforementioned code: `https://github.com/PacktPublishing/Python-Web-Development-with-Sanic/tree/main/Chapter07/accesstoken`.

A bit of self-promotion here: one of the first libraries I built for Sanic was a package to handle authentication and authorization for Sanic using JWTs. It allows you to handle this split token approach and includes all other sorts of goodies and protection. If you do not want to roll out a solution, then this is a good place to start as it has become widely adopted within the community. Check out my personal GitHub page for more details: `https://github.com/ahopkins/sanic-jwt`.

Summary

This chapter has covered a *lot* of material. Still, it has only scratched the surface of web security. To truly raise the security bar, you should continue to do some research. There are some other common headers such as `Content-Security-Policy`, `X-Content-Type-Options`, and `X-Frame-Options` that we did not have a chance to cover. Nonetheless, with the information you have gathered here, and with your self-ingenuity, you should be able to implement, for example, a `Content-Security-Policy` header that works for your application. The first place I look for this kind of material is Mozilla's MDN website: `https://developer.mozilla.org/en-US/`. I highly recommend that you visit it to learn about web standards and practices.

So, what did we cover?

You should now be familiar with the concept of same-origin and how to develop a CORS policy to defeat both CSRF and XSS attacks. We also looked at three common schemes for authenticating users: API keys, session tokens, and JWT. And, of course, by looking through all of the examples, you should have learned how to use the Sanic toolbelt to craft your own unique and *obvious* patterns to serve your applications' needs. At this point in this book, we have covered most of what you will need to build a web application. You should be familiar with all of the basic building blocks and start to have some ideas on how to piece them together to build solutions.

What we are missing now is knowledge on how to deploy our applications and run them. This is what we will cover in the next chapter.

8
Running a Sanic Server

In the time that I have been involved with the Sanic project—and specifically, in trying to assist other developers by answering their support questions—there is one topic that perhaps comes up more than any other: deployment. That one word is often bundled with a mixture of confusion and dread.

Building a web application can be a lot of fun. I suspect that I am not alone in finding a tremendous amount of satisfaction in the build process itself. One of the reasons that I love software development in general—and web development in particular—is that I enjoy the almost puzzle-like atmosphere of fitting solutions to a given problem. When the build is done and it is time to launch, that is where the anxiety kicks in.

I cannot overemphasize this next point enough. One of Sanic's biggest assets is its bundled web server. This is not just a gimmick or some side feature to be ignored. The fact that Sanic comes bundled with its own web server truly does simplify the build process. Think about traditional Python web frameworks such as Django or Flask, or about some of the newer **Asynchronous Server Gateway Interface** (**ASGI**) frameworks. For them to become operational and connected to the web, you need a production-grade web server. Building the application is only one step—deploying it requires knowledge and proficiency in another tool. Typically, the web server used to deploy your application built with one of those frameworks is not the same web server that you develop upon. For that, you have a development server.

Not only is this an added complexity and dependency, but it also means you are not developing against the actual server that will be running your code in production. Is anyone else thinking what I am thinking? Bugs.

In this chapter, we will look at what is required to run Sanic. We will explore different ways to run Sanic both in development and production to make the deployment process as easy as possible. We will start by looking at the server life cycle. Then, we will discuss setting up both a local and a production-grade scalable service. We will cover the following topics:

- Handling the server life cycle

- Configuring an application

- Running Sanic locally

- Deploying to production

- Securing your application with **Transport Layer Security** (**TLS**)

- Deployment examples

When we are done, your days of deployment-induced anxiety should be a thing of the past.

Technical requirements

We will, of course, continue to build upon the tools and knowledge from previous chapters. Earlier, in *Chapter 3, Routing and Intaking HTTP Requests*, we saw some implementations that used Docker. Specifically, we were using Docker to run an Nginx server for static content. While it is not required for deploying Sanic, knowledge of Docker and (to a lesser extent) Kubernetes will be helpful. In this chapter, we will be exploring the usage of Docker with Sanic deployments. If you are not a black-belt Docker or Kubernetes expert, don't worry. There will be examples on the GitHub repository at `https://github.com/PacktPublishing/Python-Web-Development-with-Sanic/tree/main/Chapter08`. All that we hope and expect is some basic understanding of and familiarity with these tools.

If you do not have these listed tools installed, you will need them to follow along with this chapter:

- `git`
- `docker`

- `doctl`
- `kubectl`

Handling the server life cycle

Throughout this book, we have spent a lot of time talking about the life cycle of an incoming **HyperText Transfer Protocol** (**HTTP**) request. In that time, we have seen how we can run code at different points in that cycle. Well, the life cycle of the application server as a whole is no different.

Whereas we had middleware and signals, the server life cycle has what are called "listeners". In fact, listeners are in effect (with one small exception) signals themselves. Before we look at how to use them, we will take a look at which listeners are available.

Server listeners

The basic premise of a **listener** is that you are attaching some function to an **event** in the server's life cycle. As the server progresses through the startup and shutdown process, Sanic will trigger these events and therefore allow you to easily plug in your own functionality. Sanic triggers events at both the startup and shutdown phases. For any other event during the life of your server, you should refer to the *Leveraging signals for intra-worker communication* section of *Chapter 6, Operating Outside the Response Handler*.

The order of events goes like this:

1. `before_server_start`: This event naturally begins runs before the server is started. It is a great place to connect to a database or perform any other operations that need to happen at the beginning of your application life cycle. Anything that you might be inclined to do in the global scope would almost always be better off done here. The only caveat worth knowing about is that if you are running in ASGI mode, the server is already running by the time Sanic is even triggered. In that case, there is no difference between `before_server_start` and `after_server_start`.

2. `after_server_start`: A common misconception about this event is that it could encounter a race condition where the event runs *while* your server begins responding to HTTP requests. That is *not* the case. What this event means is that there was an HTTP server created and attached to the **operating system** (**OS**). The infrastructure is in place to begin accepting requests, but it has not happened yet. Only once all of your listeners for `after_server_start` are complete will Sanic begin to accept HTTP traffic.

3. `before_server_stop`: This is a good place to start any cleanup you need to do. While you are in this location, Sanic is still able to accept incoming traffic, so anything that you might need to handle should still be available (such as database connections).

4. `after_server_stop`: Once the server has been closed, it is now safe to start any cleanup that is remaining. If you are in ASGI mode, as with `before_server_start`, this event is not actually triggered after the server is off because Sanic does not control that. It will instead immediately follow any `before_server_stop` listeners to preserve their ordering.

Two more listeners are available to you—however, these additional listeners are *only* available with the Sanic server since they are specific to the Sanic server life cycle. This is due to how the server works. When you run Sanic with multiple workers, what happens is that there is the main process that acts as an orchestrator, spinning up multiple subprocesses for each of the workers that you have requested. If you want to tap into the life cycle of each of those worker processes, then you already have the tools at your disposal with the four listeners we just saw.

However, what if you wanted to run some bit of code not on each worker process, but once in the main process: that orchestrator? The answer is the Sanic server's main process events—`main_process_start` and `main_process_stop`. Apart from the fact that they run inside the main process and not the workers, they otherwise work like the other listeners. Remember how I said that the listeners are themselves signals, with an exception? This is that exception. These listeners are not signals in disguise. For all practical purposes, this distinction is not important.

It is also worth mentioning that even though these events are meant to allow code to be run in the main process and not the worker process when in multi-worker mode, they are still triggered when you are running with a single worker process. When this is the case, it will be run at the extreme beginning and extreme end of your life cycle.

This raises an interesting and often-seen mistake: double execution. Before continuing with listeners, we will turn our attention to mistakenly running code multiple times.

Running code in the global scope

When you are preparing your application to run, it is not uncommon to initialize various services, clients, interfaces, and so on. You likely will need to perform some operations on your application very early in the process before the server even begins to run.

For example, let's imagine that you are looking for a solution to help you better track your exceptions. You find a third-party service where you can report all of your exceptions and tracebacks to help you to better analyze, debug, and repair your application. To get started, the service provides some documentation to use its **software development kit (SDK)**, as follows:

```
from third_party_sdk import init_error_reporting

init_error_reporting(app)
```

You get this set up and running in your multi-worker application, and you immediately start noticing that it is running multiple times and not in your worker processes as expected. What is going on?

Likely, the issue is that you ran your initialization code in the global scope. By *global scope* in Python, we mean something that is executing outside of a function or method. It runs on the outermost level in a Python file. In the preceding example, `init_error_reporting` runs in the global scope because it is not wrapped inside another function. The problem is that when multiple workers are running, you need to be aware of where and when that code is running. Since multiple workers mean multiple processes and each process is likely to run in your global scope, you need to be careful about what you put there.

As a very general rule, stick to putting *any* operable code inside a listener. This allows you to control the *where* and *when*, enabling the listener to operate in a more consistent and easily predictable manner.

Setting up listeners

Using listeners should look very familiar since they follow a similar pattern found elsewhere in Sanic. You create a listener handler (which is just a function) and then wrap it with a decorator. It should look like this:

```
@app.before_server_start
async def setup_db(app, loop):
    app.ctx.db = await db_setup()
```

What we see here is something *incredibly important* in Sanic development. This pattern should be committed to memory because attaching elements to your application `ctx` object increases your overall flexibility in development. In this example, we set up our database client so that it can be accessed from anywhere that our application can be (which is literally anywhere in the code).

One important thing to know is that you can control the order in which the listeners execute depending upon when they are defined. For the "start" time listeners (`before_server_start`, `after_server_start`, and `main_process_start`), they are executed in the order in which they are declared.

For the *stop* time listeners (`before_server_stop`, `after_server_stop`, and `main_process_stop`), the opposite is true. They are run in the reverse order of declaration.

How to decide to use a before listener or an after listener

As stated previously, there persists a common misconception that logic must be added to `before_server_start` in cases where you want to perform some operation before requests start. The fear is that using `after_server_start` might cause some kind of a race condition where some requests might hit the server moments before that event is triggered.

This is incorrect. Both `before_server_start` and `after_server_start` run to completion before any requests are allowed to come in.

So, then the question becomes: When should you favor one over the other? There are, of course, some personal and application-specific preferences that could be involved. Generally, however, I like to use the `before_server_start` event to set up my application context. If I need to initialize some object and persist it to `app.ctx`, then I will reach for `before_server_start`. For any other use case (such as performing any other types of external calls or configuration, I like to use `after_server_start`. This is by no means a hard and fast rule, and I often break it myself.

Now that we understand the life cycle of the server, there is one more missing bit of information that we need before we can run the application: configuration.

Configuring an application

Sanic tries to make some reasonable assumptions out of the box about your application. With this in mind, you can certainly spin up an application, and it should already have some reasonable default settings in place. While this may be acceptable for a simple prototype, as soon as you start to build your application, you will realize that you need to configure it.

And this is where Sanic's configuration system comes into play.

Configuration comes in two main flavors: tweaking the Sanic runtime operation, and declaring a state of global constants to be used across your application. Both types of configuration are important, and both follow the same general principles for applying values.

We will take a closer look at what the configuration object is, how we can access it, and how it can be updated or changed.

What is the Sanic configuration object?

When you create a Sanic application instance, it will create a configuration object. That object is really just a fancy `dict` type. As you will see, it does have some special properties. Do not let that fool you. You should remember: *it is* a `dict` object, and you can work with it like you would any other `dict` object. This will come in handy shortly when we explore how we can use the object.

If you do not believe me, then pop the following code into your application:

```
app = Sanic(__name__)
assert isinstance(app.config, dict)
```

This means that getting a configuration value with a default is no different than any other `dict` in Python, as illustrated in the following snippet:

```
environment = app.config.get("ENVIRONMENT", "local")
```

The configuration object is—however—much more important than any other `dict` object. It contains a lot of settings that are critical to the operation of your application. We have, of course, already seen in *Chapter 6, Operating Outside the Response Handler,* that we can use it to modify our default error handling, as illustrated here:

```
app.config.FALLBACK_ERROR_FORMAT = "text"
```

To understand the full scope of settings that you can tweak, you should take a look at the Sanic documentation at `https://sanic.dev/en/guide/deployment/configuration.html#builtin-values`.

How can an application's configuration object be accessed?

The best way to access the configuration object is to first get access to the application instance. Depending upon the scenario you are tackling at the moment, there are three main ways to get access to an application instance, as outlined here:

- Accessing the application instance using a request object (`request.app`)
- Accessing applications from a Blueprint instance (`bp.apps`)
- Retrieving an application instance from the application registry (`Sanic.get_app()`)

Perhaps the most common way to obtain the application instance (and, therefore, the configuration object by extension) is to grab it from the request object inside of a handler, as illustrated in the following code snippet:

```
@bp.route("")
async def handler(request):
    environment = request.app.config.ENVIRONMENT
```

If you are outside of a route handler (or middleware) where the request object is easily accessible, then the next best choice is probably to use the application registry. Rarely will it make sense to use the Blueprint `apps` property which is a set of applications that the Blueprint has been applied to. However, because it only exists *after* registration and it could be ambiguous as to which application you need, I usually will not reach for that as a solution. It is, nonetheless, good to know that it exists.

You may have seen us using the third option already. As soon as an application is instantiated, it is part of a global registry that can be looked up using the following code:

```
from sanic import Sanic


app = Sanic.get_app()
```

Whenever I am not in a handler, this is the solution I usually reach for. The two caveats that you need to be aware of are these:

1. Make sure that the application instance has already been instantiated. Using `app = Sanic.get_app()` in the global scope can be tricky if you are not careful with your import ordering. Later on, in *Chapter 11, A Complete Real-World Example,* when we build out a complete application, I will show you a trick I use to get around this.

2. If you are building a runtime with multiple application instances, then you will need to differentiate them using the application name, as follows:

```
main_app = Sanic("main")
side_app = Sanic("side")

assert Sanic.get_app("main") is main_app
```

Once you have the object, you will usually just access the configuration value as a property—for example, `app.config.FOOBAR`. As shown previously, you can also use a variety of Python accessors, as illustrated here:

```
app.config.FOOBAR
app.config.get("FOOBAR")
app.config["FOOBAR"]
getattr(app.config, "FOOBAR")
```

How can the configuration object be set?

If you go to the Sanic documentation, you will see that there are a bunch of default values already set. These values can be updated in a variety of methods as well. Of course, you can use the `object` and `dict` setters, like this:

```
app.config.FOOBAR = 123
setattr(app.config, "FOOBAR", 123)
app.config["FOOBAR"] = 123
app.config.update({"FOOBAR": 123})
```

You will usually set values like this right after creating your application instance. For example, throughout this book, I have repeatedly used `curl` to access endpoints that I created. The easiest method to see an exception is to use the text-based exception renderer. Therefore, in most cases, I have used the following pattern to make sure that when there is an exception, it is easily formatted for display in this book:

```
app = Sanic(__name__)
app.config.FALLBACK_ERROR_FORMAT = "text"
```

This is *not* usually ideal for a fully built application. If you have been involved in web application development before, then you probably do not need me to tell you that configuration should be easily changeable depending upon your deployment environment. Therefore, Sanic will load environment variables as configuration values if they are prefixed with SANIC_.

This means that the preceding FALLBACK_ERROR_FORMAT value could also be set outside of the application with an environment variable, like this:

```
$ export SANIC_FALLBACK_ERROR_FORMAT=text
```

The best method to do this will obviously depend upon your deployment strategy. We go deeper into those strategies later in this chapter, but the specifics of how to set those variables will differ and are outside the scope of this book.

Another option that you may be familiar with is centralizing all of your configurations in a single location. Django does this with settings.py. While I am personally not a fan of this pattern, you might be. You can easily duplicate it, like this:

1. Create a settings.py file by running the following code:

   ```
   FOO = "bar"
   ```

2. Apply the configuration to the application instance, like this:

   ```
   import settings
   app.update_config(settings)
   ```

3. Access the values as needed, as follows:

   ```
   print(app.config.FOO)
   ```

   There is nothing special about the settings.py filename. You just need a module with a whole bunch of properties that are uppercased. In fact, you could replicate this with an object.

4. Put all of your constants into an object now, like this:

   ```
   class MyConfig:
   FOO = "bar"
   ```

5. Apply the configuration from that object, as follows:

   ```
   app.update_config(MyConfig)
   ```

The result will be the same.

Some general rules about configuration

I have some general rules that I like to follow regarding configuration and reproduce these here. I encourage you to adopt them since they have evolved from years of making mistakes, but I just as strongly encourage you to break them when necessary:

- **Use simple values**: If you have some sort of a complex object such as a `datetime` object, perhaps configuration is not the best location for it. Part of the flexibility of configuration is that it can be set in many different ways, including outside of your application in environment variables. While Sanic will be able to convert things such as Booleans and integers, everything else will be a string. Therefore, for the sake of consistency and flexibility, try to avoid anything but simple value types.

- **Treat them as constants**: Yes, this is Python. That means everything is an object and everything is subject to runtime changes. But do not do this. If you have a value that needs to be changed *during* the running of your application, use `app.ctx` instead. In my opinion, once `before_server_start` has completed, your configuration object should be considered locked in stone.

- **Don't hardcode values**: Or, at least try really hard not to. When building out your application, you will undoubtedly encounter the need to create some sort of constant value. It is hard to guess a scenario that this might come up in without knowing your specific application, but when you realize that you are about to create a constant or some value, ask yourself whether the configuration is more appropriate. Perhaps the most concrete example of this is the settings that you might use to connect to a database, a vendor integration, or any other third-party service.

Configuring your application is almost certainly something that will change over the lifetime of your application. As you build it, run it, and add new features (or fix broken features), it is not uncommon to return to configuration often. One marker of a professional-grade application is that it relies heavily upon this type of configuration. This is to provide you with the flexibility to run the application in different environments. You may, for example, have some features that are only beneficial in local development, but not in production. It may also be the other way around. Configuration is, therefore, almost always tightly coupled with the environment where you will be deploying your application.

We now turn our attention to those deployment options to see how Sanic will behave when running in development and production environments.

Running Sanic locally

We finally are at the point where it is time to run Sanic—well, locally, that is. However, we also know we have been doing that all along since *Chapter 2*, *Organizing a Project*. The Sanic **command-line interface** (**CLI**) is already probably a fairly comfortable and familiar tool, but there are some things that you should know about it. Other frameworks have only development servers. Since we know that Sanic's server is meant for both development and production environments, we need to understand how these environments differ.

How does running Sanic locally differ from production?

The most common configuration change for local production is turning on debug mode. This can be accomplished in three ways, as follows:

1. It could be enabled directly on the application instance. You would typically see this inside of a factory pattern when Sanic is being run programmatically from a script (as opposed to the CLI). You can directly set the value, as shown here:

    ```
    def create_app(..., debug: bool = False) -> Sanic:
        app = Sanic(__name__)
        app.debug = debug
        ...
    ```

2. It is perhaps more common to see it set as an argument of app.run. A common use case for this might be when reading environment variables to determine how Sanic should initialize. In the following example, an environment value is read and applied when the Sanic server begins to run:

    ```
    from os import environ
    from path.to.somewhere import create_app

    def main():
        app = create_app()

        debug = environ.get("RUNNING_ENV", "local") !=
    "production"

        app.run(..., debug=debug)
    ```

3. The final option is to use the Sanic CLI. This is generally my preferred solution, and if you have been following along with the book, it is the one that we have been using all along. This method is straightforward, as shown here:

```
$ sanic path.to:app --debug
```

The reason that I prefer this final option is that I like to keep the operational aspects of the server distinct from other configurations.

> **Important Note**
>
> As of v22.3, the `debug` argument has changed slightly. Whereas debug used to both enable debug mode and automatic server reloading, starting in v22.3, you will need to use the `dev` argument instead.

For example, timeouts are configuration values that are closely linked to the operation of the framework and not the server itself. They impact how the framework responds to requests. Usually, these values are going to be the same, regardless of where the application is deployed.

Debug mode, on the other hand, is much more closely linked to the deployment environment. You will want to set it to `True` locally but `False` in production. Therefore, since we will be controlling how Sanic is deployed with tools such as Docker, controlling the server's operational capacity outside of the application makes sense.

"Okay," you say, *"turning on debug mode is simple, but why should I?"* When you run Sanic in debug mode, it makes a couple of important changes. The most noticeable is that you begin to see debug logs and access logs dispatched from Sanic. This is, of course, very helpful to see while developing.

> **Tip**
>
> When I sit down to work on a web application, I always have three windows in my view at all times, comprising the following:
>
> - My **integrated development environment** (IDE)
>
> - An **application programming interface** (API) client such as Insomnia or Postman
>
> - A Terminal showing me my Sanic logs (in debug mode)
>
> The Terminal with debug level logging is your window into what is happening with your application as you build it.

Perhaps the biggest change that debug mode brings is that any exception will include its traceback in the response. In the next chapter, we will look at some examples of how you can make the most of this exception information.

This is hugely important and useful while you are developing. It is also a huge security issue to accidentally leave it on in production. *DO NOT leave debug mode on in a live web application.* This includes any instance of your application that is *not* on a local machine. So, for example, if you have a staging environment that is hosted somewhere on the internet, it may not be your "production" environment. However, it still *MUST NOT* run in debug mode. At best, it will leak details about how your application was built, and at worst, it will make sensitive information available. Make sure to turn off debug mode in production.

Speaking of production, let's move on over to what it takes to deploy Sanic into the wild world of production environments.

Deploying to production

We have finally made it. After working your way through the application development process, there finally is a product to launch out into the ether of the **World Wide Web** (**WWW**). The obvious question then becomes: What are my options? There are really two sets of questions that need to be answered, as follows:

- **First question**: Which server should run Sanic?

  There are three options: Sanic server, an ASGI server, or Gunicorn.

- **Second question**: Where do you want to run the application?

  Some typical choices include a *bare-metal* **virtual machine** (**VM**), a containerized image, a **platform-as-a-service** (**PaaS**), or a self-hosted or fully managed orchestrated container cluster. Perhaps these choices might make more sense if we put some of the commonly used product names to them, as follows:

Deployment type	Potential vendors
VM	Amazon Elastic Compute Cloud (EC2), Google Cloud, Microsoft Azure, DigitalOcean, Linode
Container	Docker
PaaS	Heroku
Orchestrated cluster	Kubernetes

Table 8.1 – Examples of common hosting providers and tools

Choosing the right server option

As we stated, there are three main ways to run Sanic: the built-in server, with an ASGI compatible server, or with Gunicorn. Before we decide which server to run, we will take a brief look at the pros and cons for each option, starting with the least performant option.

Gunicorn

If you are coming to Sanic from the **Web Server Gateway Interface** (**WSGI**) world, you may already be familiar with Gunicorn. Indeed, you may even be surprised to learn that Sanic can be run with Gunicorn since it is built for WSGI applications, not asynchronous applications such as Sanic. Because of this, the biggest downside to running Sanic with Gunicorn is the *substantial* decrease in performance. Gunicorn effectively unravels much of the work done to leverage *concurrency* with the `asyncio` module. It is by far the slowest way to run Sanic, and in most use cases is not recommended.

It still could be a good choice in certain circumstances. Particularly, if you need a feature-rich set of configuration options and cannot use something such as Nginx, then this might be an approach. Gunicorn has a tremendous amount of options that can be leveraged for fine-tuning server operation. In my experience, however, I typically see people reaching for it out of habit and not out of necessity. People will use it simply because it is what they know. People transitioning to Sanic from the Flask/Django world may be used to a particular deployment pattern that was centered on tools such as Supervisor and Gunicorn. That's fine, but it is a little old-fashioned and should not be the go-to pattern for Sanic deployments.

For those people, I urge you to look at another option. You are building with a new framework, so why not deploy it with a new strategy as well?

If, however, you do find yourself needing some of the more fine-tuned controls offered by Gunicorn, I would recommend you take a look at Nginx, which has an equally (if not more) impressive set of features. Whereas Gunicorn would be set up to actually run Sanic, the Nginx implementation would rely upon Sanic running via one of the other two strategies and placing an Nginx proxy in front of it (more on Nginx proxying later in this chapter). This option will allow you to retain a great deal of server control without sacrificing performance. It does, however, require some more complexity since you need to essentially run two servers instead of just one.

If, in the end, you still decide to use Gunicorn, then the best way to do so is to use Uvicorn's worker shim. Uvicorn is an ASGI server, which we will learn more about in the next section. In this context, however, it also ships with a worker class that allows Gunicorn to integrate with it. This effectively puts Sanic into ASGI mode. Gunicorn still runs as the web server, but it will pass traffic off to Uvicorn, which will then reach into Sanic as if it were an ASGI application. This will retain much of the performance offered by Sanic and asynchronous programming (although still not as performant as the Sanic server by itself). You can accomplish this as shown next:

1. First, make sure both Gunicorn and Uvicorn are installed by executing the following command:

    ```
    $ pip install gunicorn uvicorn
    ```

2. Next, run the application like this:

    ```
    $ gunicorn \
        --bind 127.0.0.1:7777 \
        --worker-class=uvicorn.workers.UvicornWorker \
        path.to:app
    ```

You should now have the full span of Gunicorn configurations at your fingertips.

ASGI server

We visited ASGI briefly in *Chapter 1, Introduction to Sanic and Async Frameworks*. ASGI is a design specification for how servers and frameworks can communicate with each other asynchronously. It was developed as a replacement methodology for the older WSGI standard that is incompatible with modern asynchronous Python practices. This standard has given rise to three popular ASGI web servers: Uvicorn, Hypercorn, and Daphne. All three of them follow the ASGI protocol and can therefore run any framework that adheres to that protocol. The goal, therefore, is to create a common language that allows one of these ASGI servers to run any ASGI framework.

And this is where to discuss Sanic with regard to ASGI, we must have a clear distinction in our mind of the difference between the server and the framework. *Chapter 1, Introduction to Sanic and Async Frameworks,* discussed this difference in detail. As a quick refresher, the web server is the part of the application that is responsible for connecting to the OS's socket protocol and handling the translation of bytes into usable web requests. The framework takes the digested web requests and provides the application developer with the tools needed to respond and construct an appropriate HTTP response. The server then takes that response and sends the bytes back to the OS for delivery back to the client.

Sanic handles this whole process, and when it does so, it operates outside the ASGI since that interface is not needed. However, it also has the ability to speak the language of an ASGI framework and thus can be used with any ASGI web server.

One of the benefits of running Sanic as an ASGI application is that it standardizes the runtime environment with a broader set of Python tools. There is, for example, a set of ASGI middleware that could be implemented to add a layer of functionality between the server and the application.

However, some of the standardization does come at the expense of performance.

Sanic server

The default mechanism is to run Sanic with its built-in web server. It should come as no surprise that it is built with performance in mind. Therefore, what the Sanic server gives up by forfeiting the standardization and interoperability of ASGI, it makes up for in its ability to optimize itself as a single-purpose server.

We have touched on some of the potential downsides of using the Sanic server, one of which was static content. No Python server will be able to match the performance of Nginx in handling static content. If you are already using Nginx as a proxy for Sanic and you have a known location of static assets, then it might make sense to also use it for those assets. However, if you are not using it, then you need to determine whether the performance difference warrants the additional operational expense. In my opinion, if you can easily add this to your Nginx configuration: great. However, if it would take a lot of complicated effort, or you are exposing Sanic directly, then the benefit might not be as great as just leaving it as is and serving that content from Sanic. Sometimes, for example, the easiest thing to do is to run your entire frontend and backend from a single server. This is certainly a case where I would suggest learning about the competing interests and making an appropriate decision instead of trying to make a *perfect* decision.

With this knowledge, you should now be able to decide which server is the right fit for your needs. We will assume for the remainder of this book that we are still deploying with the Sanic server, but since it is mainly a matter of changing the command-line executable, this should not make a difference.

How to choose a deployment strategy?

The last section laid out three potential web servers to use for Sanic applications, but that web server needs to run on a web host. But before deciding on which web-hosting company to use, there is still a very important missing component: how are you going to get your code from your local machine to the web host? In other words: how are you going to deploy your application? We will now look through some options for deploying Sanic applications.

There is some assumed knowledge, so if some of the technologies or terms here are unfamiliar, please feel free to stop and go look them up.

VM

This is perhaps the easiest option—well, the easiest besides a PaaS. Setting up a VM is super simple these days. With just a few clicks of a button, you can have a custom configuration for a VM. The reason this then becomes a simple option is that you just need to run your Sanic application the same way you might on your local machine. This is particularly appealing when using the Sanic server since it literally means that you can run Sanic in production with the same commands that you use locally. However, getting your code to the VM, maintaining it once it is there, and then ultimately scaling it will make this option the hardest. To be blunt, I almost would never recommend this solution. It is appealing to new beginners since it looks so simple from the outside, but looks can be deceiving.

There may in fact be times when this is an appropriate solution. If that is the case, then what would deployment look like? Really, not that much different than running it locally. You run the server and bind it to an address and port. With the proliferation of cloud computing, **service providers** (**SPs**) have made it such a trivial experience to stand up a VM. I personally find platforms such as DigitalOcean and Linode to be super user-friendly and excellent choices. Other obvious choices include **Amazon Web Services** (**AWS**), Google Cloud, and Microsoft Azure. In my opinion, however, they are a little less friendly to someone new to cloud computing. Armed with their good documentation, with DigitalOcean and Linode it is relatively inexpensive and painless to click a few buttons and get an instance running. Once they provide you with an **Internet Protocol** (**IP**) address, it is now your responsibility to get your code to the machine and run the application.

You might be thinking the simplest way to move your code to the server would be to use Git. Then, all you need to do is launch the application, and you are done. But what happens if you need more instances or redundancy? Yes—Sanic comes with the ability to spin up multiple worker processes, but what if that is not enough? Now, you need another VM and some way to manage load-balancing your incoming web traffic between them. How are you going to handle redeployments of bug patches or new features? What about changes to environment variables? These complexities could lead to a lot of sleepless nights if you are not careful.

This is also somewhat ignoring the other fact that not all environments are equal. VMs could be built with different dependencies, leading to wasteful time maintaining servers and packages.

That is not to say this cannot or should not be a solution. Indeed, it might be a great solution if you are creating a simple service for your own use. Perhaps you need a web server for connecting to a smart home network, but it is certainly a case of *developer beware*. Running a web server on a *bare-metal* VM is rarely as simple as it appears at first glance.

Containers with Docker

One solution to the previous set of problems is to use a Docker container. For those that have used Docker, you can probably skip to the next section because you already understand the power that it provides. If you are new to containers, then I highly recommend you learn about them.

In brief, you write a simple manifest called a Dockerfile. That manifest describes an intended OS and some instructions needed to build an ideal environment for running your application. An example manifest is available in the GitHub repository here: `https://github.com/PacktPublishing/Python-Web-Development-with-Sanic/blob/main/Chapter08/k8s/Dockerfile`.

This might include installing some dependencies (including Sanic), copying source code, and defining a command that will be used to run your application. With that in place, Docker then builds a single image with everything needed to run the application. That image can be uploaded to a repository and used to run irrespective of the environment. You could, for example, opt to use this instead of managing all those separate VM environments. It is much simpler to bundle all that together and simply run it.

There is still some complexity involved in building our new versions and deciding where to run the image, but having consistent builds is a huge gain. This should really become a focal point of your deployment. So, although containers are part of the solution, there is still the problem of where to run it and the maintenance costs required to keep it running and up to date.

I almost *always* would recommend using Docker as part of your deployment practices, and if you know about Docker Compose, you might be thinking that is a great choice for managing deployments. I would agree with you, so long as we are talking about deployments on your local machine. Using Docker Compose for production is not something I would usually consider. The reason is simple: horizontal scaling. Just as with the issue when running Sanic on a VM, or a single container on a VM, running Docker Compose on a single VM carries the same problem: horizontal scaling. The fix is orchestration.

Container orchestration with Kubernetes

The problem with containers is that they only solve environmental problems by creating a consistent and repeatable strategy for your application—they still suffer from scalability problems. What happens when your application needs to scale past the resources that are available on a single machine? Container orchestrators such as **Kubernetes** (aka **K8s**) are a dream come true for anyone that has done **development-operations** (**DevOps**) work in the past. By creating a set of manifests, you will describe to Kubernetes what your ideal application will look like: the number of replicas, the number of resources they need, how traffic should be exposed, and so on. That is it! All you need to do is describe your application with some **YAML Ain't Markup Language** (**YAML**) files. Kubernetes will handle the rest. It has the added benefit of enabling rolling deployments where you can roll out new code with zero downtime for your application. It sounds like a dream come true for application deployments.

The downside, of course, is that this option is the most complex to set up. It is suitable for more serious applications where the level of complexity is acceptable. It may, however, be overkill for a lot of projects. This is a go-to deployment strategy for any application that will have more than a trivial amount of traffic. Of course, the complexity and scale of a Kubernetes cluster can expand based upon its needs. This dynamic quality is what makes it increasingly a standard deployment strategy that has been adopted by many industry professionals.

It is an ideal solution for platforms that consist of multiple services working together or that require scaling beyond the boundaries of a single machine.

This does bring up an interesting question, however. We know that Sanic has the ability to scale horizontally on a single host by replicating its workers in multiple processes. Kubernetes is capable of scaling horizontally by spinning up replica **pods**. You can think of a pod as encapsulating a container. Usually—especially to start—you will run Kubernetes with one container per pod. Let's say you hypothetically have decided that you need four instances of your application to handle the projected load that your application will receive. Should you have two pods each running two workers, or four pods each with one worker?

I have heard both put forth as *ideal* solutions. Some people say that you should maximize the resources per container. Other people say that you should have no more than one process per container. From a performance perspective, it is a dead heat. In all of my testing and experience, the solutions effectively perform the same. Therefore, it comes down entirely to the choice of the application builder. There is no right or wrong answer.

Later in this chapter, we will take a closer look at what it takes to launch a Sanic application with Kubernetes.

PaaS

Heroku is probably one of the most well-known **PaaS** offerings. It has been around for a while and has become an industry leader in these low-touch deployment strategies. Heroku is not the only provider—both Google and AWS have PaaS services in their respective cloud platforms, and DigitalOcean has also launched its own competing service. What makes a PaaS super convenient is that all you need to do is write the code. There is no container management, environment handling, or deployment struggles. It is intended to be a super easy low-touch solution for deploying code. Usually, deploying an application is as simple as pushing code to a Git repository.

This simple option is, therefore, ideal for **proof-of-concept** (**POC**) applications or other builds you need to deploy super quickly. I also do know plenty of people that run more robust and scalable applications through these services, and they really can be a great alternative. The huge selling point of these services is that by outsourcing the deployment, scaling, and service maintenance to the SP, you are freed up to focus on the application logic.

Because of this simplicity, and—ultimately—flexibility, we will take a closer look at launching Sanic with a PaaS vendor later in this chapter in the *Deployment examples* section. One of the things that are great about a PaaS is that it handles a lot of details such as setting up a TLS certificate and enabling an `https://` address for your application. In the next section, however, we will learn what it takes to set up an `https://` address for your application in the absence of convenience from a PaaS.

Securing your application with TLS

If you are not encrypting traffic to your web application, you are doing something wrong. In order to protect information while it is in transit between the web browser and your application, it is an absolute necessity to add encryption. The international standard for doing that is known as TLS, which is a protocol for how data can be encrypted between two sources. Often, however, it will be referred to as **SSL** (which stands for **Secure Sockets Layer** and is an earlier protocol that TLS replaces) or **HTTPS** (which stands for **HTTP Secure** and is technically an implementation of TLS, not TLS itself). Since it is not important for us *how* it works and we only care that it does what it needs to do, we will use these terms somewhat interchangeably. Therefore, it is safe for you to think about TLS and HTTPS as the same thing.

So, what is it? The simple answer is that you request a pair of keys from some reputable source on the internet. Your next step is to make them available to your web server and expose your application over a secure port—typically, that is port `443`. After that, your web server should handle the rest, and you should now be able to access your application with an `https://` address instead of `http://`.

Setting up TLS in Sanic

There are two common scenarios you should be familiar with—if you are exposing your Sanic application directly, or if you are placing Sanic behind a proxy. This will determine where you want to *terminate* your TLS connection. This simply means where you should set up your public-facing certificates. We will assume for now that Sanic is exposed directly. We will also assume that you already have certificates. If you do not know how to obtain them, don't worry—we will get to a potential solution for you in the next section.

All we need to do is to tell the Sanic server how to access those certificates. Also, since Sanic will default to port 8000, we need to make sure to set it to 443. With this in mind, these are the steps we'll take:

1. Our new runtime command (in production) will be this:

    ```
    $ sanic \
        --host=0.0.0.0 \
        --port=443 \
        --cert=/path/to/cert \
        --key=/path/to/keyfile \
        --workers=4 \
        path.to.server:app
    ```

2. It is largely the same operation if you are using app.run instead, as illustrated in the following code snippet:

    ```
    ssl = {"cert": "/path/to/cert", "key": "/path/to/
    keyfile"}
    app.run(host="0.0.0.0", port=443, ssl=ssl, workers=4)
    ```

When you are exposing your Sanic application directly and therefore terminating your TLS with Sanic, there is often a desire to add HTTP to HTTPS redirect. For your users' convenience, you probably want them to always be directed to HTTPS and for this redirection to happen *magically* for them without having to think about it.

The Sanic user guide provides us with a simple solution that involves running a second Sanic application inside our main application. Its only purpose will be to bind to port 80 (which is the default HTTP non-encrypted port) and redirect all traffic. Let's quickly examine that solution and step through it, as follows:

1. First, in addition to our main application, we need a second that will be responsible for the redirects. So, we will set up two applications and some configuration details, as follows:

```
main_app = Sanic("MyApp")
http_app = Sanic("MyHTTPProxy")


main_app.config.SERVER_NAME = "example.com"
http_app.config.SERVER_NAME = "example.com"
```

2. We add only one endpoint to the http_app application that will be responsible for redirecting all traffic to the main_app application, as follows:

```
@http_app.get("/<path:path>")
def proxy(request, path):
    url = request.app.url_for(
        "proxy",
        path=path,
        _server=main_app.config.SERVER_NAME,
        _external=True,
        _scheme="https",
    )
    return response.redirect(url)
```

In *Chapter 10, Implementing Common Use Cases with Sanic*, there is a more complete working example of how to accomplish HTTP to HTTPS redirection: https://github.com/PacktPublishing/Python-Web-Development-with-Sanic/tree/main/Chapter10/httpredirect

3. To make running the HTTP redirect application easier, we will just piggyback off of the main application's life cycle so that there is no need to create another executable. Therefore, when the main application starts up, it will also create and bind the HTTP application. The code is illustrated in the following snippet:

```
@main_app.before_server_start
async def start(app, _):
    app.ctx.http_server = await http_app.create_server(
        port=80, return_asyncio_server=True
    )
    app.ctx.http_server.app.finalize()
```

You should note how we are assigning that server to the `ctx` object for our main application so that we can use it again.

4. Finally, when the main application shuts down, it will also be responsible for shutting down the HTTP application, as illustrated in the following code snippet:

```
@main_app.before_server_stop
async def stop(app, _):
    await app.ctx.http_server.close()
```

With this in place, any request to `http://example.com` should be automatically redirected to the `https://` version of the same page.

Back in *Step 1* and *Step 2*, this example sort of skipped over the fact that you need to obtain actual certificate files to be used to encrypt your web traffic. This is largely because you need to bring your own certificates to the table. If you are not familiar with *how* to do that, the next section provides a potential solution.

Getting and renewing a certificate from Let's Encrypt

Back in the olden days of the internet, if you wanted to add HTTPS protection to your web application, it was going to cost you. Certificates were not cheap, and they were somewhat cumbersome and complicated to manage. Actually, certificates are still not cheap if you are to buy one yourself, especially if you want to buy a certificate that covers your subdomains. However, this is no longer your only option since several players came together looking for a method to create a safer online experience. The solution: free TLS certificates. These free (and reputable) certificates are available from *Let's Encrypt* and are the reason that *every* production website should be encrypted. Expense is no longer an excuse. At this point in time, if I see a website still running `http://` in a live environment, a part of me cringes as I go running for the hills.

If you do not currently have a TLS certificate for your application, head over to `https://letsencrypt.org` to get one. The process to obtain a certificate from *Let's Encrypt* requires you to follow some basic steps and then prove that you own the domain. Because there are a lot of platform specifics and it is outside the scope of this book, we will not really dive into the details of how to obtain one. Later on, this chapter does go through a step-by-step process to obtain a Let's Encrypt certificate for use in a Kubernetes deployment in the *Kubernetes (as-a-service)* section.

I do, however, highly encourage you to use Let's Encrypt if the budget for your project does not allow for you to go out and purchase a certificate.

With a certificate in hand, it is finally time to look at some actual code and decide which deployment strategy is right for your project.

Deployment examples

Earlier, when discussing the various choices for deployment strategies, two options rose above the others: PaaS and Kubernetes. When deploying Sanic into production, I would almost always recommend one of these solutions. There is no hard and fast rule here, but I generally think of Kubernetes as being the go-to solution for platforms that will be running multiple services, have the need for more controlled deployment configurations, and have more resources and a team of developers. On the other hand, a PaaS is more appropriate for single developer projects or projects that do not have resources to devote to maintaining a richer deployment pipeline. We will now explore what it takes to get Sanic running in these two environments.

PaaS

As we stated before, Heroku is a well-known industry leader in deploying applications via PaaS. This is for good reason as it has been in business providing these services since 2007 and has played a critical role in popularizing the concept. It has made the process super simple for both new and experienced developers. However, in this section, we are going to instead take a look at deploying a Sanic application with DigitalOcean's PaaS offering. The steps should be nearly identical and applicable to Heroku or any of the other services that are out there, and we look at them here:

1. First, you need to—of course—go to DigitalOcean's website and sign up for an account if you do not have one. DigitalOcean's PaaS is called **Apps**, which you can find on the left-hand side of the main dashboard once you are logged in.

2. You will next be taken through a series of steps that will ask you to connect a Git repository.

3. You will next need to configure the app through their **user interface** (**UI**). Your
    screen will probably look something like this:

Step 2 of 4

Configure your app

Python detected

Source Directory [?]   Edit

/

Type   Edit

Web Services

Environment Variables   Edit

No environment variables defined yet.

Build Command   Edit

No build command defined

Run Command   Edit

$ sanic server:app --host=0.0.0.0 --port=7777 --workers=4

HTTP Port   Edit

7777

HTTP Request Routes   Edit

/

Figure 8.1 – Example settings for PaaS setup

A very important thing to note here is that we have set `--host=0.0.0.0`. This means that we are telling Sanic that it should bind itself to any IP address that DigitalOcean provides it. Sanic will bind itself to the `127.0.0.1` address without this configuration. As anyone who has done web development knows, the `127.0.0.1` address maps to `localhost` on most computers. This means that Sanic will be accessible only to web traffic on that specific computer. This is no good. If you ever deploy an application and cannot access it, one of the first things to check is that the port and host are set up properly. One of the easiest options is to just use `0.0.0.0`, which is the equivalent of a wildcard IP address.

4. Next, you will be asked to select a location for which data center it will live in. Usually, you want to pick one that will be close to where your intended audience will be to reduce latency.

5. You will then need to select an appropriate package. If you do not know what to choose, start small and then scale it up as needed.

6. The only thing left to do is to set up the files in our repository. There is a sample in GitHub for you to follow, at `https://github.com/PacktPublishing/Python-Web-Development-with-Sanic/tree/main/Chapter08/paas`.

7. Finally, we need a `requirements.txt` file that lists out our dependencies: Sanic and a `server.py` file, just as with every other build we have done so far.

Once that is done, every time you push to the repository, your application should be rebuilt and available to you. One of the nice benefits of this is that you will get a TLS certificate with HTTPS out of the box. No configuration is needed.

Seems simple enough? Let's look at a more complex setup with Kubernetes.

Kubernetes (as-a-service)

We are going to turn our attention to Kubernetes: one of the most widely adopted and utilized platforms for orchestrating the deployment of containers. You could, of course, spin up some VMs, install Kubernetes on them, and manage your own cluster. However, I find a much more worthwhile solution is to just take one of the Kubernetes-as-a-service solutions. You still have all of the power of Kubernetes but none of the maintenance headaches. Most of the major cloud providers offer Kubernetes as a service, so you should be able to use your provider of choice.

We will again look at DigitalOcean and use their Kubernetes platform for our example. Here are the steps:

1. In our local directory, we will need a few files, as follows:

    -   `Dockerfile` to describe our Docker container

    -   `app.yml`, a Kubernetes config file described next

    -   `ingress.yml`, a Kubernetes config file described next

    -   `load-balancer.yml`, a Kubernetes config file described next

    -   `server.py`, which is again a Sanic application

    You can follow along with the files in the GitHub repository at `https://github.com/PacktPublishing/Python-Web-Development-with-Sanic/tree/main/Chapter08/k8s`.

2. Our Dockerfile is the set of instructions to build our container. We will take a shortcut and use one of the Sanic community's base images that has both Python and Sanic pre-installed, as follows:

    ```
    FROM sanicframework/sanic:3.9-latest


    COPY . /srv
    WORKDIR /srv
    EXPOSE 7777


    ENTRYPOINT ["sanic", "server:app", "--port=7777",
    "--host=0.0.0.0"]
    ```

    Just as we saw with the PaaS solution, we are binding to host `0.0.0.0` for the same reason. We are *not* adding multiple workers per container here. Again, this is something you could do if you prefer.

3. Next, we will need to build an image, as follows:

    ```
    $ docker build -t admhpkns/my-sanic-example-app .
    ```

4. Let's try running it locally to make sure it works. Here's the command to do this:

    ```
    $ docker run -p 7000:7777 --name=myapp admhpkns/my-sanic-
    example-app
    ```

    Once it is running, you should be able to access the API at `http://localhost:7000`.

5. Don't forget to clean up your environment, and remove the container when you are done, like this:

```
$ docker rm myapp
```

6. And you will, of course, need to push your container to some accessible repository. For ease of use and demonstration purposes, I will be pushing it to my public Docker Hub repository, like this:

```
$ docker push admhpkns/my-sanic-example-app:latest
```

If you are not familiar with Docker repositories, they are cloud-hosted locations for storing container images. Docker Hub is a great resource that provides a free tier. Other popular locations include GitLab, Google, and AWS.

7. For this next part, we will interact with DigitalOcean through their CLI tool. If you do not have it installed, head to https://docs.digitalocean.com/reference/doctl/how-to/install/. You will want to make sure you log in by running the following command:

```
$ doctl auth init
```

8. We next need a DigitalOcean Kubernetes cluster. Log in to their web portal, click on **Kubernetes** on the main dashboard, and set up a cluster. For now, the default settings are fine.

9. We next need to enable kubectl (the tool to interact with Kubernetes) to be able to talk to our DigitalOcean Kubernetes cluster. If kubectl is not installed, check out the instructions here: https://kubernetes.io/docs/reference/kubectl/overview/. The command you need will look something like this:

```
$ doctl kubernetes cluster kubeconfig save afb87d0b-9bbb-
43c6-a711-638bc4930f7a
```

Once your cluster is available and kubectl is set up, you can verify it is running by checking the following:

```
$ kubectl get pods
```

Since we have not set up any pods yet, there should not be anything to see.

10. When configuring Kubernetes, we need to start by running kubectl apply on our app.yml file.

> **Tip**
>
> Before going any further, you will see a lot of online tutorials that use this style of command:
>
> ```
> $ kubectl create ...
> ```
>
> I generally try to avoid that in favor of this:
>
> ```
> $ kubectl apply ...
> ```
>
> They essentially do the same thing, but the convenience is that Kubernetes resources that are created with `apply` can be continually modified by "applying" the same manifest over and over again.

What is in `app.yml`? Check out the GitHub repository for the full versions. It is rather lengthy and includes some boilerplate that is not relevant to the current discussion, so I will show only relevant snippets here. This goes for all of the Kubernetes manifests in our example.

The file should contain the Kubernetes primitives needed to run the application: a **service** and a **deployment**. A service is a stability layer on top of pods. Because Kubernetes pods can be easily created and destroyed, services exist to have a consistent internal IP address to point to those pods. A deployment is an abstraction that defines how pods are to be created, which containers should they contain, how many there should be, and so on.

The service should look something like this:

```
spec:
  ports:
    - port: 80
      targetPort: 7777
  selector:
    app: ch08-k8s-app
```

Notice how we are mapping port 7777 to 80. This is because we will be terminating TLS in front of Sanic, and our ingress controller will talk to Sanic over HTTP unencrypted. Because it is all in a single cluster, this is acceptable. Your needs might be more sensitive, and then you should look into encrypting that connection as well.

The other thing in `app.yml` is the deployment, which should look something like this:

```
spec:
  selector:
    matchLabels:
```

```
      app: ch08-k8s-app
  replicas: 4
  template:
    metadata:
      labels:
        app: ch08-k8s-app
    spec:
      containers:
        - name: ch08-k8s-app
          image: admhpkns/my-sanic-example-app:latest
          ports:
            - containerPort: 7777
```

Here, we are defining the number of replicas we want, as well as pointing the container to our Docker image repository.

11. After creating that file, we will apply it, and you should see a result similar to this:

```
$ kubectl apply -f app.yml
service/ch08-k8s-app created
deployment.apps/ch08-k8s-app created
```

You can now check out to see that it worked, as follows:

```
$ kubectl get pods
$ kubectl get svc
```

12. We will next use an off-the-shelf solution to create an Nginx ingress. This will be the proxy layer that terminates our TLS and feeds HTTP requests into Sanic. We will install it as follows:

```
$ kubectl apply -f https://raw.githubusercontent.com/
kubernetes/ingress-nginx/controller-v1.0.0/deploy/static/
provider/do/deploy.yaml
```

Note, at the time of writing, v1.0.0 is the latest. That probably won't be true by the time you are reading this, so you may need to change that. You can find the latest version on their GitHub page at https://github.com/kubernetes/ingress-nginx.

13. Next, we will set up our ingress. Create an `ingress.yml` file following the pattern in our GitHub repository example, like this:

```
$ kubectl apply -f ingress.yml
```

You will notice there are *intentionally* some lines commented out. We will get to that in a minute. Let's just quickly verify that it worked by executing the following command:

```
$ kubectl get pods -n ingress-nginx
```

14. We should take a step back and jump over to the **DigitalOcean** dashboard. On the left is a tab called **Networking**. Go there, and then in the tab for **Domains**, follow the procedure to add your own domain there. In that example, in `ingress.yml` we added `example.com` as the ingress domain. Whichever domain you add to DigitalOcean's portal should match your ingress. If you need to go back and update and re-apply the `ingress.yml` file with your domain, do that now.

15. Once that is all configured, we should be able to see our application working, as in the following example:

```
$ curl http://example.com
Hello from 141.226.169.179
```

This is, of course, not ideal because it is still on `http://`. We will now get a Let's Encrypt certificate and set up TLS.

16. The easiest method for this is to set up a tool called `cert-manager`. It will do all of the interfacing we need with Let's Encrypt. Start by installing it, as follows:

```
$ kubectl apply --validate=false -f https://github.com/
jetstack/cert-manager/releases/download/v1.5.3/cert-
manager.yaml
```

Again, please check to see what the most up-to-date version is and update this command accordingly.

We can verify its installation here:

```
$ kubectl get pods --namespace cert-manager
```

17. Next, create a `load-balancer.yml` file following the example in the GitHub repository. It should look something like this:

```
apiVersion: v1
kind: Service
metadata:
```

```
  annotations:
    service.beta.kubernetes.io/do-loadbalancer-hostname:
example.com
  name: ingress-nginx-controller
  namespace: ingress-nginx
spec:
  type: LoadBalancer
  externalTrafficPolicy: Local
  ports:
    - name: http
      port: 80
      protocol: TCP
      targetPort: http
    - name: https
      port: 443
      protocol: TCP
      targetPort: https
  selector:
    app.kubernetes.io/name: ingress-nginx
    app.kubernetes.io/instance: ingress-nginx
    app.kubernetes.io/component: controller
```

18. Apply that manifest and confirm that it worked, as follows:

```
$ kubectl apply -f load-balancer.yml
service/ingress-nginx-controller configured
```

19. Your Kubernetes cluster will now start the process of obtaining a certificate.

> **Tip**
>
> One thing that you might encounter is that the process gets stuck while
> requesting the certificate. If this happens to you, the solution is to turn on
> **Proxy Protocol** in your **DigitalOcean** dashboard. Go to the following setting
> and turn this on if you need to:
>
> **Networking** > **Load Balancer** > **Manage Settings** > **Proxy Protocol** > **Enabled**

20. We're almost there! Open up that `ingress.yml` file and uncomment those few lines that were previously commented out. Then, apply the file, as follows:

```
$ kubectl apply -f ingress.yml
```

Done! You should now automatically have a redirect from `http://` to `https://`, and your application is fully protected.

Better yet, you now have a deployable Sanic application with all the benefits, flexibility, and scalability that Kubernetes container orchestration provides.

Summary

Building a great Sanic application is only half of the job. Deploying it to make our application usable out in the wild is the other half. In this chapter, we explored some important concepts for you to consider. It is never too early to think about deployment either. The sooner you know which server you will use and where you will host your application, the sooner you can plan accordingly.

There are of course many combinations of deployment options, and I only provided you with a small sample. As always, you will need to learn what works for your project and team. Take what you have learned here and adapt it.

However, if you were to ask me to boil all of this information down and ask for my personal advice on how to deploy Sanic, I would tell you this:

- Run your applications using the built-in Sanic server.

- Terminate TLS outside of your application.

- For personal or smaller projects, or if you want a simpler deployment option, use a PaaS provider.

- For larger projects that need to scale and have more developer resources, use a hosted Kubernetes solution.

There you have it. You should now be able to build a Sanic application and run it on the internet. Our work is done, right? You should have the skills and knowledge you need now to go out and build something great, so go ahead and do that now. In the remainder of this book, we will start to look at some more practical issues that arise while building web applications and look at some best-practice strategies for how to solve them.

Part 3: Putting It All together

After spending time looking at specific issues individually, we start to look at how some issues relate to one another. This part will include a lot more examples and code snippets. A lot of the code will only be available online, so you will be highly encouraged to also flip through the source code repository on GitHub: `https://github.com/PacktPublishing/Python-Web-Development-with-Sanic`. The goal is once again to provide you with knowledge, not snippets.

This Part contains the following chapters:

- *Chapter 9, Best Practices to Improve Your Web Applications*
- *Chapter 10, Implementing Common Use Cases with Sanic*
- *Chapter 11, A Complete Real-World Example*

9
Best Practices to Improve Your Web Applications

From *Chapter 1, Introduction to Sanic and Async Frameworks*, through *Chapter 8, Running a Sanic Server*, we learned how to build a web application from conception through deployment. Pat yourself on the back and give yourself a round of applause. Building and deploying a web application is not a simple feat. So, what have we learned? We, of course, spent time learning about all of the fundamental tools that Sanic provides: route handlers, blueprints, middleware, signals, listeners, decorators, exception handlers, and so on. More importantly, however, we spent some time thinking about how HTTP works and how we can use these tools to design and build applications that are secure, scalable, maintainable, and easily deployable.

There have been a lot of specific patterns in this book for you to use, but also, quite intentionally, I have left a lot of ambiguity. You have continually read statements such as *it depends upon your application's needs*. After all, one of the goals of the Sanic project is to remain *unopinionated*.

That's all well and good, and flexibility is great. But what if you are a developer that has not yet determined what patterns work and which do not? The difference between writing a *Hello, world* application and a production-ready, real-world application is huge. If you only have limited experience in writing applications, then you also only have had limited experience in making mistakes. It is through those mistakes (whether made by yourself or from lessons learned by others who have made them) that I truly believe we become better developers. Like so many other things in life, failure leads to success.

The purpose of this chapter, therefore, is to include several examples and *preferences* that I have learned from my 25+ years of building web applications. That means for every best practice you will learn in this chapter, there is probably some *mistake* that I made to go along with it. These are a set of base-level *best practices* that I think are critical for any professional-grade application to include from the beginning.

In this chapter, we are going to look at the following:

- Implementing practical real-world exception handlers
- Setting up a testable application
- Gaining insight from logging and tracing
- Managing database connections

Technical requirements

There are no new technical requirements that you have not already seen. By this point, you should hopefully have a nice environment available for building Sanic, along with all the tools, such as Docker, Git, Kubernetes, and cURL, that we have been using all along. You can follow along with the code examples on the GitHub repository: `https://github.com/PacktPublishing/Python-Web-Development-with-Sanic/tree/main/Chapter09`.

Implementing practical real-world exception handlers

Exception handling is not a new concept at this point. We explored the topic in the *Implementing proper exception handling* section in *Chapter 6, Operating Outside the Response Handler*. I emphasized the importance of creating our own set of exceptions that include default status messages and response codes. This useful pattern was meant to get you up and running very quickly to be able to send *useful* messages to your users.

For example, imagine we are building an application for travel agents to book airline tickets for customers. You can imagine one of the steps of the operation might be to assist in matching flights through connecting airports.

Well, what if the customer selected two flights where the time between the flights was too short? You might do something like this:

```
from sanic.exceptions import SanicException


class InsufficientConnection(SanicException):
    status_code = 400
    message = "Selected flights do not leave enough time for
connection to be made."
```

I love this pattern because it makes it super easy for us to now repeatably raise an InsufficientConnection exception and have a known response for the user. But responding properly to the user is only half of the battle. When something goes wrong in our applications in the *real world*, we want to know about it. Our applications need to be able to report back so that if there is indeed a problem, then we can fix it.

So, how do we go about solving this problem? Logging is, of course, essential (we will look at that in the *Gaining insight from logging and tracing* section later). Having a reliable way to get to your system logs is an absolute must for a lot of reasons. But do you want to monitor your logs all day long, every day, looking for a traceback? Of course not!

Somehow, in some way, you need to set up some alerts to notify you that an exception happened. Creating proper notifications is an important part of maintaining a web application since they tell you when something is not operating as you intended. However, receiving a notification on every issue may become very noisy and overwhelming. In some applications, you easily become lost and stop paying attention to notifications if there are too many, or if they are difficult to consume. Luckily, not all exceptions are created equal, and only sometimes will you actually want to be notified. Some errors are fine to just display to the user and to remain ignorant of their existence. If a customer forgets to input valid data, you do not need your mobile phone waking you up at 3 a.m. while you are on call. While setting up system monitoring and alerting tools is outside the scope of this book, the point that I am trying to make is that your application should be proactive about warning you when certain things happen and ignoring the issues that you do not care about. Sometimes bad things will happen, and you want to make sure that you are able to sift through the noise and not miss out on the issues that really matter. A simple form of this might be to send an email when something particularly bad happens.

Knowing what you do about Sanic so far, if I came to you and asked you to build a system that sent me an email whenever `PinkElephantError` is raised, how would you do it?

I hope this is not your answer:

```
if there_is_a_pink_elephant():
    await send_adam_an_email()
    raise PinkElephantError
```

"*Why?*" you might ask. For starters, what if this needs to be implemented in a few locations, and then we need to change the notification from `send_adam_an_email()` to `build_a_fire_and_send_a_smoke_signal()`? You now need to go searching through all of your code to make sure it is done consistently and hope you did not miss anything.

What else could you do? How can you simply write the following code in your application and have it know that it needs to send me an email?

```
if there_is_a_pink_elephant():
    raise PinkElephantError
```

Let's learn that next.

Catching errors with middleware

Adding the notification mechanism right next to where we raise the exception, as in the preceding example, would work, but is not the best solution. The goal is to run `send_adam_an_email()` at the same time that we raise `PinkElephantError`. One solution would be to catch the exception with response middleware and send out the alert from there. The problem with this is that the response is not likely to have an easily parseable exception. If `PinkElephantError` raises a 400 response, how would you be able to distinguish it from any other 400 response? You could, of course, have JSON formatting and check the exception type, or try and read the exception message. But that will only work in `DEBUG` mode because in `PRODUCTION` mode, you may not have that information available.

One creative solution I have seen is to attach an arbitrary exception code and rewrite it in the middleware as follows:

```
class PinkElephantError(SanicException):
    status_code = 4000
    message = "There is a pink elephant in the room"
```

```
@app.on_response
async def exception_middleware(request: Request, response:
HTTPResponse):
    if response.status == 4000:
        response.status = 400
        await send_adam_an_email()
```

This solution will likely become very tedious to maintain and it is not at all obvious to anyone (including your future self) what is happening. It reminds me of the old-school style of error coding. I am talking about those errors where you need a lookup table to translate a number to an error description, which undoubtedly will be incomprehensible because of a lack of standardization or documentation. Just thinking about seeing *E19* on my coffee machine as I race around to find the owner's manual to look up what that means is enough to raise my stress levels. What I am trying to say is: *Save yourself the hassle and try to find a nicer solution for identifying exceptions than attaching some otherwise hard-to-understand error codes that you later need to translate.* We need a better solution.

Catching errors with signals

Remember our old friend signals from way back in the *Leveraging signals for intra-worker communication* section in *Chapter 6, Operating Outside the Response Handler*? If you recall, Sanic can dispatch event signals when certain things occur. One of them is when an exception is raised. Better yet, the signal context includes the exception instance, making it *much* easier to identify which exception occurred.

A cleaner and more maintainable solution to the aforementioned code would look like this:

```
@app.signal("http.lifecycle.exception")
async def exception_signal(request: Request, exception:
Exception):
    if isinstance(exception, PinkElephantError):
        await send_adam_an_email()
```

I think you can already see this is a much classier and fitting solution. For a lot of use cases, this might very well be the best solution for you. Therefore, I suggest you commit this simple four-line pattern to memory. Now, when we need to change send_adam_an_email() to build_a_fire_and_send_a_smoke_signal(), that will be a super-simple change to our code.

Long-time builders of Sanic applications may be looking at this example and wondering whether we can just use `app.exception`. This is certainly an acceptable pattern, but not without its potential pitfalls. Let's look at that next.

Catching the error and responding manually

When an exception is raised, Sanic stops the regular route handling process and moves it over to an `ErrorHandler` instance. This is a single object that exists throughout the lifespan of your application instance. Its purpose is to act as a sort of mini-router to take incoming exceptions and make sure they are passed off to the proper exception handler. If there is none, then it uses the default exception handler. As we have seen already, the default handler is what we can modify by using the `error_format` argument.

Here is a quick example of what an exception handler looks like in Sanic:

```
@app.exception(PinkElephantError)
async def handle_pink_elephants(request: Request, exception:
Exception):

    ...
```

The problem with this pattern is that because you took over the actual handling of the exception, it is now your job to respond appropriately. If you build an application with 10, 20, or even more of these exception handlers, keeping their responses consistent becomes a chore.

It is for this reason that I genuinely try to avoid custom exception handling unless I need to. In my experience, I get much better results by controlling formatting, as discussed in the *Fallback handling* section in *Chapter 6*, *Operating Outside the Response Handler*. I try to avoid one-off response customizations that only target a single use case. While building an application, we likely need to build error handlers for many types of exceptions and not just `PinkElephantError`. Therefore, I tend to disfavor using exception handlers when I need to do something with the error—such as sending an email—and not just deal with how it is an output for the user.

Okay, okay, I give in. I will let you in on a secret: you can still use the `app.exception` pattern to intercept the error and still use the built-in error formatting. You can even still trigger some action with it—such as sending an email, lighting a smoke signal, or triggering your coffee machine—so you can get to work fixing the bug. If you like the exception handler pattern better than the signal, then it is possible to use it without my concern of formatting too many custom error responses.

Let's see how we can take an action with the error handler and still retain a consistent error formatting experience:

1. First, let's make a simple endpoint to throw our error and report back in text format:

```
class PinkElephantError(SanicException):
    status_code = 400
    message = "There is a pink elephant in the room"
    quiet = True


@app.get("/", error_format="text")
async def handler(request: Request):
    raise PinkElephantError
```

I have added quiet = True to the exception because that will suppress the traceback from being logged. This is a helpful technique when the traceback is not important to you and it just gets in the way.

2. Next, create an exception handler to send the email, but still use the default error response:

```
async def send_adam_an_email():
    print("EMAIL ADAM")


@app.exception(PinkElephantError)
async def handle_pink_elephants(request: Request,
exception: Exception):
    await send_adam_an_email()
    return request.app.error_handler.default(request,
exception)
```

We can access the default ErrorHandler instance using our application instance, as shown in the preceding code.

I would like you to hit that endpoint using curl so you can see that this works as expected. You should get the default text response and see that a mock email was sent to me as faked in the logs.

As you can also see, we are using the error_handler object that exists application-wide. In our next section, we will look at modifying that object.

Modifying ErrorHandler

When Sanic starts up, one of the first things that it does is create an `ErrorHandler` instance. We saw in the previous example that we can access it from the application instance. Its purpose is to make sure that when you define an exception handler, the request is responded to from the proper location.

One of the other benefits of this object is that it is easily customizable and is triggered on every single exception. Therefore, in the days before Sanic introduced signals, it was the easiest way to get some arbitrary code to run on every exception, such as our error-reporting utility.

Modifying the default `ErrorHandler` instance might have looked something like this:

1. Create `ErrorHandler` and inject the reporting code:

    ```
    from sanic.handlers import ErrorHandler


    class CustomErrorHandler(ErrorHandler):
        def default(self, request: Request, exception:
    Exception):

            ...
    ```

2. Instantiate your application using your new handler:

    ```
    from sanic import Sanic


    app = Sanic(..., error_handler=CustomErrorHandler())
    ```

That's it. Personally, I would almost *always* go for the signals solution when dealing with alerting or other error reporting. Signals have the benefit of being a much more succinct and targeted solution. It does not require me to subclass or monkey patch any objects. However, it is helpful to know how to create a custom `ErrorHandler` instance, as we have just seen, as you will see it out there in the wild.

For example, you will see them in third-party error-reporting services. These services are platforms that you can subscribe to that will aggregate and track exceptions in your application. They can be incredibly helpful in identifying and debugging problems in production applications. Usually, they work by hooking into your normal exception handling process. Since overriding `ErrorHandler` used to be the best method for low-level access to all exceptions in Sanic, many of these providers will provide sample code or libraries that implement this strategy.

Whether you use a custom `ErrorHandler` or signals is still a matter of personal taste. The biggest benefit, however, of signals is that they are run in a separate `asyncio` task. This means that Sanic will efficiently manage the *concurrent* response to the user with the reporting (provided you do not introduce other blocking code).

Does this mean that subclassing `ErrorHandler` is not a worthwhile effort? Of course not. In fact, if you are unhappy with the default error formats that Sanic uses, I would recommend that you change it using the previous example with `CustomErrorHandler`.

With this in mind, you now have the ability to format all of your errors as needed. An alternative strategy to this would be to manage this with exception handlers like in the `app.exception` pattern. The problem with that method is that you potentially lose out on Sanic's built-in auto-formatting logic. As a reminder, one of the great benefits of the default `ErrorHandler` is that it will attempt to respond with an appropriate format, such as HTML, JSON, or plain text, depending upon the circumstances.

Exception handling is an incredibly important component of any professional-grade web application. Make sure to put some thought into your application needs when designing a strategy. You very well may find that you need a mixture of signals, exception handlers, and a custom `ErrorHandler`.

We'll now turn our attention to another important aspect of professional-grade application development that may also not be exciting for some people to build: testing.

Setting up a testable application

Imagine this scenario: inspiration strikes you and you have a great application idea. Your excitement and creative juices are flowing as you start formulating ideas in your head about what to build. Of course, you do not rush straight into building it because you have read all the earlier chapters in this book. You take some time to plan it out, and in a caffeine-induced marathon, you start hacking away. Slowly, you start to see the application take shape and it is working beautifully. Hours go by, maybe it's days or weeks—you are not sure because you are in the zone. Finally, after all that work, you have a **minimum viable product** (**MVP**). You deploy it and go for some much-deserved sleep.

The problem is that you never set up testing. Undoubtedly, when you now come online and check out the error-handling system that you set up with advice from the previous section, you notice that it is swamped with errors. Uh oh! Users are doing things in your application that you did not anticipate. Data is not behaving as you thought it might. Your application is broken.

I would venture to guess that most people that have developed a web application or done any software development can relate to this story. We have all been there before. For many newcomers and experienced developers alike, testing is not fun. Maybe you are one of those rare breeds of engineers that completely love setting up a testing environment. If so, with all honesty, I tip my hat to you. For the rest of us, suffice it to say that if you want to build a professional application, you need to find the patience in you to develop a test suite.

Testing is a *huge* field, and I will not cover it here. There are plenty of testing strategies out there, including the often-celebrated **test-driven design** (**TDD**). If you know what this is and it works for you, great! If not, I will not judge you. If you are unfamiliar with it, I do suggest that you take some time and do some internet research on the topic. TDD is a fundamental part of many professional development workflows and many companies have adopted it.

Similarly, there are a lot of testing terms, such as **unit testing** and **integration testing**. We will use my simplified definitions of these terms: unit testing is when you test a single component or endpoint and integration testing is when you test the component or endpoint interacting with another system (such as a database).

What we care about in this book is how you can test your Sanic application in both unit and integration tests. Therefore, while I hope the general idea and approaches here are useful, to truly have a well-tested application, you will need to go beyond the pages of this book.

The last ground rule that we need to get out of the way is that the tests here will all assume that you are using `pytest`. It is one of the most widely used testing frameworks with many plugins and resources.

Getting started with sanic-testing

The Sanic Community Organization (the community of developers that maintain the project) also maintains a testing library for Sanic. Although its primary utility is by the Sanic project itself to achieve a high level of test coverage, it nonetheless has found a home and use case for developers working with Sanic. We will use it extensively because it provides a convenient interface for interacting with Sanic.

To start, we will need to install it in your virtual environment. While we are at it, we will install `pytest` too:

```
$ pip install sanic-testing pytest
```

So, what does `sanic-testing` do? It provides an HTTP client that you can use to reach your endpoints.

A typical barebones implementation would look like this:

1. First, you will have your application defined in some module or factory. For now, it will be a global-scoped variable, but later in the chapter, in the *Testing a full application* section, we will start working with factory pattern applications where the application instance is defined inside of a function:

    ```python
    # server.py
    from sanic import Sanic

    app = Sanic(__name__)


    @app.get("/")
    async def handler(request: Request):
        return text("...")
    ```

2. Then, in your testing environment, you initialize a test client. Since we are using `pytest`, let's set that up in a `conftest.py` file as a fixture so we can easily access it:

    ```python
    # conftest.py
    import pytest
    from sanic_testing.testing import SanicTestClient
    from server import app


    @pytest.fixture
    def test_client():
        return SanicTestClient(app)
    ```

3. You will now have access to the HTTP client in your unit tests:

    ```python
    # test_sample.py
    def test_sample(test_client):
        request, response = test_client.get("/")

        assert response.status == 200
    ```

4. Running your tests now is a matter of executing the `pytest` command. It should look something like this:

```
$ pytest
================= test session starts =================
platform linux -- Python 3.9.7, pytest-6.2.5, py-1.11.0,
pluggy-1.0.0
rootdir: /path/to/testing0
plugins: anyio-3.3.4
collected 1 item


test_sample.py . [100%]


================= 1 passed in 0.09s =================
```

So, what just happened here? What happened is that the test client took your application instance and actually ran it locally on your operating system. It initiated the Sanic server, binding it to a host and port address on your operating system, and ran whatever event listeners were attached to your application. Then, once the server was running, it used `httpx` as an interface to send an actual HTTP request to the server. It then bundled up both the `Request` and `HTTPResponse` objects and provided them as the return value.

The code for this example can be found in the GitHub repository: `https://github.com/PacktPublishing/Python-Web-Development-with-Sanic/tree/main/Chapter09/testing0`.

This is something that I cannot stress enough. Just about every time that someone has come to me with a question about or problem using `sanic-testing`, it is because the person failed to understand that the test client is *actually* running your application. This happens on every single call.

For example, consider the following:

```
request, response = test_client.get("/foo")
request, response = test_client.post("/bar")
```

When you run this, it will first start up the application and send a GET request to `/foo`. The server then goes through the full shutdown. Next, it stands up the application again and sends a POST request to `/bar`.

For most test cases, this starting and stopping of the server is preferred. It will make sure that your application runs in a clean environment every time. It happens very quickly, and you can still whip through a bunch of unit tests without feeling this as a performance penalty.

There are some other options that we will explore later in the following sections.

A more practical test client implementation

Now that you have seen how the test client works, I am going to let you in on a little secret: you do not actually need to instantiate the test client. In fact, other than the previous example, I have *never* used sanic-testing like this in a real application.

The Sanic application instance has a built-in property that can set up the test client for you if sanic-testing has been installed. Since we already installed the package, we can just go ahead and start using it. All that you need is access to your application instance.

Setting up an application fixture

Before going further, we will revisit the pytest fixtures. If you are unfamiliar with them, they might seem somewhat magical to you. In brief, they are a pattern in pytest to declare a function that will return a value. That value can then be used to inject an object into your individual tests.

So, for example, in our last use case, we defined a fixture in a special file called conftest.py. Any fixtures that are defined there will be available anywhere in your testing environment. That is why we were able to inject test_client as an argument in our test case.

I find it almost always beneficial to do this with the application instance. Whether you are using a globally defined instance or a factory pattern, you will make testing much easier with fixtures.

Therefore, I will always do something like this in my conftest.py:

```
import pytest
from server import app as application_instance


@pytest.fixture
def app():
    return application_instance
```

I now have access to my application instance everywhere in the test environment without importing it:

```
def test_sample(app):
    ...
```

> **Tip**
>
> There is one more quick trick you should know about fixtures. You can use the `yield` syntax here to help you inject code before and after your test. This is particularly helpful with an application if you need to do any sort of cleanup after the test runs. To achieve this, do the following:
>
> ```
> @pytest.fixture
> def app():
>     print("Running before the test")
>     yield application_instance
>     print("Running after the test")
> ```

With access to our app instance using fixtures, we can now rewrite the previous unit test like this:

```
def test_sample(app: Sanic):
    request, response = app.test_client.get("/")

    assert response.status == 200
```

To make our lives a little simpler, I added the type annotation for the fixture so that my **integrated development environment** (IDE) knows that it is a Sanic instance. Even though the main purpose of type hinting is to catch mistakes early, I also like to use it in cases like this to just make my IDE experience nicer.

This example shows that access to the test client is simply a matter of using the `app.test_client` property. By doing that, Sanic will automatically instantiate the client for you as long as the package is installed. This makes it super simple to write unit tests like this.

Testing blueprints

Sometimes, you may run across a scenario where you want to test some functionality that exists on a blueprint alone. In this case, we are assuming that any application-wide middleware or listeners that run before the blueprint are not relevant to our test. This means that we are testing some functionality that is entirely contained within the boundaries of the Blueprint.

I love situations like this and actively seek them out. The reason is that they are super easy to test, as we will see in a minute. These types of testing patterns are probably best understood as they contrast to what we will do in the *Testing a full application* section. The main difference is that in these tests, our endpoints do not rely upon the existence of a third-party system, such as a database. Perhaps more accurately, I should say that they do not rely upon the impacts that a third-party system might have. The functionality and business logic are self-contained, and therefore very conducive to unit testing.

When I find a situation like this, the first thing that I do is add a new fixture to my `conftest.py` file. It will act as a dummy application that I can use for testing. Each unit test I create can use this dummy application with my target blueprint attached and nothing else. This allows my unit test to be more narrowly focused on my single example. Let's see how that looks next:

1. Here, we will create a new fixture that creates a new application instance:

```
# conftest.py
import pytest
from sanic import Sanic


@pytest.fixture
def dummy_app():
    return Sanic("DummyApp")
```

2. We can now stub out a test in our blueprint tests:

```
# test_some_blueprint.py
import pytest
from path.to.some_blueprint import bp


@pytest.fixture
def app_with_bp(dummy_app):
    dummy_app.blueprint(bp)
    return dummy_app
```

```
def test_some_blueprint_foobar(app_with_bp):
    ...
```

In this example, we see that I created a fixture that is localized to this one module. The point of this is to create a reusable application instance that has my target blueprint attached to it.

A simple use case for this kind of testing might be input validation. Let's add a blueprint that does some input validation. The blueprint will have a simple POST handler that looks at the incoming JSON body and just checks that the key exists and the type matches the expectation:

1. First, we will create a schema that will be the keys and the value type that we expect our endpoint to be able to test:

    ```
    from typing import NamedTuple


    class ExpectedTypes(NamedTuple):
        a_string: str
        an_int: int
    ```

2. Second, we will make a simple type checker that responds with one of three values depending upon whether the value exists, and is of the expected type:

    ```
    def _check(
        exists: bool,
        value: Any,
        expected: Type[object],
    ) -> str:
        if not exists:
            return "missing"
        return "OK" if type(value) is expected else "WRONG"
    ```

3. Finally, we will create our endpoint that will take the request JSON and respond with a dictionary about whether the passed data was valid:

    ```
    from sanic import Blueprint, Request, json


    bp = Blueprint("Something", url_prefix="/some")
    ```

```
@bp.post("/validation")
async def check_types(request: Request):
    valid = {
        field_name: _check(
            field_name in request.json,
            request.json.get(field_name), field_type
        )
        for field_name, field_type in
        ExpectedTypes.__annotations__.items()
    }
    expected_length = len(ExpectedTypes.__annotations__)
    status = (
        200
        if all(value == "OK" for value in valid.values())
        and len(request.json) == expected_length
        else 400
    )
    return json(valid, status=status)
```

As you can see, we have now created a very simplistic data checker. We loop over the definitions in the schema and check each to see whether it is as expected. All of the values should be "OK" and the request data should be the same length as the schema.

We can now test this out in our test suite. The first thing that we could test is to make sure that all the required fields are present. There are three potential scenarios here: the input has missing fields, the input has only the correct fields, and the input has extra fields. Let's take a look at these scenarios and create some tests for them:

1. First, we will create a test to check that there are no missing fields:

    ```
    def test_some_blueprint_no_missing(app_with_bp):
        _, response = app_with_bp.test_client.post(
            "/some/validation",
            json={
                "a_string": "hello",
                "an_int": "999",
            },
    ```

```
        )

        assert not any(
            value == "MISSING"
            for value in response.json.values()
        )
        assert len(response.json) == 2
```

In this test, we sent some bad data. Notice how the `an_int` value is actually a string. But we do not care about that right now. What this is meant to test is that all the proper fields were sent.

2. Next up is a test that should contain all of the inputs, of the correct types, but nothing more:

```
    def test_some_blueprint_correct_data(app_with_bp):
        _, response = app_with_bp.test_client.post(
            "/some/validation",
            json={
                "a_string": "hello",
                "an_int": 999,
            },
        )

        assert response.status == 200
```

Here, all we need to assert is that the response is a 200 since we know that it will be a 400 if it is bad data.

3. Lastly, we will create a test that checks that extraneous information is not sent:

```
    def test_some_blueprint_bad_data(app_with_bp):
        _, response = app_with_bp.test_client.post(
            "/some/validation",
            json={
                "a_string": "hello",
                "an_int": 999,
                "a_bool": True,
            },
```

```
    )

    assert response.status == 400
```

In this final test, we are sending known bad data since it contains the exact same payload as the previous test, except for the additional `"a_bool": True`. Therefore, we should assert that the response will be 400.

Looking at these tests, it seems very repetitive. While the general rule of **don't repeat yourself** (**DRY**) is often cited as a reason to abstract logic, be careful with this in testing. I would prefer to see repetitive testing code over some highly abstracted, beautiful, shiny factory pattern. In my experience—yes, I have been burned by this many times in the past—adding fancy abstraction layers to testing code is a recipe for disaster. Some abstraction might be helpful (creating the `dummy_app` fixture is an example of good abstraction), but too much will become difficult to maintain and update as your application needs to change. Testing code should be simple to read and easy to edit. This is certainly one of those areas where development straddles the line between science and art. Creating a powerful testing suite with a proper balance of repetition and abstraction will take some practice and is highly subjective.

With that warning out of the way, there is an abstraction layer that I do really like. It makes use of `pytest.parametrize`. This is a super-helpful feature that allows you to create a test and run it against multiple inputs. We are not abstracting our tests, per se, but instead are testing the same code with a variety of inputs.

Using `pytest.mark.parametrize`, we can actually condense those three tests into a single test:

1. We create a decorator that has two arguments: a string containing a comma-delimited list of argument names and an iterable that contains values to be injected into the test:

    ```
    @pytest.mark.parametrize(
    "input,has_missing,expected_status",
    (
        (
            {
                "a_string": "hello",
            }, True, 400,
        ),
        (
            {
    ```

```
                "a_string": "hello",
                "an_int": "999",
            }, False, 400,
        ),
        (
            {
                "a_string": "hello",
                "an_int": 999,
            }, False, 200,
        ),
        (
            {
                "a_string": "hello","an_int": 999,
                "a_bool": True,
            }, False, 400,
        ),
    ),
)
```

We have three values that we are going to inject into our test: input, has_
missing, and expected_status. The test is going to run multiple times, and
each time it will pull one of the tuples of arguments to inject into the test function.

2. Our test function can now be abstracted to use these arguments:

```
def test_some_blueprint_data_validation(
    app_with_bp,
    input,
    has_missing,
    expected_status,
):
    _, response = app_with_bp.test_client.post(
        "/some/validation",
        json=input,
    )

    assert any(
        value == "MISSING"
```

```
        for value in response.json.values()
    ) is has_missing
    assert response.status == expected_status
```

In this way, it is much easier for us to write multiple unit tests across different use cases. You may have noticed that I actually just created a fourth test. Since it was so simple to add more tests using this method, I included one use case that we had not previously tested. I hope you see the huge benefit that this creates and come to learn to love testing with @pytest.mark.parametrize.

In this example, we are defining the inputs and what our expected outcome should be. By parametrizing the single test, it actually turns this into multiple tests inside pytest.

The code for these examples can be found in the GitHub repository: https://github.com/PacktPublishing/Python-Web-Development-with-Sanic/tree/main/Chapter09/testing2.

Mocking out services

The sample blueprint that we were testing against is obviously not something we would ever use in real life. In that example, we were not actually doing anything with the data. The oversimplified example removed the need to worry about how to handle interactions with services such as a database access layer. What if we are testing a real endpoint? And, by a real endpoint, I mean one that is meant to interface with a database. For example, how about a registration endpoint? How can we test that the registration endpoint actually does what it is supposed to do and injects data as expected? Mocking is the answer. We will look at how we can use Python's mocking utilities to pretend we have a real database layer. We will also still use the dummy_app pattern for testing. Let's see what that will look like now:

1. First, we will need to refactor our blueprint so that it looks like something you might actually encounter in the wild:

    ```
    @bp.post("/")
    async def check_types(request: Request):
        _validate(request.json, RegistrationSchema)
        connection: FakeDBConnection = request.app.ctx.db
        service = RegistrationService(connection)
        await service.register_user(request.json["username"],
    request.json["email"])
        return json(True, status=201)
    ```

We are still doing the input validation. However, instead of simply storing the registration details to memory, we will send them off to a database for writing to disk. You can check out the full code at `https://github.com/PacktPublishing/Python-Web-Development-with-Sanic/tree/main/Chapter09/testing3` to see the input validation. The important things to note here are that we have `RegistrationService`, which is calling a `register_user` method.

2. Since we still have not looked at the usage of **object relationship mapping (ORM)**, our database storage function will ultimately just call some raw SQL queries. We will look at ORM in more detail in the *Managing database connections* section later in the chapter, but for now, let's create the registration service:

```
from .some_db_connection import FakeDBConnection


class RegistrationService:
    def __init__(self, connection: FakeDBConnection) ->
None:
        self.connection = connection

    async def register_user(
        self, username: str, email: str
    ) -> None:
        query = "INSERT INTO users VALUES ($1, $2);"
        await self.connection.execute(query, username,
email)
```

3. The registration service calls into our database to execute some SQL. We will also need a connection to our database. For the sake of the example, I am using a fake class, but this would (and should) be the actual object that your application uses to connect to the database. Therefore, imagine that this is a proper database client:

```
from typing import Any


class FakeDBConnection:
    async def execute(self, query: str, *params: Any):
        ...
```

4. With this in place, we can now create a new fixture that will take the place of our data access layer. Normally, you would create something like this to instantiate the client:

```
from sanic import Sanic
from .some_db_connection import FakeDBConnection


app = Sanic.get_app()



@app.before_server_start
async def setup_db_connection(app, _):
    app.ctx.db = FakeDBConnection()
```

Imagine that this code exists on our *actual* application. It initiates the database connection and allows us to access the client within our endpoints, as shown in the preceding code, because our connection uses the application ctx object. Since our unit tests will not have access to a database, we need to create a *mock* database instead and attach that to our dummy application.

5. To do that, we will create our dummy_app and then import the actual listener used by the real application to instantiate the fake client:

```
@pytest.fixture
def dummy_app():
    app = Sanic("DummyApp")

    import_module("testing3.path.to.some_startup")
    return app
```

6. To force our client to use a mocked method instead of actually sending a network request to a database, we are going to monkeypatch the database client using a feature of pytest. Set up a fixture like this:

```
from unittest.mock import AsyncMock


@pytest.fixture
def mocked_execute(monkeypatch):
    execute = AsyncMock()
    monkeypatch.setattr(
```

```
            testing3.path.to.some_db_connection.
    FakeDBConnection, "execute", execute
        )

        return execute
```

We now have a mock object in place of the real `execute` method, and we can proceed to build out a test on our registration blueprint. One of the great benefits of using the `unittest.mock` library is that it allows us to assert that the database client would have been called. We will see what that looks like next.

7. Here, we create a test with some assertions that help us to know that the correct data will make its way to the data access layer:

```
@pytest.mark.parametrize(
    "input,expected_status",
    (
        (
            (
                {
                    "username": "Alice",
                    "email": "alice@bob.com",
                },
                201,
            ),
        ),
    )
)
def test_some_blueprint_data_validation(
    app_with_bp,
    mocked_execute,
    input,
    expected_status,
):
    _, response = app_with_bp.test_client.post(
        "/registration",
        json=input,
    )

    assert response.status == expected_status

    if expected_status == 201:
```

```
        mocked_execute.assert_awaited_with(
            "INSERT INTO users VALUES ($1, $2);",
    input["username"], input["email"]
            )
```

Just like before, we are using `parametrize` so that we can run multiple tests with different inputs. The key takeaway is the usage of the mocked `execute` method. We can ask `pytest` to provide that mocked object to us so that our test can make assertions upon it and we know that it was executed as expected.

This is certainly helpful for testing isolated issues, but what if there needs to be application-wide testing? We will look at that next.

Testing a full application

As an application progresses from its infancy, there is likely to be a network of middleware, listeners, and signals that process requests that are not just limited to the route handler. In addition, there are likely to be connections to other services (such as databases) that complicate the entire process. A typical web application cannot be run in a vacuum. When it starts up, it needs to connect to other services. These connections are critical to the proper performance of the application, and therefore if they do not exist, then the applications cannot start. Testing these can be very troublesome. Do not just throw your hands up and give up. Resist the temptation. In the previous tests, there was a glimpse of how this can be achieved quite simply. We did in fact successfully test against our database. But what if that is not enough?

Sometimes testing against `dummy_app` is not sufficient.

This is why I really like applications that are created by a factory pattern. The GitHub repository for this chapter provides an example of a factory pattern that I use a lot. It has some very helpful features in it. Essentially, the end result is a function that returns a Sanic instance with everything attached to it. Through the implementation of the Sanic standard library, the function crawls through your source code looking for things to attach to it (routes, blueprints, middleware, signals, listeners, and much more) and is set up to avoid circular import issues. We talked about factory patterns and their benefits back in *Chapter 2, Organizing a Project*.

What is particularly important right now is that the factory in the GitHub repository can selectively choose what to instantiate. This means we can use our actual application with targeted functionality. Let me provide an example.

Once, I was building an application. It was critical to know exactly how it was performing in the real world. Therefore, I created middleware that would calculate some performance metrics and then send them off to a vendor for analysis. Performance was critical—which was part of my decision to use Sanic to begin with. When I tried to do some testing, I realized that I could not run the application in my test suite if it did not connect to the vendor. Yes, I could have mocked it out. However, a better strategy was to just skip the operation altogether. Sometimes, there really is no need to test every bit of functionality.

To make this concrete, here is a real quick explanation of what I am talking about. Here is a middleware code snippet that calculates runtime at the beginning and end of the request and sends it off:

```python
from time import time
from sanic import Sanic


app = Sanic.get_app()


@app.on_request
async def start_timer(request: Request) -> None:
    request.ctx.start_time = time()


@app.on_response
async def stop_timer(request: Request, _) -> None:
    end_time = time()
    total = end_time - request.ctx.start_time
    async send_the_value_somewhere(total)
```

One solution to my problem of contrasting testing versus production behavior could be to change the application code to only run in production:

```python
if app.config.ENVIRONMENT == "PRODUCTION":
    ...
```

But in my opinion, a better solution is to skip this middleware altogether. Using the factory pattern shown in the repo, I could do this:

```python
from importlib import import_module
from typing import Optional, Sequence
```

```python
from sanic import Sanic


DEFAULT = ("path.to.some_middleware.py",)


def create_app(modules: Optional[Sequence[str]] = None) ->
Sanic:
    app = Sanic("MyApp")

    if modules is None:
        modules = DEFAULT

    for module in modules:
        import_module(module)

    return app
```

In this factory, we are creating a new application instance and looping through a list of known modules to import them. In normal usage, we would create an application by calling create_app(), and the factory would import the DEFAULT known modules. By importing them, they will attach to our application instance. More importantly, however, this factory allows us to send an arbitrary list of modules to load. This allows us the flexibility to create a fixture in our tests that uses the actual factory pattern for our application but has the control to pick and choose what to load.

In our use case, we decided that we do not want to test the performance middleware. We can skip it by creating a test fixture that simply ignores that module:

```python
from path.to.factory import create_app


@pytest.fixture
def dummy_app():
    return create_app(modules=[])
```

As you can see, this opens up the ability for me to create tests that are specifically targeting parts of my actual application, and not just a dummy application. Using a factory through the use of inclusion and exclusion, I can create unit tests with only the functionality that I need and avoid the unneeded functionality.

I hope your mind is now racing with possibilities that this opens up for you. Testing becomes so much easier when the application is itself composable. This awesome trick is one way you can really take your application development to the next level. An easily composable application becomes an easily testable application. This leads to the application being well tested and now you are truly on your way to becoming a next-level developer.

If you have not already begun, I highly suggest that you use a factory like mine. Go ahead and copy it. Just promise me that you will use it to create some unit tests.

Using ReusableClient for testing

Up until this point, we have been using a test client that starts and stops a service on every call to it. The `sanic-testing` package ships with it another test client that can be manually started and stopped. Therefore, it is possible to reuse it between calls, or even tests. In the next subsection, we will learn about this reusable test client.

Running a single test server per test

You may sometimes need to have multiple calls to your API running on the same instance. For example, this could be useful if you were storing some temporary state in between calls in memory. This is obviously not a good solution in most use cases because storing the state in memory makes horizontal scaling difficult. Leaving that issue aside, let's take a quick look at how you might implement this:

1. We will first create an endpoint that just spits out a counter:

    ```python
    from sanic import Sanic, Request, json
    from itertools import count


    app = Sanic("test")


    @app.before_server_start
    def setup(app, _):
    ```

```
            app.ctx.counter = count()


    @app.get("")
    async def handler(request: Request):
        return json(next(request.app.ctx.counter))
```

In this simplified example, every time that you hit the endpoint, it will increment a number.

2. We can test this endpoint that maintains an internal state by using a ReusableClient instance, as follows:

```
    from sanic_testing.reusable import ReusableClient


    def test_reusable_context(app):
        client = ReusableClient(app, host="localhost",
    port=9999)

        with client:
            _, response = client.get("/")
            assert response.json == 0

            _, response = client.get("/")
            assert response.json == 1

            _, response = client.get("/")
            assert response.json == 2
```

As long as you are using the client inside that with context manager, then you will be hitting the exact same instance of your application in each call.

3. We can simplify the preceding code by using fixtures:

```
    from sanic_testing.reusable import ReusableClient
    import pytest


    @pytest.fixture
```

```
def test_client(app):
    client = ReusableClient(app, host="localhost",
port=9999)
    client.run()
    yield client
    client.stop()
```

Now, when you set up a unit test, it will keep the server running for as long as the test function is executing.

4. This unit test could be written as follows:

```
def test_reusable_fixture(test_client):
    _, response = test_client.get("/")
    assert response.json == 0

    _, response = test_client.get("/")
    assert response.json == 1

    _, response = test_client.get("/")
    assert response.json == 2
```

As you can see, this is a potentially powerful strategy if you want to run only a single server for the duration of your test function.

What if you want to keep the instance running for the entire duration of your testing? The simplest way would be to change the scope of the fixture to session:

```
@pytest.fixture(scope="session")
def test_client():
    client = ReusableClient(app, host="localhost",
port=9999)
    client.run()
    yield client
    client.stop()
```

With this setup, no matter where you are running tests in pytest, it will be using the same application. While I personally have never felt the need for this pattern, I can definitely see its utility.

The code for this example can be found in the GitHub repository: `https://github.com/PacktPublishing/Python-Web-Development-with-Sanic/tree/main/Chapter09/testing4`.

With both proper exception management and testing out of the way, the next critical addition of any true professional application is logging.

Gaining insight from logging and tracing

When it comes to logging, I think that most Python developers fall into three main categories:

- People that always use `print` statements

- People that have extremely strong opinions and absurdly complex logging setups

- People that know they should not use `print` but do not have the time or energy to understand Python's `logging` module

If you fall into the second category, you might as well skip this section. There is nothing in it for you except if you want to criticize my solutions and tell me there is a better way.

If you fall into the first category, then you really need to learn to change your habits. Don't get me wrong, `print` is fantastic. However, it does not have a place in professional-grade web applications because it does not provide the flexibility that the `logging` module offers. "*Wait a minute!*" I hear the first-category people shouting already. "*If I deploy my application with containers and Kubernetes, it can pick up my print output and redirect it from there.*" If you are deadset against using `logging`, then I suppose I might not be able to change your mind. But I am still going to try.

I used to fall into the third category and taking the time to learn about the `logging` module changed the way I develop. If you are like me, then I hope to finally convince you to make the switch as we break down the mystery of Python logging.

Leaving aside the configuration complexity, consider that the `logging` module provides a rich API to send messages at different levels and with meta context. If you want to take a giant leap forward from an amateur to a professional, then I suggest that you change from `print` to `logging`.

Let's examine the standard Sanic access logs. The message that the access logger sends out is actually blank. Take a look for yourself in the Sanic codebase if you want. The access log is this:

```
access_logger.info("")
```

The message is an empty string. What you actually see is something more like this:

```
[2021-10-21 09:39:14 +0300] - (sanic.access)[INFO]
[127.0.0.1:58388]: GET http://localhost:9999/  200 13
```

How does the logged message have all data from an empty string? Embedded in that line is a bunch of metadata that is both machine-friendly and human-readable, thanks to the `logging` module. In fact, you can store arbitrary data with logs that some logging configurations will store for you, something like this:

```
log.info("Some message", extra={"arbitrary": "data"})
```

If I have convinced you and you want to learn more about how to use logging in Sanic, let's continue.

Types of Sanic loggers

Sanic ships with three loggers. You can access all of them in the `log` module:

```
from sanic.log import import access_logger, error_logger, logger
```

Feel free to use these in your own applications. Especially in smaller projects, I will often use the Sanic `logger` object for convenience. These are, of course, actually intended for use by Sanic itself, but nothing is stopping you from using them. In fact, it might be convenient as you know that all of your logs are formatted consistently. My only word of caution is that it's best to leave the `access_logger` object alone since it has a highly specific job.

Why would you want to use both `error_logger` and a regular logger? I think the answer depends upon what you want to happen to your logs. There are many options to choose from. The simplest form is obviously just to output to the console. This is not a great idea for error logs, however, since you have no way to persist the message and review them when something bad happens. Therefore, you might take the next step and output your `error_logger` to a file. This, of course, could become cumbersome, so you might decide instead to use a third-party system to ship off your logs to another application to store and make them accessible. Whatever setup you desire, using multiple loggers may play a particular role in how the logging messages are handled and distributed.

Creating your own loggers, my first step in application development

When I approach a new project, one of the things I ask myself is what will happen with my production logs? This is, of course, a question highly dependent upon your application, and you will need to decide this for yourself. Asking the question, though, highlights a very important point: there is a distinction between development logs and production logs. More often than not, I have no clue what I want to do with them in production yet. We can defer that question for another day.

Before I even begin writing my application, I will create a logging framework. I know that the goal is to have two sets of configurations, so I begin with my development logs.

I want to emphasize this again: the *very first step* in building an application is to make a super-simple framework for standing up an application with logging. So, let's go through that setup process now:

1. The first thing we are going to do is make a super-basic scaffold following the patterns that we established in *Chapter 2, Organizing a Project*:

    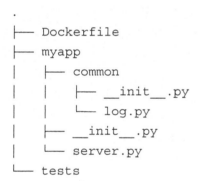

    ```
    .
    ├── Dockerfile
    ├── myapp
    │   ├── common
    │   │   ├── __init__.py
    │   │   └── log.py
    │   ├── __init__.py
    │   └── server.py
    └── tests
    ```

    This is the application structure that I like to work with because it makes it very easy for me to develop on. Using this structure, we can easily create a development environment focused upon running the application locally, testing the application, logging, and building images. Here, we obviously are concerned with running the application locally with logging.

2. The next thing I like to create is my application factory with a dummy route on it
    that I will remove later. This is how we can begin `server.py`. We will continue to
    add to it:

```python
from sanic import Sanic, text
from myapp.common.log import setup_logging, app_logger


def create_app():
    app = Sanic(__name__)
    setup_logging()

    @app.route("")
    async def dummy(_):
        app_logger.debug("This is a DEBUG message")
        app_logger.info("This is a INFO message")
        app_logger.warning("This is a WARNING message")
        app_logger.error("This is a ERROR message")
        app_logger.critical("This is a CRITICAL message")
        return text("")

    return app
```

There is a very important reason that I call `setup_logging` after creating my
app instance. I want to be able to use the configuration logic from Sanic to load
environment variables that may be used in creating my logging setup.

Here's a quick aside that I want to point out before continuing. There are two
different camps when it comes to creating a Python `logger` object. One side says
that it is best practice to create a new `logger` in every module. In this scenario, you
would put the following code at the top of *every single Python file*:

```python
from logging import getLogger


logger = getLogger(__name__)
```

The benefit of this approach is having the module name of where it was created closely related to the logger name. This is certainly helpful in tracking down where a log came from. The other camp, however, says that it should be a single global variable that is imported and reused since that may be easier to configure and control. Besides, we can specifically target filenames and line numbers quickly with proper log formatting, so it is unnecessary to include the module name in the logger name. While I do not discredit the localized, per-module approach, I too prefer the simplicity of importing a single instance like this:

```
from logging import getLogger


logger = getLogger("myapplogger")
```

If you dive really deep into logging, this also provides you with a much greater ability to control how different logger instances operate. Similar to the conversation about exception handlers, I would rather limit the number of instances I need to control. In the example that I just showed for `server.py`, I chose the second option to use a single global `logging` instance. This is a personal choice and there is no wrong answer in my opinion. There are benefits and detriments of both strategies, so choose which makes sense to you.

3. The next step is to create the basic `log.py`. For now, let's keep it super simple, and we will build from there:

```
import logging


app_logger = logging.getLogger("myapplogger")



def setup_logging():
    ...
```

4. With this in place, we are ready to run the application and test it out. But wait! Where is the app that we pass to our `sanic` command?

    We previously used this to run our application:

    ```
    $ sanic src.server:app -p 7777 --debug --workers=2
    ```

    Instead, we will tell the Sanic CLI the location of the `create_app` function and let it run that for us. Change your startup to this:

    ```
    $ sanic myapp.server:create_app --factory -p 7777 --debug
    --workers=2
    ```

You should now be able to hit your endpoint and see some basic messages output to your terminal. You likely will not have the DEBUG message since the logger is still probably set to only INFO and above. You should see something basic like this:

```
This is a WARNING message
This is a ERROR message
This is a CRITICAL message
```

Awesome, we now have the basics of logging down. Next, we will look to see how we can inject some more helpful information into our logs.

Configuring logging

The preceding logging messages are exactly what using print could provide. The next thing that we need to add is some configuration that will output some metadata and format the messages. It is important to keep in mind that some logging details may need to be customized to suit the production environment:

1. We, therefore, will start by creating a simple configuration:

    ```python
    DEFAULT_LOGGING_FORMAT = "[%(asctime)s] [%(levelname)s]
    [%(filename)s:%(lineno)s] %(message)s"


    def setup_logging(app: Sanic):
        formatter = logging.Formatter(
            fmt=app.config.get("LOGGING_FORMAT", DEFAULT_
    LOGGING_FORMAT),
            datefmt="%Y-%m-%d %H:%M:%S %z",
        )


        handler = logging.StreamHandler()
        handler.setFormatter(formatter)
        app_logger.addHandler(handler)
    ```

    Make sure to note that we changed the signature function of setup_logging to now take the application instance as an argument. Make sure to go back to update your server.py file to reflect this change.

As a side note, sometimes you might want to simplify your logging to force Sanic to use the same handlers. While you can certainly go through the process of updating the Sanic logger configuration (see `https://sanic.dev/en/guide/best-practices/logging.html#changing-sanic-loggers`), I find that to be much too tedious. A simpler approach is to set up the logging handlers and then simply apply them to the Sanic loggers, as follows:

```
from sanic.log import logger, error_logger


def setup_logging(app: Sanic):
    ...
    logger.handlers = app_logger.handlers
    error_logger.handlers = app_logger.handlers
```

It is good practice to always have `StreamHandler`. This will be used to output your logs to the console. But what if we want to add some additional logging utilities for production? Since we are not 100% sure yet what our production requirements will be, we will set up logging to a file for now. This can always be swapped out at another time.

2. Change your `log.py` to look like this:

```
def setup_logging(app: Sanic):
    formatter = logging.Formatter(
        fmt=app.config.get("LOGGING_FORMAT", DEFAULT_
LOGGING_FORMAT),
        datefmt="%Y-%m-%d %H:%M:%S %z",
    )

    handler = logging.StreamHandler()
    handler.setFormatter(formatter)
    app_logger.addHandler(handler)

    if app.config.get("ENVIRONMENT", "local") ==
"production":
        file_handler = logging.FileHandler("output.log")
        file_handler.setFormatter(formatter)
        app_logger.addHandler(file_handler)
```

You can easily see how this could be configured with a different kind of logging handler or formatting that might more closely match your needs in different environments.

All of the configurations shown used programmatic controls of the logging instance. One of the great flexibilities of the `logging` library is that all of this can be controlled with a single `dict` configuration object. You, therefore, will find it a very common practice to keep YAML files containing logging configurations. These files are easy to update and swap in and out of build environments to control production settings.

Adding color context

The preceding setup is entirely functional, and you could stop there. However, to me, this is not enough. When I am developing a web application, I always have my terminal open spitting out logs. In a sea of messages, it might be hard to sift through all of the text. How can we make this better? We will achieve this through the appropriate use of color.

Because I generally do not need to add color to my production output, we will go through adding color formatting in my local environment only:

1. We will begin by setting up a custom logging formatter that will add colors based upon the logging level. Any debug messages are blue, warnings are yellow, errors are red, and a critical message will be red with a white background to help them stand out (in a dark-colored terminal):

    ```python
    class ColorFormatter(logging.Formatter):
        COLORS = {
            "DEBUG": "\033[34m",
            "WARNING": "\033[01;33m",
            "ERROR": "\033[01;31m",
            "CRITICAL": "\033[02;47m\033[01;31m",
        }

        def format(self, record) -> str:
            prefix = self.COLORS.get(record.levelname)
            message = super().format(record)

            if prefix:
                message = f"{prefix}{message}\033[0m"

            return message
    ```

We are using the standard color escape codes that most terminals understand to apply the colors. This will color the entire message. You, of course, could get much fancier by coloring only parts of your messages, and if that interests you, I suggest you play around with this formatter to see what you can achieve.

2. After we create this, we will make a quick internal function to decide which formatter to use:

```python
import sys


def _get_formatter(is_local, fmt, datefmt):
    formatter_type = logging.Formatter
    if is_local and sys.stdout.isatty():
        formatter_type = ColorFormatter

    return formatter_type(
        fmt=fmt,
        datefmt=datefmt,
    )
```

If we are in a local environment, that is, a TTY terminal, then we use our color formatter.

3. We need to change the start of our `setup_logging` function to account for these changes. We will also abstract some more details to our configuration for easy access to change them per environment:

```python
DEFAULT_LOGGING_FORMAT = "[%(asctime)s] [%(levelname)s]
[%(filename)s:%(lineno)s] %(message)s"
DEFAULT_LOGGING_DATEFORMAT = "%Y-%m-%d %H:%M:%S %z"


def setup_logging(app: Sanic):
    environment = app.config.get("ENVIRONMENT", "local")
    logging_level = app.config.get(
        "LOGGING_LEVEL", logging.DEBUG if environment ==
"local" else logging.INFO
    )
    fmt = app.config.get("LOGGING_FORMAT", DEFAULT_
LOGGING_FORMAT)
    datefmt = app.config.get("LOGGING_DATEFORMAT",
DEFAULT_LOGGING_DATEFORMAT)
```

```
        formatter = _get_formatter(environment == "local",
    fmt, datefmt)


        . . .
```

Besides dynamically getting a formatter, this example adds another new piece to the puzzle. It is using a configuration value to determine the logging level of your logger.

Adding some basic tracing with request IDs

A common problem with logs is that they can become noisy. It might be tough to correlate a specific log with a specific request. For example, you might be handling multiple requests at the same time. If there is an error, and you want to look back at earlier messages, how do you know which logs should be grouped together?

There are entire third-party applications that add what is known as **tracing**. This is particularly helpful if you are building out a system of inter-related microservices that work together to respond to incoming requests. While we're not necessarily diving into microservice architecture, it is worth mentioning here that tracing is an important concept that should be added to your application. This is true regardless of whether your application architecture uses microservices or not.

For our purpose, what we want to achieve is to add a request identifier to every single request. Whenever that request attempts to log something, that identifier will automatically be injected into our request format. In order to accomplish this goal, we do the following:

1. First, we need a mechanism to inject the request object into every logging operation.

2. Second, we need a way to show the identifier whether it exists or ignore it if it does not.

Before we get to the code implementation, I would like to point out that the second part could be handled in a couple of ways. The simplest might be to create a specific logger that will only be used inside of a request context. This means that you would have one logger that is used in startup and shutdown operations, and another that is used only for requests. I have seen this approach used well.

The problem is that we are again using multiple loggers. To be entirely honest, I really do prefer the simplicity of having just a single instance that works for all of my use cases. This way, I do not need to bother thinking about which logger I should reach for. Therefore, I will show you here how to build option two: an omni-logger that can be used anywhere in your application. If you instead prefer the more targeted types, then I challenge you to take my concepts here and build out two loggers instead of one.

We will get started by tackling the issue of passing the request context. Remember, because Sanic operates asynchronously, there is no way to guarantee which request will be handled in what order. Luckily, the Python standard library has a utility that works great with `asyncio`. It is the `contextvars` module. What we will do to start is create a listener that sets up a context that we can use to share our request object and pass it to the logging framework:

1. Create a file called `./middleware/request_context.py`. It should look like this:

```python
from contextvars import ContextVar

from sanic import Request, Sanic

app = Sanic.get_app()


@app.after_server_start
async def setup_request_context(app: Sanic, _):
    app.ctx.request = ContextVar("request")


@app.on_request
async def attach_request(request: Request):
    request.app.ctx.request.set(request)
```

What is happening here is that we are creating a context object that can be accessed from anywhere that has access to our app. Then, on every single request, we will attach the current request to the context variable to make it accessible from anywhere the application instance is accessible.

2. The next thing that needs to happen is to create a logging filter that will grab the request (if it exists) and add it to our logging record. In order to do this, we will actually override Python's function that creates logging records in our `log.py` file:

```
old_factory = logging.getLogRecordFactory()


def _record_factory(*args, app, **kwargs):
    record = old_factory(*args, **kwargs)
    record.request_info = ""

    if hasattr(app.ctx, "request"):
        request = app.ctx.request.get(None)
        if request:
            display = " ".join([str(request.id), request.method, request.path])
            record.request_info = f"[{display}] "

    return record
```

Make sure you notice that we need to stash the default record factory because we want to make use of it. Then, when this function is executed, it will check to see whether there is a current request by looking inside that request context we just set up.

3. We also need to update our format to use this new bit of information. Make sure to update this value:

```
DEFAULT_LOGGING_FORMAT = "[%(asctime)s] [%(levelname)s] [%(filename)s:%(lineno)s] %(request_info)s%(message)s"
```

4. Finally, we can inject the new factory as shown:

```
from functools import partial


def setup_logging(app: Sanic):
    ...


    logging.setLogRecordFactory(partial(_record_factory, app=app))
```

Feel free to check this book's GitHub repository to make sure that your `log.py` looks like mine: `https://github.com/PacktPublishing/Python-Web-Development-with-Sanic/tree/main/Chapter09/tracing/myapp`.

5. With all of this in place, it is time to hit our endpoint. You should now see some nice, pretty colors in your terminal, and some request information inserted:

```
[2021-10-21 12:22:48 +0300] [DEBUG] [server.py:12]
[b5e7da51-68b0-4add-a850-9855c0a16814 GET /] This is a
DEBUG message
[2021-10-21 12:22:48 +0300] [INFO] [server.py:13]
[b5e7da51-68b0-4add-a850-9855c0a16814 GET /] This is a
INFO message
[2021-10-21 12:22:48 +0300] [WARNING] [server.py:14]
[b5e7da51-68b0-4add-a850-9855c0a16814 GET /] This is a
WARNING message
[2021-10-21 12:22:48 +0300] [ERROR] [server.py:15]
[b5e7da51-68b0-4add-a850-9855c0a16814 GET /] This is a
ERROR message
[2021-10-21 12:22:48 +0300] [CRITICAL] [server.py:16]
[b5e7da51-68b0-4add-a850-9855c0a16814 GET /] This is a
CRITICAL message
```

After running through these examples, one thing you might have noticed and not seen before is `request.id`. What is this and where does it come from?

Using X-Request-ID

It is a common practice to use **Universally Unique Identifier** (UUIDs) to track requests. This makes it very easy for client applications to also track requests and *correlate* them to specific instances. This is why you will often hear them called correlation IDs. If you hear the term, they are the exact same thing.

As a part of the practice of correlating requests, many client applications will send an X-Request-ID header. If Sanic sees that header in an incoming request, then it will grab that ID and use it to identify the request. If not, then it will automatically generate a UUID for you. Therefore, you should be able to send the following request to our logging application and see that ID populated in the logs:

```
$ curl localhost:7777 -H 'x-request-id: abc123'
```

For the sake of simplicity, I am not using a UUID.

Your logs should now reflect this:

```
[2021-10-21 12:36:00 +0300] [DEBUG] [server.py:12] [abc123 GET
/] This is a DEBUG message
[2021-10-21 12:36:00 +0300] [INFO] [server.py:13] [abc123 GET
/] This is a INFO message
[2021-10-21 12:36:00 +0300] [WARNING] [server.py:14] [abc123
GET /] This is a WARNING message
[2021-10-21 12:36:00 +0300] [ERROR] [server.py:15] [abc123 GET
/] This is a ERROR message
[2021-10-21 12:36:00 +0300] [CRITICAL] [server.py:16] [abc123
GET /] This is a CRITICAL message
```

Logging is a critical component of professional-grade web applications. It really does not need to be that complicated. I have seen super lengthy and overly verbose configurations that quite honestly scared me away. With a little bit of attention to detail, however, you can make a truly fantastic logging experience without much effort. I encourage you to grab the source code for this and hack it until it meets your needs.

We'll next turn our attention to another critical component of web applications: database management.

Managing database connections

This book above all else is really hoping to provide you with confidence to build applications *your* way. This means we are actively looking to stomp out copy/paste development. You know what I mean. You go to *Stack Overflow* or some other website, copy code, paste it, and then move on with your day without thinking twice about it.

This sort of copy/paste mentality is perhaps most prevalent when it comes to database connections. Time for a challenge. Go start up a new Sanic app and connect it to a database. Some developers might approach this challenge by heading to some other codebase (from another project, an article, documentation, or a help website), copying some basic connection functions, changing the credentials, and calling it a day. They may never have put much thought into what it means to connect to a database: if it works, then it must be okay. I know I certainly did that for a long time.

This is not what we are doing here. Instead, we will consider a couple of common scenarios, think through our concerns, and develop a solution around them.

To ORM or not to ORM, that is the question

For the benefit of anyone that does not know what ORM is, here is a quick definition.

ORM is a framework used to build Python-native objects. Those objects are related directly to a database schema and are also used to build queries to fetch data from the database to be used when building the Python objects. In other words, they are a data access layer that has the capability of two-way translation from Python and to the database. When people talk about ORM, they are typically referring to one that is intended to be used with an SQL-based database.

The question about whether to use ORM or not is answered with some strong opinions. In some contexts, people might think you are living in the Stone Age if instead of using it you are hand-writing your SQL queries. On the other hand, some people will think ORM is a nuisance and leads to both overly simplistic yet grotesquely complicated and inefficient queries. I suppose to an extent both groups are correct.

Ideally, I cannot tell you what you should or should not do. The implementation details and the use case are highly relevant to any decision. In my projects, I tend to shy away from it. I like to use the `databases` project (`https://github.com/encode/databases`) to build custom SQL queries, and then map the results to `dataclass` objects. After handcrafting my SQL, I use some utilities to hydrate them from raw, unstructured values into schema-defined Python objects. I have also, in the past, made extensive use of ORM tools such as peewee (`https://github.com/coleifer/peewee`) and SQLAlchemy (`https://github.com/sqlalchemy/sqlalchemy`). And, of course, since I developed in Django for many years, I have done a lot of work in its internal ORM tool.

When should you use ORM? First and foremost, for most projects, using ORM should probably be the default option. ORM tools are great at adding the required safety and security to make sure that you do not accidentally introduce a security bug. By enforcing types, they can be extremely beneficial in maintaining data integrity. And, of course, there is the benefit of abstracting away a lot of the database knowledge. Where ORM falls short, perhaps, is in its ability to handle complexity. As a project grows in the number of tables and interconnected relationships, it may be more difficult to continue using ORM. There also are a lot of more advanced options in SQL languages, such as *PostgreSQL*, that you simply cannot accomplish by using an ORM tool to build your queries. I find them to really shine in more simplistic **create/read/update/delete** (**CRUD**) applications, but actually get in the way of more complex database schemas.

Another potential downside to ORM is that it makes it super easy to sabotage your own project. A little mistake in building an inefficient query could be the difference between absurdly long response times and super-fast responses. Speaking from experience as someone who was bit by this bug, I find that applications that are built with ORM tools tend to over fetch data and inefficiently run more network calls than are needed. If you feel comfortable with SQL and know that your data will become fairly complicated, then perhaps you are better off writing your own SQL queries. The biggest benefit of using hand-crafted SQL is that it overcomes the complexity-scaling issue of ORM.

Even though this book is not about SQL, after much consideration, I think the best use of our time is to build a custom data layer and not use an off-the-shelf ORM tool. This option will force us into making good choices about maintaining our connection pools and developing secure and practical SQL queries. Moreover, anything that is discussed here in regard to implementation can easily be swapped out to a more fully featured ORM tool. If you are more familiar and comfortable with SQLAlchemy (which now has async support), then feel free to swap out my code accordingly.

Creating a custom data access layer in Sanic

When deciding upon which strategy to use for this book, I explored a lot of the options out there. I looked at all of the popular ORM tools that I see people using with Sanic. Some options, such as SQLAlchemy, have so much material out there that I could not possibly do it justice. Other options encouraged lower-quality patterns. Therefore, we turn to one of my favorites, the `databases` package, using `asyncpg` to connect to Postgres (my relational database of choice). The goal will be to implement good connection management, provide a simple and intuitive pattern to query the data, and output a set of models that will make building applications easier and more consistent.

I highly encourage you to look at the code in the repository at `https://github.com/PacktPublishing/Python-Web-Development-with-Sanic/tree/main/Chapter09/hikingapp`. This is one of the first times that we have created a *complete* application. By that, I mean an example of an application that goes out to fetch some data. Going back to the discussion from *Chapter 2*, *Organizing a Project*, about project layout, you will see an example of how we might structure a real-world application. There is also a lot going on in there that is somewhat outside of the scope of the discussion here (which is much more narrowly focused on database connections), so we will not dive too deeply into it right now. But do not worry, we will come back to the application's patterns again in *Chapter 11*, *A Complete Real-World Example*, when we build out a full application.

In the meantime, it might be a good opportunity for you to review that source code now. Try to understand how the project is structured, run it, and then test out some of the endpoints. Instructions are in the repository: `https://github.com/PacktPublishing/Python-Web-Development-with-Sanic/blob/main/Chapter09/hikingapp/README.md`.

I also would like to point out that since our applications are growing with the addition of another service, I am going to start running the application using `docker-compose` and Docker containers locally. All the build materials are in the GitHub repository for you to copy for your own needs. But, of course, you would not dare to just copy and paste the code without actually understanding it, so let's make sure that you do.

The application we are talking about is a web API for storing details about hiking. It connects its database of known hiking trails to users who can keep track of the total distance they have hiked and when they hiked certain trails. When you spin up the database, there should be some information prepopulated for you.

The first thing that we must do is make sure that our connection details are coming from environment variables. Never store them in the project files. Besides the security concerns associated with this, it is super helpful to make changes by redeploying your application with different values if you need to change the size of your connection pool or rotate your passwords. Let's begin:

1. Store your connection settings using `docker-compose`, Kubernetes, or any other tool you are using to run your containers. If you are not running Sanic in a container (for example, you plan to deploy to a PaaS solution that offers environment variables for you through a **graphical user interface** (**GUI**)), an option that I like to use for local development is `dotenv` (`https://github.com/theskumar/python-dotenv`).

    The config values that we care about right now are the **data source name** (**DSN**) and the **connection pool** settings. If you are not familiar with a DSN, it is a string that contains all of the information needed to connect to a database in a form that might look familiar to you as a URL.

A connection pool is an object that holds open a set number of connections to your database, and then allows your application to make use of those connections as needed. Imagine a scenario without a connection pool where a web request comes in and then your application goes and opens a network socket to your database. It fetches information, serializes it, and sends it back to the client. But it also closes that connection because there is no common pool to draw from. The next time that a request comes to your application, it will need to reopen a connection to the same database. This is hugely inefficient. Instead, your application can warm up several connections by opening them and holding them in reserve in the common pool.

2. Then, when the application needs a connection, instead of opening a new connection, it can simply connect to your database using a connection pool and store that object in your application ctx:

```python
# ./application/hiking/worker/postgres.py
@app.before_server_start
async def setup_postgres(app: Sanic, _):
    app.ctx.postgres = Database(
        app.config.POSTGRES_DSN,
        min_size=app.config.POSTGRES_MIN,
        max_size=app.config.POSTGRES_MAX,
    )


@app.after_server_start
async def connect_postgres(app: Sanic, _):
    await app.ctx.postgres.connect()


@app.after_server_stop
async def shutdown_postgres(app: Sanic, _):
    await app.ctx.postgres.disconnect()
```

As you can see, three main things are happening:

i.   The first is we are creating the Database object, which stores our connection pool and acts as the interface for querying. We store it in the app.ctx object so that it will be easily accessible from anywhere in the application. This was placed inside of the before_server_start listener since it alters the state of our application.

ii. The second is that the listener actually opens up the connections to the database and holds them at the ready until they are needed. We are warming up the connection pool prematurely so that we do not need to spend the overhead at query time.

iii. Lastly, of course, the important step we do is to make sure that our application properly shuts down its connections.

3. The next thing we need to do is create our endpoints. In this example, we will use class-based views:

```
from sanic import Blueprint, json, Request
from sanic.views import HTTPMethodView
from .executor import TrailExecutor


bp = Blueprint("Trails", url_prefix="/trails")



class TrailListView(HTTPMethodView, attach=bp):
    async def get(self, request: Request):
        executor = TrailExecutor(request.app.ctx.
postgres)
        trails = await executor.get_all_trails()
        return json({"trails": trails})
```

Here, the GET endpoint on the root level of the /trails endpoint is meant to provide a list of all trails in the database (forgetting about pagination). TrailExecutor is one of those objects that I do not want to dive too deeply into right now. But, as you can probably guess from this code, it takes the instance of our database (which we initiated in the last step) and provides methods to fetch data from the database.

One of the reasons that I really like the databases package is that it makes it incredibly easy to handle connection pooling and session management. It basically does it all for you under the hood. But one thing that is a good habit to get into (regardless of what system you are using) is to wrap multiple consecutive writes to your database in a transaction.

Imagine that you needed to do something like this:

```
executor = FoobarExecutor(app.ctx.postgres)
await executor.update_foo(value=3.141593)
await executor.update_bar(value=1.618034)
```

Often, when you have multiple database writes in a single function, you want either all of them to succeed or all of them to fail. Having a mixture of success and failures might, for example, leave your application in a bad state. When you identify situations like this it is almost always beneficial to nest your functions inside of a single transaction. To implement such a transaction within our sample, it would look something like this:

```
executor = FoobarExecutor(app.ctx.postgres)
async with app.ctx.postgres.transaction():
    await executor.update_foo(value=3.141593)
    await executor.update_bar(value=1.618034)
```

Now, if one of the queries fails for whatever reason, the database state will be rolled back to where it was before the change. I highly encourage you to adopt a similar practice no matter what framework you use for connecting to your database.

Of course, a discussion of databases is not necessarily limited to SQL databases. There are plenty of NoSQL options out there, and you, of course, should figure out what works for your needs. We will next take a look at connecting my personal favorite database option to Sanic: Redis.

Connecting Sanic to Redis

Redis is a blazingly fast and simple database to work with. Many people think of it simply as a key/value store, which is something that it does extremely well. It also has a lot of other features that could be thought of as a sort-of shared primitive data type. For example, Redis has hashes, lists, and sets. These correspond nicely to Python's `dict`, `list`, and `set`. It is for this reason that I often recommend this as a solution to someone that needs to share data across a horizontal scale-out.

In our example, we will use Redis as a caching layer. For this, we are relying upon its hashmap capability to store a `dict`-like structure with details about a response. We have an endpoint that might take several seconds to generate a response. Let's simulate that now:

1. First, create a route that will take a while to generate a response:

    ```python
    @app.get("/slow")
    async def wow_super_slow(request: Request):
        wait_time = 0
        for _ in range(10):
            t = random.random()
            await asyncio.sleep(t)
            wait_time += t
        return text(f"Wow, that took {wait_time:.2f}s!")
    ```

2. Check to see that it works:

    ```
    $ curl localhost:7777/slow
    Wow, that took 5.87s!
    ```

The response took 5.87 seconds, which is very slow for a response time. To make this faster on subsequent requests, we will create a decorator that will serve precached responses if existing:

1. First, we will install `aioredis`:

    ```
    $ pip install aioredis
    ```

2. Create a database connection pool similar to what we did in the previous section:

    ```python
    from sanic import Sanic
    import aioredis


    app = Sanic.get_app()


    @app.before_server_start
    async def setup_redis(app: Sanic, _):
        app.ctx.redis_pool = aioredis.BlockingConnectionPool.
    from_url(
    ```

```
        app.config.REDIS_DSN, max_connections=app.config.
REDIS_MAX
    )
    app.ctx.redis = aioredis.Redis(connection_pool=app.
ctx.redis_pool)


@app.after_server_stop
async def shutdown_redis(app: Sanic, _):
    await app.ctx.redis_pool.disconnect()
```

3. Next, we will create a decorator to use with our endpoints:

```
def cache_response(build_key, exp: int = 60 * 60 * 72):
    def decorator(f):
        @wraps(f)
        async def decorated_function(request: Request,
*handler_args, **handler_kwargs):
            cache: Redis = request.app.ctx.redis
            key = make_key(build_key, request)

            if cached_response := await get_cached_
response(request, cache, key):
                response = raw(**cached_response)
            else:
                response = await f(request, *handler_
args, **handler_kwargs)
                await set_cached_response(response,
cache, key, exp)

            return response

        return decorated_function

    return decorator
```

What is happening here is pretty simple. First, we generate some keys that will be used to look up and store values. Then, we check to see whether anything exists for that key. If yes, then use that to build a response. If no, then execute the actual route handler (which we know takes some time).

4. Let's see what we have accomplished in action. First, we will hit the endpoint again. To emphasize my point, I will include some stats from `curl`:

```
$ curl localhost:7777/v1/slow
Wow, that took 5.67s!
status=200   size=21 time=5.686937 content-type="text/
plain; charset=utf-8"
```

5. Now, we will try it again:

```
$ curl localhost:7777/v1/slow
Wow, that took 5.67s!
status=200   size=21 time=0.004090 content-type="text/
plain; charset=utf-8"
```

Wow! It returned almost instantly! In the first attempt, it took just under 6 seconds to respond. In the second, because the information has been stored in Redis, we got an identical response in about 4/1,000 of a second. And, don't forget that in those 4/1,000 of a second, Sanic went to fetch data from Redis. Amazing!

Using Redis as a caching layer is incredibly powerful as it can be used to significantly boost your performance. The flip side—as anyone that has worked with caching before knows—is that you need to have an appropriate use case and a mechanism for invalidating your cache. In this example, it is accomplished in two ways. If you check the source code at GitHub (https://github.com/PacktPublishing/Python-Web-Development-with-Sanic/blob/main/Chapter09/hikingapp/application/hiking/common/cache.py#L43), you will see that we are expiring the value automatically after 72 hours, or if someone sends a ?refresh=1 query argument to the endpoint.

Summary

Since we are past the point of talking about basic concepts in application development, we have graduated to the level of exploring some best practices that I have learned over the years of developing web applications. This is clearly just the tip of the iceberg, but they are some very important foundational practices that I encourage you to adopt. The examples from this chapter could become a great foundation for starting your next web application process.

First, we saw how you can use smart and repeatable exception handling to create a consistent and thoughtful experience for your users. Second, we explored the importance of creating a testable application, and some techniques to make it easily approachable. Third, we discussed implementing logging in both development and production environments, and how you could use those logs to easily debug and trace requests through your application. Finally, we spent time learning how databases could be integrated into your application.

In the next chapter, we will continue to expand upon the basic platform that we have built. You will continue to see a lot of the same patterns (such as logging) in our examples as we look at some common use cases of Sanic.

10
Implementing Common Use Cases with Sanic

Years ago when I was in law school, I set out to build a web application to help me with my studies. I wanted to create something that would both help me organize my notes and make it easy for my classmates to share outlines and study materials. I had been building websites at that point for a number of years, so I thought I knew what I was doing. I sat down to begin by creating an endpoint to store notes. Next, I created a database to persist the notes. I realized that I also needed to tie those notes to specific courses, so I added course management. Little by little I started adding features as I saw the need for them. The end result was a mess. I failed to establish good basics in my project, and it snowballed as new features crept into scope. With an idea in my head, I jumped straight to the implementation, skipped over all of the planning, and created none of the application infrastructures that set up a project for success. All of my logic was in the various route handlers, there was no project organization, and no consistency in things such as logging, exception handling, or access control. Like many amateur web applications, there was absolutely zero unit testing.

This book has been thoughtfully organized to build an understanding of HTTP and Sanic first. We tried to establish patterns of good practices and learn how to wield all of the tools that Sanic provides us. Only after gaining this knowledge and leveling up our technical skills along the way can we dig into implementation details. We know that before we can go ahead and build the next chatbot, we have some infrastructure details to take care of first. The goal of this chapter is to look at some practical features that you may be asked to build. With our foundation already established, we want to use what we have learned so far and see how we could solve these common problems.

While reviewing these use cases, we will look at some implementation details, talk about some of the considerations that factor into our planning decisions, and describe the general approach you might take to the problem. I hope to show you some insight into my own thought process when tasked with a project like this. The insight and the approach to building are much more important takeaways than the code we use here. Therefore, just like the previous chapter, there will be a lot of code in the repository that will not be in the book. It simply is not all relevant to the conversation. I will point out specific design decisions and include some choice code snippets that are worth mentioning. However, to have a complete picture of how these applications work, you should follow along with the full code.

So, what are we going to build?

- Synchronizing and scaling websocket feeds
- Powering a progressive web application
- Designing a GraphQL API
- Building a Discord bot and running Sanic from another service
- Creating an HTTP to HTTPS proxy and nesting Sanic inside Sanic

Technical requirements

Since this chapter builds upon the previous chapters, you should have all of the technical needs already fulfilled. We will start using some additional third-party packages, so make sure you have `pip` handy.

If you would like to jump ahead to make sure your environment is set up, here are the `pip` packages that we plan to use:

```
$ pip install aioredis ariadne "databases[postgresql]" nextcord
```

Furthermore, if you recall, back in *Chapter 2*, *Organizing a Project*, we discussed using factory patterns. Because we are now starting to build what could become the base of a *real-world* application, I feel it is much better to use a factory pattern here that can be expanded. Therefore, for the remainder of this book, you will see more and more usage of the factory pattern that we already have established and used.

To see the full application code, please follow along with the GitHub repository at `https://github.com/PacktPublishing/Python-Web-Development-with-Sanic/tree/main/Chapter10`.

Synchronizing and scaling websocket feeds

Earlier in this book, we explored websockets in the *Websockets for two-way communication* section of *Chapter 5*, *Building Response Handlers*. If you have not read that section yet, I encourage you to do that now. At this time, we are going to take our websocket implementation and create a horizontally scalable websocket feed. The basic premise of the code here will be the same as in that section, which is why you should have an understanding of what we build there before moving on to the example here.

The purpose of the feed we will build is to share events that happen in one browser across to another browser. Building upon the example from *Chapter 5*, *Building Response Handlers*, we are going to add a third-party broker that will allow us to run multiple application instances. This means that we can horizontally scale our application. The previous implementation suffered from the fact that it stored client information in memory. With no mechanism to share state or broadcast messages across multiple application instances, there was no way for one websocket connection to guarantee that it would be able to push messages to every other client. At best it would only be able to push messages to clients that happened to be routed and connected to the same application instance. Ultimately, this made it impossible to scale the application with multiple workers.

The goal now will be to create what is known as a **pubsub**. This is a term that means *publish and subscribe* since the pattern relies upon multiple sources subscribing to a central broker. When one of those sources publishes a message to the broker, all of the other sources that are subscribed receive that message. The term pubsub is a simple description of this push and pull between the broker and the sources. We will use this concept when building our feed.

The simplest way to handle pubsub in my opinion is with Redis, which has a very simple built-in pubsub mechanism. The idea is simple: every application instance will be a source. At startup, the application instance will subscribe to a specific channel on the Redis instance. With this connection established, it has the ability to push and pull messages from that broker on a specific channel. By pushing this off to a third-party service, all of our applications will be able to access the same information through the push and pull of the pubsub.

In the websockets example in the *Websockets for two-way communication* section of *Chapter 5*, *Building Response Handlers*, when a message was received, the server would push that message out to other clients that were connected to the same application instance. What we are about to build will accomplish something similar. Browser clients will open a websocket with one of many web servers, which will hold onto that client connection. This again will be held in memory. When there is an incoming message from a client instance, it will publish that message not by directly distributing it to the other clients, but instead, it will push the message to the pubsub broker. Then, all of the other instances will receive that message since they are subscribed to the same broker and can push the message to any websocket clients that happen to be connected to that application instance. In this way, we can build a distributed websocket feed.

To get started, we will spin up a Redis service using `docker-compose`, as well as our development application. Take a look in the repository for the details on how to accomplish that: `https://github.com/PacktPublishing/Python-Web-Development-with-Sanic/tree/main/Chapter10/wsfeed`. We will assume that you have a Redis instance available and running:

1. We begin by creating a websocket handler and attaching it to a blueprint:

```
from sanic import Blueprint
from sanic.log import logger
from .channel import Channel

bp = Blueprint("Feed", url_prefix="/feed")


@bp.websocket("/<channel_name>")
async def feed(request, ws, channel_name):
    logger.info("Incoming WS request")
    channel, is_existing = await Channel.get(
        request.app.ctx.pubsub, request.app.ctx.redis,
channel_name
```

```
)

if not is_existing:
    request.app.add_task(channel.receiver())
client = await channel.register(ws)

try:
    await client.receiver()
finally:
    await channel.unregister(client)
```

This is the entirety of our Sanic integration in this example. We defined a websocket endpoint. The endpoint requires us to access a feed by going to channel_name, which is meant to be a unique listening location. This could be a username, a chatroom stock ticker, and so on. The point is that channel_name is meant to represent some location in your application where people will want to continuously retrieve information from your application as a feed. For example, this also could be used to build out a sort of shared editing application where multiple users are able to make changes simultaneously to the same resource.

The handler in this example works by fetching a Channel object. If it created new Channel, then we send off a receiver task to the background that is responsible for listening to our pubsub broker. The next thing in the handler is to register our current websocket connection on the channel, and then create another receiver. The point of this second client.receiver is to listen to the websocket connection and take incoming messages to push off to the pubsub broker.

2. Let's take a quick look at the Client object:

```
from dataclasses import dataclass, field
from uuid import UUID, uuid4
from aioredis import Redis
from sanic.server.websockets.impl import
WebsocketImplProtocol


@dataclass
class Client:
    protocol: WebsocketImplProtocol
    redis: Redis
```

```
        channel_name: str
        uid: UUID = field(default_factory=uuid4)


        def __hash__(self) -> int:
            return self.uid.int


        async def receiver(self):
            while True:
                message = await self.protocol.recv()
                if not message:
                    break
                await self.redis.publish(self.channel_name,
    message)


        async def shutdown(self):
            await self.protocol.close()
```

As just stated, the purpose of this object is to listen to the current websocket connection and to send a message to the pubsub broker when available. This is accomplished by the `publish` method.

3. We now will take a look at the `Channel` object. This class is a bit longer than `Client`, so we will look at the code for it in sections. It might be helpful to open the GitHub repository to see the class definition in full:

```
    class ChannelCache(dict):

        ...



    class Channel:
        cache = ChannelCache()


        def __init__(self, pubsub: PubSub, redis: Redis,
    name: str) -> None:
            self.pubsub = pubsub
            self.redis = redis
            self.name = name
            self.clients: Set[Client] = set()
            self.lock = Lock()
```

```python
    @classmethod
    async def get(cls, pubsub: PubSub, redis: Redis,
name: str) -> Tuple[Channel, bool]:
        is_existing = False

        if name in cls.cache:
            channel = cls.cache[name]
            await channel.acquire_lock()
            is_existing = True
        else:
            channel = cls(pubsub=pubsub, redis=redis,
name=name)
            await channel.acquire_lock()

            cls.cache[name] = channel

        await pubsub.subscribe(name)

        return channel, is_existing
```

A channel is created and cached in each application instance in memory. This means that for every single application instance where an incoming request asks to join a specific channel, there is only one instance of that channel created. Even if we have 10 application instances, it does not matter that we have 10 instances of the channel. What we care about is that on any *single* application instance, there is never more than one `Channel` subscribing to a single Redis pubsub channel. Having multiple channels on the same application instance could get messy and lead to a memory leak.

4. Therefore, we also want to add a mechanism to clean up the cache when a channel is no longer needed. It can be done with the following code:

```python
    async def destroy(self) -> None:
        if not self.lock.locked():
            logger.debug(f"Destroying Channel {self.
name}")
            await self.pubsub.reset()
            del self.__class__.cache[self.name]
```

```
        else:
               logger.debug(f"Abort destroying Channel
    {self.name}. It is locked")
```

The reason we are using Lock is to try and avoid race conditions where multiple requests make an attempt to destroy a channel instance. Without this, you might end up in a case where concurrent requests try to destroy the channel out of order, leading to an otherwise very difficult bug to replicate consistently.

If you recall from earlier in our review of this example, after the channel is created (or fetched from the cache), we register the websocket connection on the Channel instance, which looks like this:

```
    async def register(self, protocol: WebsocketImplProtocol)
 -> Client:
        client = Client(protocol=protocol, redis=self.redis,
    channel_name=self.name)
        self.clients.add(client)
        await self.publish(f"Client {client.uid} has joined")
        return client
```

We simply create the Client object, add it to the known clients that need to be notified from this instance of an incoming message, and send off a message to let other clients know that someone has just joined. The publish message method simply looks like this:

```
    async def publish(self, message: str) -> None:
        logger.debug(f"Sending message: {message}")
        await self.redis.publish(self.name, message)
```

Once a client has been registered, it also needs to have the ability to unregister. A method to unregister is as follows:

```
    async def unregister(self, client: Client) -> None:
        if client in self.clients:
            await client.shutdown()
            self.clients.remove(client)

            await self.publish(f"Client {client.uid} has left")

        if not self.clients:
```

```
        self.lock.release()
        await self.destroy()
```

Here, we remove the current client from the known clients on `Channel`. If there are no longer any more clients listening to this channel, then we can close it and clean up after ourselves.

This is a super simple pattern that provides an immense amount of flexibility. In fact, in my course of providing support and helping people with their Sanic applications, I have provided assistance in building applications using a similar pattern to this on numerous occasions. I keep a Gist on GitHub that I share with people whenever they ask about websockets. You can find it here: `https://gist.github.com/ahopkins/5b6d380560d8e9d49e25281ff964ed81`. Using this, you could create some truly incredible frontend applications. Speaking of which, in the next section, we are going to start looking at the interplay between Sanic and frontend web applications that run in the browser.

Powering a progressive web application

A lot of use cases for building web APIs are to power a **progressive web application** (**PWA**—also known as a **single-page application**, or **SPA**). Like many other web developers out there, the real draw to web development was for the purpose of building a usable application or website in the browser. Let's be honest, not many of us are out there writing `curl` commands to use our favorite APIs. The real power of a web API is when it powers something else.

What does a PWA need in order to run? Well, when you build a PWA, the final product is a bunch of static files. Okay, so we put those files into a directory called `./public` and then we serve them:

```
  app.static("/", "./public")
```

There you go—we now are running a PWA. We're finished.

Well, not so fast. Being able to serve the static content is important, but it is also not the only factor you need to consider. Let's look at some considerations when building PWAs.

Dealing with subdomains and CORS

In *Chapter 7, Dealing with Security Concerns*, we spent a significant amount of time looking into CORS from a security lens. I would venture a guess that by far the biggest rationale for requiring CORS protection is the need to serve content to a PWA. These types of applications are ubiquitous on the internet, and usually have to tackle. The reason this usually happens is that, oftentimes, the frontend of a PWA and the backend are on different subdomains. This usually is because they are running on different servers. The static content might be served with a CDN, and the backend is on a VPS or PaaS offering (see *Chapter 8, Running a Sanic Server*, for more on Sanic deployment options).

CORS is a big topic. As we saw in *Chapter 7, Dealing with Security Concerns*, it is also something easy to get wrong. I participated in a conversation once with the maintainer of one of the most widely used pieces of CORS software on the web. Even he joked around that he does not understand CORS and cannot implement it properly on his own. Luckily, there is a simple method for getting CORS up and running in your Sanic applications using Sanic Extensions. If you are not familiar with Sanic Extensions, it is a package that is developed and maintained by the Sanic team to add extra features to Sanic. Sanic Extensions focuses on all of the more opinionated and use-case-specific implementations that are inappropriate for the core project. CORS is one of those features.

So, how do we get going out of the box?

```
$ pip install "sanic[ext]"
```

or

```
$ pip install sanic sanic-ext
```

That's it. Just install the `sanic-ext` package in your environment and you will get CORS protection out of the box. As of version 21.12, if you have `sanic-ext` in your environment, Sanic will auto-instantiate it for you.

The only thing we need to do now is to configure it. Usually, to get started with CORS configuration, we need to set the allowed origins:

```
app.config.CORS_ORIGINS = "http://foobar.com,http://bar.com"
```

"Well, hang on a minute," you say, *"Can't I just serve the frontend assets from Sanic and avoid CORS because the front and back are on the same server?"* Yes. If that approach works for you, go for it. Let's see what that might look like (from a development perspective).

Running a development server

What happens when you decide that you want both frontend and backend applications to run on the same server? Or, when you want to use the `app.static` method shown above to serve your project files? Building this locally could be very tricky.

The reason this is the case is, when building a frontend application, you need a frontend server. Most frameworks have some sort of build requirement. That is to say that you type some code, hit save, then some package such as `webpack` or `rollup` compiles your JS and serves it to you from a local development web server. Your frontend development server might run on port `5555`, so you go to `http://localhost:5555`.

But, you want to access your locally running backend from that frontend application to populate content. The backend is running on `http://localhost:7777`. Uh oh, do you see where this is going? We are right back to CORS all over again. As long as your frontend application is being run by a different server than your backend, you will continue to run into CORS issues.

Ultimately, we are trying to get a single server to run both the backend and frontend. Since we are talking about local development, we also want auto-reload capabilities for both our Python files and our Javascript files. We also need to trigger a rebuild of the JavaScript, and finally, we need to serve this all from one location. Luckily, Sanic can do all of this for us. Let's now use Sanic as a local development server for a frontend project.

This will work with any frontend tools you want since we will essentially be calling those tools from within Python. My frontend development framework of choice is Svelte, but feel free to try this with React, Vue, or any of the other many alternatives. I will not walk you through the steps of setting up a frontend project since that is not important here. Imagine that you have already done it. If you would like to follow along with the code, please see the GitHub repository: `https://github.com/PacktPublishing/Python-Web-Development-with-Sanic/tree/main/Chapter10/pwa`.

To accomplish our goals, we will set up the Sanic server to add auto-reload capabilities to the build directory of the frontend application. For Svelte projects using `rollup` (a popular JS build tool), it will put the compiled assets into a `./public` directory. We want to serve that directory as static content, and then tell the Sanic server that it should auto-reload on the contents of that directory. Let's dive in to see how we can accomplish that:

1. We start by declaring the location of the static files and serving them with `static`:

```
app = Sanic("MainApp")
app.config.FRONTEND_DIR = Path(__file__).parent /
"my-svelte-project"
app.static("/", app.config.FRONTEND_DIR / "public")
```

2. When we run Sanic, make sure to add that directory to the auto-reloader like this:

```
$ sanic server:app -d -p7777 -R ./my-svelte-project/src
```

3. The next thing we want to do is define a custom signal. We are going to use this later, so all it needs to do now is define it. It just needs to exist so that we can later await the event:

```
@app.signal("watchdog.file.reload")
async def file_reloaded():
    ...
```

4. We are now ready to build something that will check the files that were reloaded and decide whether or not we need to trigger the `rollup` build process. We will look at this in two parts. First, we create a startup listener that checks the file extensions to determine the server start was triggered by a reload from any `.svelte` or `.js` file extensions:

```
@app.before_server_start
async def check_reloads(app, _):
    do_rebuild = False
    if reloaded := app.config.get("RELOADED_FILES"):
        reloaded = reloaded.split(",")
        do_rebuild = any(
            ext in ("svelte", "js")
```

```
        for filename in reloaded
        if (ext := filename.rsplit(".", 1)[-1])
)
```

As of version 21.12, the files that triggered a reload are stashed in a SANIC_
RELOADED_FILES environment variable. Since any environment variables starting
with the SANIC_ prefix are loaded into our app.config, we can simply read that
value if it exists and check the file extensions.

Assuming there is a rebuild required, we next want to trigger a subprocess call to
our shell to run the build command:

```
if do_rebuild:
    rebuild = await create_subprocess_shell(
        "yarn run build",
        stdout=PIPE,
        stderr=PIPE,
        cwd=app.config.FRONTEND_DIR,
    )

    while True:
        message = await rebuild.stdout.readline()

        if not message:
            break
        output = message.decode("ascii").rstrip()
        logger.info(f"[reload] {output}")

    await app.dispatch("watchdog.file.reload")
```

Finally, when this is all done, we are going to dispatch that custom event that we
created earlier.

Up until now, we have the auto-reload and auto-rebuilding working as expected. The only thing we are missing now is the ability to trigger the web browser to refresh the page. This can be accomplished using a tool called `livereload.js`. You can access `livereload.js` by searching for it and installing JavaScript. Essentially, what it will do is create a websocket connection to a server on port 35729. Then, from that websocket, you can send messages prompting the browser to perform a refresh:

1. To do this from Sanic, we are going to run nested applications. Add a second application definition:

   ```
   livereload = Sanic("livereload")
   livereload.static("/livereload.js", app.config.FRONTEND_
   DIR / "livereload.js")
   ```

2. We will also need to declare a few more constants. These are mainly to run the two types of messages that `livereload` needs to send from the server:

   ```
   INDEX_HTML = app.config.FRONTEND_DIR / "public" / "index.
   html"
   HELLO = {
       "command": "hello",
       "protocols": [
           "http://livereload.com/protocols/official-7",
       ],
       "serverName": app.name,
   }
   RELOAD = {"command": "reload", "path": str(INDEX_HTML)}
   ```

3. Next, set up the necessary listeners to run the nested server:

   ```
   @app.before_server_start
   async def start(app, _):
       app.ctx.livereload_server = await livereload.create_
   server(
           port=35729, return_asyncio_server=True
       )
       app.add_task(runner(livereload, app.ctx.livereload_
   server))


   @app.before_server_stop
   ```

```
async def stop(app, _):
    await app.ctx.livereload_server.close()
```

The `runner` task used in the code above should look like this:

```
async def runner(app, app_server):
    app.is_running = True
    try:
        app.signalize()
        app.finalize()
        await app_server.serve_forever()
    finally:
        app.is_running = False
        app.is_stopping = True
```

4. It is time to add the websocket handler:

```
@livereload.websocket("/livereload")
async def livereload_handler(request, ws):
    global app
    logger.info("Connected")
    msg = await ws.recv()
    logger.info(msg)
    await ws.send(ujson.dumps(HELLO))

    while True:
        await app.event("watchdog.file.reload")
        await ws.send(ujson.dumps(RELOAD))
```

As you can see, the handler accepts an initial message from `livereload`, and then sends a `HELLO` message back. Afterward, we are going to run a loop and wait until the custom signal we created is triggered. When it is, we send off the `RELOAD` message, which triggers the browser to refresh the web page.

Voila! We now have a full JavaScript development environment running inside of Sanic. This is a perfect setup for those PWAs where you want to serve the frontend and backend content from the same location. Visit the site in your browser at `http://localhost:7777`. Make sure to make some changes to the Svelte app to see how it updates the page in the browser for you automatically.

Figure 10.1 – Screenshot showing a Sanic-powered PWA built with JavaScript

Since we are already talking about frontend content, we will next visit another important topic for frontend developers: GraphQL.

Designing a GraphQL API

In 2015, Facebook released a project meant to rival traditional web APIs and flip the concept of a RESTful web application on its head. This project is what we now know as GraphQL. This book has so far assumed that we are building out endpoints using the traditional method of combining HTTP methods with thoughtful paths to point to specific resources. In this approach, web servers are responsible for being the interface between a client and the source of data (for example, a database). The concept of GraphQL pushes all of that aside and allows the client to directly request what information it wants to receive. There is a single endpoint (usually `/graphql`) and a single HTTP method (usually `POST`). The single route definition is meant to be used for both retrieving data and causing state changes in the application. This all happens through a set of queries that are sent as the body on that single endpoint. GraphQL was meant to revolutionize the way we build the web and to take over as the standard practice of the future. At least, that is what many people said was going to happen.

This has not actually come to pass. At the time of writing, the popularity of GraphQL has seemingly peaked and is now on the decline. Nevertheless, I do believe that GraphQL fulfills a necessary niche in the web application world, and it will continue to live on as an alternative implementation for years to come (just not as a replacement). We, therefore, do need to know how to integrate it with Sanic for the instances where you may be asked to deploy one of these servers.

Before we can answer the question of *Why use GraphQL?*, we must understand what it is. As the name seemingly implies, **GraphQL** is a sort of query language. A query is a JSON-like request for information to be delivered in a specific format. A client looking to receive information from a web server might send a POST request with a body that includes a query like this:

```
{
  countries (limit: 3, offset:2) {
    name
    region
    continent
    capital {
      name
      district
    }
    languages {
      language
      isofficial
      percentage
    }
  }
}
```

In return, a server would go and fetch whatever data it needed and compile a return JSON document matching that description:

```
{
  "data": {
    "countries": [
      {
        "name": "Netherlands Antilles",
        "region": "Caribbean",
```

```
        "continent": "North America",
        "capital": {
          "name": "Willemstad",
          "district": "Curaçao"
        },
        "languages": [
            {
              "language": "Papiamento",
              "isofficial": true,
              "percentage": 86.19999694824219
            },
            {
              "language": "English",
              "isofficial": false,
              "percentage": 7.800000190734863
            },
            {
              "language": "Dutch",
              "isofficial": true,
              "percentage": 0
            }
          ]
        },
        ...
      ]
    }
  }
```

As you might be able to tell, this becomes a very powerful tool for the client as it can bundle what might otherwise be multiple network calls into a single operation. It also allows a client (for example a PWA) to specifically retrieve the exact data that it needs in the format that it needs it.

Why would I want to use GraphQL?

I believe that GraphQL is the best friend of the frontend developer, but the bane of existence for the backend developer. It is certainly true that web applications using GraphQL will generally issue fewer HTTP calls to web servers than their counterparts. It is also certainly true that a frontend developer will have an easier time manipulating responses from a web server using GraphQL since they get to be the architect of how that data is structured.

GraphQL provides a very easy method for data retrieval. Because it is a strongly typed specification, it makes it possible to have tools that make the whole process of generating a query very elegant. For example, many GraphQL implementations come with an out-of-the-box web UI that can be used for development. See *Figure 10.2* for an example. The UI usually provides an area on the left for writing queries. Once ready, you can execute that query and see the results.

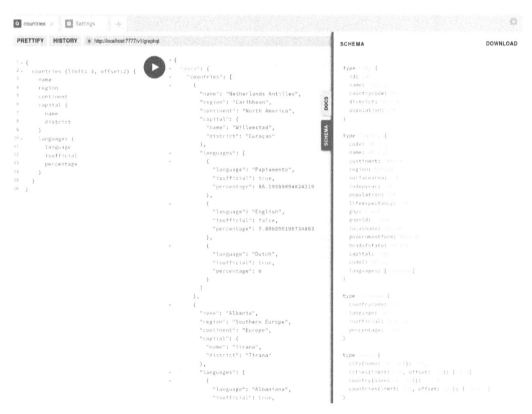

Figure 10.2 – Example of a GraphQL UI showing the SCHEMA tab that
displays all of the available information

One of the most important features that these UIs include is the ability to navigate the schema. You can see a closeup of that section in *Figure 10.3*. This **SCHEMA** display provides an easy-to-navigate overview of the types of queries that can be made, and what information is retrievable for different object types. There is certainly a fun factor that goes into these tools as you play with them to craft exactly the information you want. Simply put: *GraphQL is simple to use and implement*. It also has a very satisfying *coolness* factor to it when you start building ad hoc custom queries.

```
SCHEMA                                    DOWNLOAD


 type City {
   id: Int
   name: String
   countrycode: String
   district: String
   population: Int
   languages: [Language]
 }

 type Country {
   code: String
   name: String
   continent: String
```

Figure 10.3 – Closeup view of the schema navigator of the GraphQL UI

Except, GraphQL is a nightmare in the backend. For all of the simplification from the client perspective, the web server needs to deal with a much greater level of complexity. For this reason, when someone tells me that they want to build a GraphQL application, I usually ask them: why? If they are building it as a public-facing API, then it may be wonderful. GitHub is a great example of a public-facing GraphQL API that is a treat to work in. Querying the GitHub API is simple and intuitive. If, however, they are building the API for their own internal purposes, then there is a set of trade-offs that must be considered.

GraphQL is not in totality any easier or simpler than REST. Instead, it represents the shifting of complexity almost entirely to the web server. This may be acceptable, but it is a trade-off that you must consider. I generally find that the overall increase in complexity of the backend outweighs any benefits of implementation.

I know it may sound like I am not a fan of GraphQL. This is not true. I do think that GraphQL is a great concept, and I think there are some amazing tools out there (including in the Python world) to help build these applications. If you want to include GraphQL in your Sanic application, I would highly recommend tools such as `Ariadne` (`https://ariadnegraphql.org/`) and `Strawberry` (`https://strawberry.rocks/`). Even with these tools, a good GraphQL application, in my opinion, is more difficult to build with a few pitfalls waiting to swallow you up. As we look into how we can build a Sanic GraphQL application, I will try and point out these issues so that we can work around them.

Adding GraphQL to Sanic

I have built a small GraphQL application for this section. All of the code is, of course, on the GitHub repository for this book: `https://github.com/PacktPublishing/Python-Web-Development-with-Sanic/tree/main/Chapter10/graphql`. I highly suggest you have the code available while reading. Quite frankly, the code in its entirety is much too complex and lengthy to include it all here. So, instead, we will talk through it in general, and I will refer you back to the repository for specifics. For your convenience, I have also added a number of comments and further discussion points in the code base itself.

When we discussed database access in the *To ORM or not to ORM* section of *Chapter 9, Best Practices to Improve Your Web Applications*, we talked about whether you should or should not implement an ORM. The discussion was about whether you should use a tool to help you build the SQL queries or to build them yourself. There are very good arguments on both sides: pro-ORM versus anti-ORM. I opted for a somewhat hybrid approach to build the SQL queries by hand, and then build a lightweight utility to hydrate the data to a usable model. Let's call it semi-ORM.

A similar question *could* be posed here: should I build a GraphQL implementation myself or use a package? My answer is that you should absolutely use a package. I cannot see any reason to try and build a custom implementation yourself. There are several good options in Python; my personal preference is Ariadne. I particularly like the schema-first approach that the package takes. Using it allows me to define the GraphQL parts of my application in `.gql` files, therefore enabling my IDE to add syntax highlighting and other language-specific conveniences.

Let's begin:

1. Since we are using Ariadne in our example here, we begin by installing it in our virtual environment:

   ```
   $ pip install ariadne
   ```

2. To get up and running with Ariadne's *hello world* application does not take much. First, we will add our imports and setup Ariadne:

   ```python
   from ariadne import QueryType, graphql, make_executable_
   schema
   from ariadne.constants import PLAYGROUND_HTML
   from graphql.type import GraphQLResolveInfo
   from sanic import Request, Sanic, html, json


   app = Sanic(__name__)
   query = QueryType()



   type_defs = """
       type Query {
           hello: String!
       }
   """



   @query.field("hello")
   async def resolve_hello(_, info: GraphQLResolveInfo):
       user_agent = info.context.headers.get("user-agent",
   "guest")
       return "Hello, %s!" % user_agent
   ```

3. Next, we will add the two endpoints that we need: one to handle the GraphQL requests, and the other to display the GraphQL UI like the one seen in Figure 10.2:

   ```python
   @app.post("/graphql")
   async def graphql_handler(request: Request):
       success, result = await graphql(
           request.app.ctx.schema,
   ```

```
            request.json,
            context_value=request,
            debug=app.debug,
        )
        status_code = 200 if success else 400
        return json(result, status=status_code)
```

```
    @app.get("/graphql")
    async def graphql_playground(request: Request):
        return html(PLAYGROUND_HTML)
```

As you can see, there are two endpoints:

. A GET that displays the GraphQL query builder

. A POST that is the ingress to the GraphQL backend

4. Finally, we bring it all together by executing the schema and making it available in our application ctx:

```
    @app.before_server_start
    async def setup_graphql(app, _):
        app.ctx.schema = make_executable_schema(type_defs,
    query)
```

From this humble beginning, you can build from Sanic and Ariadne to meet the needs of your application. Let's take a look at a potential strategy you might take.

5. Scrapping the above, we begin with an app that looks very similar in structure to what we have seen before. Create ./blueprints/graphql/query.py and place your root-level GraphQL object in it as shown:

```
    from ariadne import QueryType


    query = QueryType()
```

6. Now, we create the two endpoints needed inside of a CBV on our GraphQL
    Blueprint instance:

```python
from sanic import Blueprint, Request, html, json
from sanic.views import HTTPMethodView
from ariadne.constants import PLAYGROUND_HTML


bp = Blueprint("GraphQL", url_prefix="/graphql")



class GraphQLView(HTTPMethodView, attach=bp, uri=""):
    async def get(self, request: Request):
        return html(PLAYGROUND_HTML)

    async def post(self, request: Request):
        success, result = await graphql(
            request.app.ctx.schema,
            request.json,
            context_value=request,
            debug=request.app.debug,
        )

        status_code = 200 if success else 400
        return json(result, status=status_code)
```

As you can see, this is nearly identical to the simple version from before.

7. On this same Blueprint instance, we are going to place all of our startup logic. This
    keeps it all in a convenient location and allows us to attach it to our application
    instance all at once:

```python
from ariadne import graphql, make_executable_schema
from world.common.dao.integrator import RootIntegrator
from world.blueprints.cities.integrator import
CityIntegrator
from world.blueprints.countries.integrator import
CountryIntegrator
from world.blueprints.languages.integrator import
LanguageIntegrator
```

```
@bp.before_server_start
async def setup_graphql(app, _):
    integrator = RootIntegrator.create(
        CityIntegrator,
        CountryIntegrator,
        LanguageIntegrator,
        query=query,
    )
    integrator.load()
    integrator.attach_resolvers()
    defs = integrator.generate_query_defs()
    additional = integrator.generate_additional_schemas()
    app.ctx.schema = make_executable_schema(defs, query,
*additional)
```

You may be wondering, *what is an integrator*, and *what is all of that code doing*? This is where I am going to refer you to the repository for the specifics, but we will walk through the concept here.

In my application example, an `Integrator` is an object that lives inside of a domain and is the conduit to setting up a GraphQL schema that Ariadne can use.

In the GitHub repository, you will see that the simplest `Integrator` is for the `languages` module. It looks like this:

```
from world.common.dao.integrator import BaseIntegrator


class LanguageIntegrator(BaseIntegrator):
    name = "language"
```

Next to it is a file called `schema.gql`:

```
type Language {
    countrycode: String
    language: String
    isofficial: Boolean
    percentage: Float
}
```

It is then the job of RootIntegrator to marshall all of the various domains together and generate the schema for Ariadne using both the dynamically generated schema and the hardcoded schema, as in the preceding snippet.

We also need to create a place for our GraphQL query to start. A query might look like this:

```
async def query_country(
    self, _, info: GraphQLResolveInfo, *, name: str
) -> Country:
    executor = CountryExecutor(info.context.app.ctx.
postgres)
    return await executor.get_country_by_name(name=name)
```

A user creates a query and we go and fetch it from the database. The Executor here works exactly as it does in hikingapp. Refer back to *Chapter 9, Best Practices to Improve Your Web Applications*, and the code in the GitHub repository: https://github.com/PacktPublishing/Python-Web-Development-with-Sanic/tree/main/Chapter09/hikingapp. Therefore, with a query like this, we can now translate the GraphQL query to an object:

```
{
  country(name: "Israel") {
    name
    region
    continent
    capital {
      name
      district
    }
    languages {
      language
      isofficial
      percentage
    }
  }
}
```

With the power of GraphQL, our response should be this:

```
{
  "data": {
    "country": {
      "name": "Israel",
      "region": "Middle East",
      "continent": "Asia",
      "capital": {
        "name": "Jerusalem",
        "district": "Jerusalem"
      },
      "languages": [
        {
          "language": "Hebrew",
          "isofficial": true,
          "percentage": 63.099998474121094
        },
        {
          "language": "Arabic",
          "isofficial": true,
          "percentage": 18
        },
        {
          "language": "Russian",
          "isofficial": false,
          "percentage": 8.899999618530273
        }
      ]
    }
  }
}
```

The way that Ariadne (and other GraphQL implementations) works is that you define a strongly typed schema. With knowledge of that schema, you might end up with nested objects. For example, the preceding `Country` schema might look like this:

```
type Country {
    code: String
    name: String
    continent: String
    region: String
    capital: City
    languages: [Language]
}
```

The `Country` type has a field called `capital`, which is a `City` type. Since this is not a simple scalar value that easily serializes to JSON, we need to tell Ariadne how to translate—or resolve—that field. Given the example on GitHub (`https://github.com/PacktPublishing/Python-Web-Development-with-Sanic/blob/main/Chapter10/graphql/application/world/blueprints/countries/integrator.py`), it would be to query our database like this:

```
class CountryIntegrator(BaseIntegrator):
    name = "country"

    async def resolve_capital(
        self,
        country: Country,
        info: GraphQLResolveInfo
    ) -> City:
        executor = CityExecutor(info.context.app.ctx.postgres)
        return await executor.get_city_by_id(country.capital)
```

This is how we can follow the path between different objects. It is then the job of Ariadne to piece all of these different queries and resolvers together to generate a final object to return. This is the power of GraphQL.

You may have also noticed a flaw. Because every resolver is meant to operate independently and to handle the conversion of a single field into a value, you can very easily overfetch data from the database. This is especially true if you have an array of objects that all resolve to the same database instance. This is known as the *n+1* problem. While it is not a unique problem to GraphQL, the design of many GraphQL systems makes it acutely prone to it. If you ignore this problem, while responding to a single request your server might ask the database for the same information over and over again even though it should otherwise already have it.

Many applications suffer from this issue. They rely upon many more database queries than should be needed. All of this overfetching adds up and reduces the performance and efficiency of web applications. While you should certainly be aware of this issue and cognizant as you develop any application, I feel it is something you must particularly plan for with GraphQL implementations since they thrive off of simplified resolvers. Therefore, the biggest piece of advice I can provide when building one of these applications is to think about in-memory, request-based caching. That is to say that caching objects on a request instance might save a ton of SQL queries.

I encourage you to take some time to review the rest of the code in the GitHub repository. There are some helpful patterns that could be usable in a real-world application. Since they are not necessarily related to Sanic or implementing GraphQL in Sanic, we will leave the discussion here for now and turn to another popular use case with Sanic: chatbots.

Building a Discord bot: running Sanic from another service

At some point early in 2021, I was convinced by a few people in the Sanic community that we needed to move our primary discussion and community-building tool. We had a somewhat underutilized chat application and also the community forums, which were mainly used for longer-style support questions. Discord is a more intimate conversation than what other options could offer. When it was suggested to me that we use Discord, I was a little hesitant to add another application to my tool belt. Nevertheless, we went forward with it. If there are Discord fans out there reading this book, then you do not need me to explain to you its benefits. For everyone else, Discord is a very easy-to-use and engaging platform that really facilitates the types of discussion helpful to our corner of the internet.

As I learned more about the platform, the biggest thing that stuck out to me was that chatbots are everywhere. There is an incredible subculture I was unaware of relating to the building of bots. The vast majority of these bots are built using SDKs, which are open source projects that wrap much of the client HTTP interactions needed to interface with Discord's API. There are entire ecosystems and frameworks built up on top of this to help developers make engaging bots.

Naturally, one of the next questions that gets asked all the time is: how can I integrate Sanic with my bot application? We are going to try and do that.

But first, I want to point out that while the example we are going to build uses Discord, the principles are in no way connected to running this on Discord. The core of what we are about to do is to run some `asyncio` process and reuse that loop for running Sanic. This means that you could in fact use this exact same technique to run nested Sanic applications. We will see what that looks like in the next section.

Building a simple Discord bot

I am not an expert with Discord. There is an entire realm of development that occurs based upon this platform and I will not pretend to be an authority. Our goal here is to integrate a bot application with Sanic. To do this, we are going to stand up a basic Discord bot using `nextcord`. If you are not familiar with `nextcord`, as of the time of the writing this book, it is an actively maintained fork of the abandoned `discord.py` project. If you are also not familiar with that, no worries. The simple explanation is that these are frameworks used to build a bot application on Discord. Similar to how Sanic provides tools for HTTP communications, these frameworks provide tools to communicate with Discord.

Let's take a minute to consider the basic hello world application from the documentation:

```
import nextcord


client = nextcord.Client()


@client.event
async def on_ready():
    print(f'We have logged in as {client.user}')


@client.event
async def on_message(message):
    if message.author == client.user:
```

```
        return

    if message.content.startswith('$hello'):
        await message.channel.send('Hello!')


client.run('your token here')
```

To be honest, this looks similar to what we built in Sanic. It starts with an application instance. Then, there are decorators that wrap handlers. The last thing we see is `client.run`.

This is the key to what we want to build. This `run` method is going to create a loop and run it until the application is shut down. Our job now is to run Sanic inside of this application. This means we will *not* be using the Sanic CLI to stand up our application. Instead, we will run the application using this:

$ python bot.py

Let's get started:

1. Start by copying the minimal bot example from the documentation into bot.py. You can grab the code here: `https://nextcord.readthedocs.io/en/latest/quickstart.html`

2. Create a simple Sanic application as a proof of concept:

```
from sanic import Sanic, Request, json


app = Sanic(__name__)



@app.get("/")
async def handler(request: Request):
    await request.app.ctx.general.send("Someone sent a
message")
    return json({"foo": "bar"})



@app.before_server_start
async def before_server_start(app, _):
```

```
    await app.ctx.general.send("Wadsworth, reporting for
duty")
```

Nothing fancy is happening so far. We have a single handler that will send off a message in a listener before the server starts. We also have a single handler that will also trigger a message to our Discord server when the route endpoint is hit.

3. To integrate this with the Discord bot, we will use the on_ready event to run our Sanic server:

```
from server import app


@client.event
async def on_ready():
    app.config.GENERAL_CHANNEL_ID = 906651165649928245
    app.ctx.wadsworth = client
    app.ctx.general = client.get_channel(app.config.
GENERAL_CHANNEL_ID)


    if not app.is_running:
        app_server = await app.create_server(port=9999,
return_asyncio_server=True)
        app.ctx.app_server = app_server
        client.loop.create_task(runner(app_server))
```

> **Important Note**
>
> For the sake of simplicity, I am just importing from the server import app. That is because it is a super simple implementation. In actuality, if I were building a proper application, I would *not* use this pattern. Instead, I would use the factory pattern discussed repeatedly throughout this book and build my application from a callable. This is to help with import management and to avoid passing global scope variables.

A few things are happening here that we need to discuss. This is the syntax used to tell nextcord to run this handler when the application starts up and is connected to Discord, and therefore "ready." But, according to their documentation, this event could be triggered multiple times. That would be a mistake to try and run Sanic multiple times since it would fail to properly bind to a socket.

To avoid this, we look at the `app.is_running` flag to determine whether we should run this again.

What happens next is that we are going to manually create a Sanic server. After that—and this part is critical—we pass that app server instance into a *new* task. Why? Because if we ran Sanic from the current task, it would block indefinitely, and the Discord bot would never actually run. Since we want them both to run concurrently, it is imperative that we run Sanic from another `asyncio` task.

4. Next, we need to create that `runner` operation. The job here is to run the created server. This means that we need to manually trigger all of the listener events. It also means that we need to perform some shutdown of connections. Because we are operating at a *much* lower level than normal, you will be required to be more hands-on:

```python
async def runner(app_server: AsyncioServer):
    app.is_running = True
    try:
        await app_server.startup()
        await app_server.before_start()
        await app_server.after_start()
        await app_server.serve_forever()
    finally:
        app.is_running = False
        app.is_stopping = True
        await app_server.before_stop()
        await app_server.close()
        for connection in app_server.connections:
            connection.close_if_idle()
        await app_server.after_stop()
        app.is_stopping = False
```

The job here looks simple. It starts the application, runs some listener events, and then will listen forever until the application shuts down. Before completely exiting, we need to run some cleanup operations inside the `finally` block.

Once you have all of this implemented, you can run it as we said before by executing the bot.py script. You should now see this message in your Discord server that was triggered by Sanic during the application startup life cycle.

Figure 10.4 – Screenshot of our Discord bot sending a message

Next, you should be able to hit your single endpoint and see another message:

Figure 10.5 – Screenshot of our Discord bot sending a message

In this example, we are running Sanic from another service: the Discord bot service. I generally do not like to recommend this approach. It is easy to mess up by leaving out critical events, or improperly handling operations. The solution that we just saw has an incomplete shutdown mechanism because it does not include any sort of handling for the graceful shutdown of existing connections. Running a web server manually is not trivial, so I would rarely recommend this. People try it often, so it is important to see some of the considerations and how it could be done. However, since I have seen even highly experienced senior developers get tripped up trying this, it might be better to stay away from this pattern unless it is an absolute requirement.

The complete example is in the GitHub repository: `https://github.com/PacktPublishing/Python-Web-Development-with-Sanic/tree/main/Chapter10/wdsbot/from_bot`.

This leads to the next question: instead of running Sanic inside the Discord bot, can we run the bot inside Sanic? We will answer this question for you in the following section.

Running the Discord bot from Sanic

Before we get started, let's consider what `client.run` is doing. It does whatever internal instantiation is needed to run its service, including making a connection to the Discord server. Then, it enters into a loop to asynchronously receive and send messages to the Discord server. This sounds very similar to what the Sanic server does. And, therefore, we can do the exact same thing that we just did, except in reverse:

1. Take the code we just built and remove the `on_ready` event from the bot. Your `./bot.py` should now look like this:

    ```
    import nextcord


    client = nextcord.Client()


    @client.event
    async def on_message(message):
        if message.author == client.user:
            return


        if message.content.startswith("$hello"):
            await message.channel.send("Hello!")
    ```

2. Add a startup time listener to your Sanic application that starts the bot in a new background task:

    ```
    @app.before_server_start
    async def startup_wadsworth(app, _):
        app.ctx.wadsworth = client
        app.add_task(client.start(app.config.DISCORD_TOKEN))


        while True:
            if client.is_ready():
                app.ctx.general = client.get_channel(app.
    config.GENERAL_CHANNEL_ID)
                await app.ctx.general.send("Wadsworth,
    reporting for duty")
                break
            await asyncio.sleep(0.1)
    ```

In this listener, we are also doing the same thing we did in the previous example. We set up `app.ctx.wadsworth` and `app.ctx.general` so that they are easily accessible for use later on in the build. Also, we want to send a message when Wadsworth is online and ready to work. Yes, we could do this from the bot using `on_ready` as before, but we can also do this from Sanic. In the preceding code, we create a loop to check for the state of the bot. Once it is ready, we will send the message and close out the loop.

3. The next thing we need to make sure to do is to properly close the bot connection. We will do that in a shutdown listener:

```
@app.before_server_stop
async def shutdown(app, _):
    await client.close()
```

Now, you have full capability to run your bot from Sanic. This should behave exactly as before, but you have the full power of running your application with the Sanic CLI as we have throughout the rest of this book. Go ahead and fire it up now:

```
$ sanic server:app -p 7777 --debug
```

You can find the source code for this example in the GitHub repository: `https://github.com/PacktPublishing/Python-Web-Development-with-Sanic/tree/main/Chapter10/wdsbot/from_sanic`.

This pattern of nesting other `asyncio` applications has broader applicability than just running Discord bots and Sanic together. It also allows us to run multiple Sanic applications in the same process, albeit on different ports. This is what we are going to do next.

Creating an HTTP to HTTPS proxy: nesting Sanic inside Sanic

Running Sanic from within Sanic seems a bit like those Russian nesting dolls. While it may initially seem like an amazing thought experiment, it does have some real-world applicability. The most obvious example of running two instances of Sanic together like this would be to create your own HTTP to HTTPS proxy. That is what we are going to do now.

The caveat that I want to add to this is that this example will use a **self-signed certificate**. That means that it is not suitable for production use. You should look at the *Securing your application with TLS* section in *Chapter 8, Running a Sanic Server*, for details on how to properly secure your application using TLS.

To begin, we will create two servers. For the sake of simplicity, one will be server.py (your main application running HTTPS over port 443) and the other will be redirect.py (the HTTP to HTTPS proxy running on port 80).

1. We will start by creating our self-signed certificate. If you are on a Windows machine, you might need to look up how to do this on your OS:

   ```
   $ openssl req -x509 -newkey rsa:4096 -keyout key.pem -out
   cert.pem -sha256 -days 365
   ```

2. Store those files in a ./certs directory.

3. Next, we start building our Sanic application in server.py with a simple factory pattern. The code for this build is available at https://github.com/PacktPublishing/Python-Web-Development-with-Sanic/tree/main/Chapter10/httpredirect:

   ```
   from sanic import Sanic
   from wadsworth.blueprints.view import bp
   from wadsworth.blueprints.info.view import bp as info_
   view
   from wadsworth.applications.redirect import attach_
   redirect_app


   def create_app():
       app = Sanic("MainApp")
       app.config.SERVER_NAME = "localhost:8443"
       app.blueprint(bp)
       app.blueprint(info_view)


       attach_redirect_app(app)


       return app
   ```

> **Tip**
>
> The first thing that I would like to point out is the usage of SERVER_NAME. This is a configuration value that is unset out of the box in Sanic. It is usually something that you should use in all of your applications. It is a helpful value used by Sanic behind the scenes in a few locations. For our purpose in this example, we want to use it to help us generate URLs further down the road with app.url_for. The value should be the domain name of your application, plus the port (if it is not using the standard 80 or 443). You should not include the http:// or https://.

What is attach_redirect_app? This is another application factory. But it will work a little bit differently since it will also act to nest the redirect app inside of MainApp.

The last thing worth pointing out is that there is the Blueprint Group bp that we will attach all of our Blueprints to. Except, info_view will be separate. More on that in just a bit.

4. We begin the second factory pattern, attach_redirect_app, at redirect. py:

```
def attach_redirect_app(main_app: Sanic):
    redirect_app = Sanic("RedirectApp")
    redirect_app.blueprint(info_view)
    redirect_app.blueprint(redirect_view)
    redirect_app.ctx.main_app = main_app
```

We are attaching two views—the same info_view that we just attached to MainApp, and the redirect_view that will do our redirection logic. We will look at that once we are done with the factory and server here in redirect.py.

Also, please notice that we are attaching main_app to redirect_app.ctx for later retrieval. As we have learned, passing objects through ctx is the preferred method for handling objects that need to be referenced throughout an application.

5. Next, we will add a few listeners to MainApp. This is going to happen inside of the attach_redirect_app factory. There are some software architects that may dislike my nesting of logical concerns together. We are going to silence the critics and do it anyway because what we are after is necessarily tightly coupled logic that will be easy for us to debug and update in the future:

```
def attach_redirect_app(main_app: Sanic):
    ...
```

```
@main_app.before_server_start
async def startup_redirect_app(main: Sanic, _):
    app_server = await redirect_app.create_server(
        port=8080, return_asyncio_server=True
    )
    if not app_server:
        raise ServerError("Failed to create redirect
server")
    main_app.ctx.redirect = app_server
    main_app.add_task(runner(redirect_app, app_
server))
```

Here, we are dropping down into some lower-level operations of the Sanic server. We basically need to mimic what the Sanic CLI and `app.run` do, except inside the confines of the already existing loop.

When you run a Sanic server instance, it will block the process until shut down. But we need to have two servers running. Therefore, the `RedirectApp` server needs to be run in a background task. We accomplish that by pushing off the work of running the server by using `add_task`. We will come back to the runner when we are done with the factory.

6. `RedirectApp` also needs to be turned down. Therefore, we attach to `MainApp` another listener to do that:

```
def attach_redirect_app(main_app: Sanic):
    ...

    @main_app.after_server_stop
    async def shutdown_redirect_app(main: Sanic, _):
        await main.ctx.redirect.before_stop()
        await main.ctx.redirect.close()
        for connection in main.ctx.redirect.connections:
            connection.close_if_idle()
        await main.ctx.redirect.after_stop()
        redirect_app.is_stopping = False
```

This includes all of the major elements you need for turning down Sanic. It is a little bit basic and if you are implementing this in the real world, you might want to take a look into how the Sanic server performs a graceful shutdown to close out any existing requests.

7. We now turn to `runner`, the function that we passed off to be run in a background task to run `RedirectApp`:

```python
async def runner(app: Sanic, app_server: AsyncioServer):
    app.is_running = True
    try:
        app.signalize()
        app.finalize()
        ErrorHandler.finalize(app.error_handler)
        app_server.init = True

        await app_server.before_start()
        await app_server.after_start()
        await app_server.serve_forever()
    finally:
        app.is_running = False
        app.is_stopping = True
```

Again, what we are accomplishing are some of the high-level steps that Sanic takes under the hood to stand up a server. It does run `before_start` slightly out of order. Typically, that would happen before `create_server`. The impact is minimal. Since our `RedirectApp` does not even use any of the event listeners, we could do without `before_start` and `after_start` (and the shutdown events too).

8. Now to the important part of the application—the redirection view:

```python
from sanic import Blueprint, Request, response
from sanic.constants import HTTP_METHODS


bp = Blueprint("Redirect")


@bp.route("/<path:path>", methods=HTTP_METHODS)
async def proxy(request: Request, path: str):
```

```
    return response.redirect(
        request.app.url_for(
            "Redirect.proxy",
            path=path,
            _server=request.app.ctx.main_app.config.
SERVER_NAME,
            _external=True,
            _scheme="https",
        ),
        status=301,
    )
```

This route is going to be fairly all-encompassing. It basically will accept every endpoint that remains unmatched, irrespective of the request's HTTP method. This is accomplished using the `path` parameter type and passing the `HTTP_METHODS` constant to the route definition.

The job is to redirect the exact same request to the `https` version. You could do this a few ways. For example, the following works:

```
f"https://{request.app.ctx.main_app.config.SERVER_NAME}
{request.path}"
```

However, for me and my brain, I like to use `url_for`. If you prefer the alternative, you do you. The redirect function is a convenient method for generating the appropriate redirection response. Since our use case calls for a redirection from `http` to `https`, we use a `301` redirect to indicate that this is a permanent (and not temporary) redirection. Let's see it in action.

9. To run our application, we need to use the TLS certificates that we generated:

```
$ sanic wadsworth.applications.server:create_app \
    --factory --port=8443 \
    --cert=./wadsworth/certs/cert.pem \
    --key=./wadsworth/certs/key.pem
```

We are running the application again using the CLI. Make sure to use `--factory` since we are passing it a callable. Also, we are telling Sanic where it can find the certificate and key that were generated for the TLS encryption.

10. Once that is running, we jump into a terminal to test with `curl`. First, we will make sure that both applications are accessible:

```
$ curl http://localhost:8080/info
{"server":"RedirectApp"}
```

That looks right.

```
$ curl -k https://localhost:8443/info
{"server":"MainApp"}
```

This also looks right. Please note that I included -k in the `curl` command. This is because of the self-signed certificate we created. Since it is not from an official trusted certificate authority, `curl` will not automatically issue the request until you specifically tell it that the certificate is okay.

Something that is really interesting about this is that the /info endpoint is *not* defined twice. If you look in the source code, you will see that it is a single blueprint that has been applied to both applications.

11. And now we come to the final test—the redirection:

```
$ curl -kiL http://localhost:8080/v1/hello/Adam
HTTP/1.1 301 Moved Permanently
Location: https://localhost:8443/v1/hello/Adam
content-length: 0
connection: keep-alive
content-type: text/html; charset=utf-8

HTTP/1.1 200 OK
content-length: 16
connection: keep-alive
content-type: application/json

{"hello":"Adam"}
```

Make sure to notice that we are hitting port 8080, which is RedirectApp. We again use -k to tell `curl` to not worry about certificate validation. We also use -L to tell `curl` to follow forward any redirections. Lastly, we add -i to output the full HTTP responses so that we can see what is going on.

As you can see from the above response, we generated an appropriate 301 redirection and sent the user on to the https version, which greeted me so nicely by my first name.

And that's it: a simple HTTP to HTTPS redirection application running Sanic inside Sanic.

Summary

What I love about building web applications is the chance to build solutions to problems. For example, earlier in this chapter, we had the problem of wanting to run a JavaScript development server from Sanic. If you put five different developers on that problem, you might end up with five different solutions. I believe that building web applications is on some level an art form. That is to say that it is not a strict field that must be solved in only one *obvious* way. Rather, what is obvious can only be determined given the unique circumstances and parameters surrounding your build.

Of course, what we have built here is just the tip of the iceberg for what is possible with Sanic. The choices displayed are both some popular use cases and also some use cases that might not be so straightforward. I hope that you can take some of the ideas and patterns and put them to good use. By reading this book and internalizing the examples in this chapter, I hope that I have helped to stimulate the creative ideas of application building for you.

If we mashed up all of the ideas from this chapter into a single application, you would end up with a PWA powered by Sanic using distributed websocket feeds and a GraphQL API that also runs a Discord bot. My point is that creating features to implement in your application cannot be done in a vacuum. You must consider other parts of your architecture when deciding on how to build something. This chapter should help you see some of my thought processes when I tackled these problems.

As we near the conclusion of this book, the last thing we need to do is actually pull together a lot of what we know into a single deployable application. That is what we'll do next, in *Chapter 11, A Complete Real-World Example*, where we'll build a fully functional, production-grade Sanic application.

11
A Complete Real-World Example

I remember clearly the feeling I used to get in college and law school when walking into the classroom on exam day. I do not mean the sense of anguish and panic someone might feel when they are unprepared. Rather, I mean the build-up of excitement to tackle the unknown and solve difficult problems. It is the exhilarating feeling after a semester of learning and studying that I would get as I prepared to face what lies ahead. Armed with the confidence from my studies and preparations, I clearly remember sitting with a pen in hand looking forward to the challenge of proving myself. I feel this way every time that I begin a brand-new web application project.

With the knowledge and practice of the previous chapters, I hope you feel this too. I hope that you are feeling excited and empowered. You should be confident that the material we have covered has equipped you with the tools that you need to build, deploy, and maintain a web application. You indeed are ready to build something awesome.

Of course, web application development is not truly a test, but if we did want to draw a comparison, then it would be more like an open-book exam. It is okay to feel anxious or nervous, or even unsure and underprepared on some topics. Anytime that you need to reference material, look it up. The most important key to success is knowing how to tackle a problem and not memorizing the actual code needed to solve it. I urge you to prepare yourself with tools at the ready: keep this book next to your keyboard, become familiar with the Sanic User Guide (`https://sanic.dev/en/`), and join the Sanic community forums or Discord server to ask questions or float ideas.

In the end, the person scoring your exam is you, so there are no right or wrong answers. Way back in *Chapter 1*, *Introduction to Sanic and Async Frameworks*, we looked at one of the maxims from the Zen of Python and redefined it: *"There should be one ... obvious way for you to do it."* This is a very important concept that I hope you can internalize and apply throughout your own careers.

Although I have been building web applications since the late 1990s, I began my career as a practicing attorney and only later changed professions to software engineering. When people learn this about me, their first question will often be, *"How?"* I agree that it does not outwardly seem like a standard career path, since it is not an obvious progression. But for me, it was. What I loved about lawyering is closely aligned with what I love about web development: problem-solving. My career has been a journey that has admittedly taken twists and turns that I could not and did not foresee from the beginning. However, from my perspective, the progression has been entirely logical, consistent, and ongoing. I encourage you to take a similar feeling and approach with you on your journey. You plan, you execute, and you adapt to the ever-changing circumstances. Most importantly, you constantly evaluate your options and choose or invent the path that makes sense for your individual needs and circumstances.

In this final chapter, I hope to bring together everything learned from the earlier chapters and present you with a final working web application. The web application presented in this chapter is not the only way to build, but it is the obvious way for me to build. With my 25 years of experience, I hope that there is some wisdom and useful patterns for you to use in building out your web applications. But my solutions may not meet your needs. Please feel free to reuse, improve, and ignore whatever code I have shared with you.

We will discuss the following main topics in this chapter:

- The process of building a web application
- Highlighting select features of the Booktracker

Technical requirements

This chapter does not introduce any new technologies. All of the source code is available at `https://github.com/PacktPublishing/Web-Development-with-Sanic/tree/main/Chapter11`. More importantly, the final working application is hosted online and fully functional at `https://sanicbook.com`. Please feel free to poke around on the application to see how it works.

The process of building a web application

Now that all of the preparations and learning is complete, it is time to write the final web application. Just like the big exam at the end of a semester of hard work and study, it is time to prove what we have been working toward. I always did best in exams when I entered the classroom with a process in my head. I may not know what challenges or problems would await me, but I knew the process that I would use to break apart the issues on my way toward a solution. For me, I found a direct correlation between having a defined process for approaching the course's subject matter and the ultimate grade that I received on the exam.

We need a process. Over the years, I have developed a process that I like to take when building a web application. It is broken down into eight steps:

1. Define the functionality and workflow.
2. Decide on the technology stack.
3. Architect the data structures.
4. Plan and build the **User Interface** (**UI**).
5. Build the application infrastructure.
6. Prototype the minimally viable backend features.
7. Create continuous integration, deployment, and automation tools.
8. Iterate, iterate, iterate.

These steps are a starting point and more of a suggested course of action. You will often find while completing one of the steps that you may need to revisit an earlier one. This is okay. For example, in *step 2*, we need to decide upon the technology stack for our application. That does not mean that we must foresee every possible component from the beginning. Of course, if a need arises to add some new technology, then repeat the step and add the technology. Similarly, I try to build as much of the frontend UI through mockups or working code as possible. This helps me to evaluate all of the use cases that may arise. I also know that I will need to come back and revise the work.

Some people might take the approach that their design stage must cover every possible edge case. They need to plan for all contingencies, and the entire scope of the project must be defined at the beginning. In my opinion, this is both impossible and entirely impractical. Even if you can anticipate every feature that you need to build and can, therefore, predetermine all of the planning needed to account for it, this inflexible approach is destined for failure. The needs of an application are constantly changing, and therefore, the design is a living and evolving organism. It must be adaptable, and you must be flexible enough to realize that, sometimes, your early design decisions will need to be reevaluated.

Of course, this also means that, sometimes, your early design decisions can have a large impact on later choices. You will often find that your choices will need to account for something that could have been decided differently: *"If only I had used X instead of Y."* My caution to you is to not second-guess these choices. Instead, embrace them as a challenge to overcome. Building within the confines of earlier decisions will help you refine your own decision-making skills for the next web application process.

We will take a walk through each step of my process to see what the major goals and milestones are that we need to pass on our way toward a completed project. The very first set of examples in this book in *Chapter 2, Organizing a Project*, relate to a discussion about how to organize a project. Our theoretical use case is an application that I am calling **the Booktracker**. The remainder of this chapter will focus on the building of this application. There is a fully developed application with the source code available at `https://github.com/ahopkins/sanicbook`. Furthermore, I have gone through and deployed this application live, meaning that you can interact with it and see how it works. It can be accessed here: `https://sanicbook.com`.

Step 1—Define the functionality and workflow

The very first step in building an application seems axiomatic: you must decide what you want to build. What I suggest that we do is dig in a little deeper to flesh out the idea. It may be tempting to skip this step, especially if you have a general idea in your head of what you want to build. You do—after all—know what you want to build, so your mind is already spinning with the specifics of *how* you will build your web application. I personally take a lot of enjoyment from the problem-solving aspect of web development. Therefore, answering the *how does it work?* question is one of the most fun parts for me. My mind likes to jump ahead to crafting solutions, which means I may skip over the part where I actually define the problem.

We will slow down and try to take a methodical approach to define the problem that our web application might solve. Let's say that you and some friends share a favorite hobby: extreme ironing (yes, it's a real thing—go search it online). You have decided that you want to enable fellow extreme-ironers from all over the world to share photos and experiences. Before we can figure out how we are going to store images, there are two basic questions that you must answer:

- What is my application?

- Why do people need or want to use it?

By asking and answering these questions, you can identify the problem that your web application will solve. It is only by clearly knowing the problem that you can develop a properly suited solution. It is now your job to stay focused on that problem as you begin to flesh out the idea with features. I would suggest that you make two lists of features right now. The first list is the set of features without which your application has no use. The second list is reserved for everything else. Since this second list is defined as the alternative to the first, we will focus upon the first list.

At the beginning of your project, you will define a hyper-focused list of features that serve as a starting point. These features are going to be the first ones that you develop to get the application off the ground and organized. I am not suggesting that you must go to market with a minimal product. You may be familiar with the concept of a **Minimally Viable Product** (**MVP**). The features required to get the application off the ground are related to the MVP but are not the same. An MVP is a concept commonly used in software development where you launch a product with the least amount of features needed to attract early users and begin a feedback loop with them. This is important and is, of course, something you should be working toward. This process helps you work toward that MVP, but the first list of features that we want to develop is only a small subset of it. The second list should really be split into two lists: items that belong in the MVP and wish-list items. There are, therefore, three lists of features that need to be created:

- Features that are minimally necessary to make the application functional

- Features that fulfill the requirements of an MVP

- Features that are for future development (often called the wish list)

The first list of features that I would like you to create will be narrowly defined and should be the most basic set of features of the MVP. Your initial feature list is merely the compilation of the items that you need to make the application operational. If you remove one feature, then the application cannot even start to run. This is what the first list should contain. There is an extreme amount of power in being able to build something, then stand it up, and actually see and interact with it. Once you achieve this level, it then becomes much easier to iterate over your feature list. After seeing an operational web server, it becomes much easier to work on it. When we get to building features in *step 6*, your first accomplishment will be to build the first list. Only after that will you get to the second list.

It is also worth mentioning that the second list should itself only be what is needed to reach your MVP. If you have an idea of something you might like to build, put that away on a third list. If it does not fulfill the need of the MVP, then you need to actually have a launched web application to decide whether it is even something worth building. For example, maybe you think it would be a really awesome feature to allow people to live-stream their extreme-ironing expeditions. That may sound nice, but it also may be a lot of work for something that does not meet the community's needs. Focus upon the simple use cases for the second list before tackling the hard use cases.

Answering questions about the Booktracker application

When building out the *Booktracker*, I needed to ask myself these two questions so that I knew what problem I wanted to solve. Let's ask them now and see how we can define the problem that the *Booktracker* is meant to solve:

- *Q: What is my application?*

  The *Booktracker* is a web-based portal and API that enables users to keep track of the books that are in their personal collection, that they would like to read, or that they have already read. Okay, so we know that the application will center around books, and the idea of tying book ownership or readership is paramount. I have also defined that the application should both be usable from a web portal and that I want users to be able to directly integrate with the API. Let's move on to the next question, and then we can start to look at some features.

- *Q: Why do people need or want to use it?*

  Users will use the *Booktracker* to maintain digital lists of the books in their personal library, books that they would like to borrow or purchase to read in the future, and books that they may have lent to their friends so that they know where all of their books are.

Aha! Now, we are starting to see the beginning of some features that we will need. We will need to have things such as the following:

- Login and authentication, perhaps with social media to make it easier.

- Adding and updating books.

- Tracking the read/unread status.

- Tracking the ownership and location of a book, or who has borrowed a book.

- Perhaps we also need some social management, since users might be interacting with other users; this might even include some discussion forums or live chat.

Our list is starting to lose a little focus. I think that it is okay to define a long list of features, but you should be realistic about what is necessary and what is required. For example, I cannot launch my application if users cannot register, log in, and add books. This would be something that is a part of my MVP. Creating a social network does not seem so critical for the first stage, and this should probably appear on the third list.

So, what is the list of features that should be on the second list? It is common to think of features within the **FRUD** system, which stands for **function**, **reliability**, **usability**, and **delightfulness**. Features will primarily fall into one of these categories. They can be arranged in a pyramid, as shown in *Figure 11.1*:

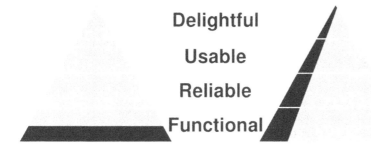

Figure 11.1 – Features displayed in the FRUD system, with the MVP shown in black

When building an MVP, you must pick features that touch all of the categories. If you instead focus just on the base layer of the pyramid (functional features), then you will have a hard time attracting users. The pyramid on the left in *Figure 11.1* shows an MVP where the application only focuses on function. This application would be completely unusable. Instead, by selecting a small number of features from each of the categories, the application can be something that attracts users.

But what are the features that an application needs to start? What are the features that should be on my first hyper-focused list of to-dos on the first list? I would suggest that those are a subset of the functional and reliable features of your MVP. They are the items needed to run the most basic iteration of the web application. Even before login and even before being able to add a book, I think that the first feature we need to build is to be able to display a list of books. This is where the development process will begin. I will start by building a database to store books and a method to display them to the user.

But what about login? A social login does seem like it might be extraneous, and perhaps that is something that should wait until *after* the MVP. It clearly is a feature that belongs in one of the top categories of the pyramid. Oftentimes, it is probably advisable to leave social sign-on features until after the MVP. However, the reason that I want to have a social login is so that I do not need to develop a system for registering, storing, and validating passwords. Therefore, for my application, that will be an item on the second list. My lists now look like this:

List one—basic stand-up	List two—MVP	List three—the wish list
• Runs the frontend Svelte UI • Pulls information about books from the database and displays them in an endpoint	• Social sign-on with GitHub • Allow users to add and update book states • Only show books relevant to users	• Track ownership/borrowing/location of books • Social media-style sharing and discussion

Table 11.1 – The list of features for the Booktracker application

Moving to the next step, it is time for me to decide on my technology stack.

Step 2—Decide on the technology stack

The next step is to decide on the **technology stack**. You must explore the technologies that will be used to build the application and loosely how they will function together. A technology stack is the list of tools that are used to build the application. For example, deciding that your backend application will be powered with a Python asynchronous web application called Sanic is part of your stack. What other technologies are needed?

You will need to decide how users are going to interact with the application. Will there be a frontend web UI? Will it power a mobile application? Or, perhaps, will it be accessed by other applications? Think about what you need to use to enable the type of user interactions you are looking to build. You also need to think about the technologies that are needed to power the set of features that you want. One of the most critical components of a web application is, of course, the database or storage engines that will be needed. Other important parts of the stack are the tools that may be used to monitor and maintain it. Of course, these features can always be added later. But if you know that you want to use a specific tool for tracking your logs or performing application tracing, perhaps you should plan for it upfront.

Building a stack for the Booktracker

I know that the application will be powered by Sanic, so that becomes the first item in my stack. I also know that I want users to be able to interact via a Web API but also by accessing a frontend browser-based application. Therefore, my stack should include a web UI, but it also probably needs an **OpenAPI Specification** (**OAS**) that helps anyone who is interacting directly with the API.

How should I build these two frontend components? My personal favorite framework for frontend applications is Svelte, so I will use that. As for the OAS, I will use Sanic extensions, which will help me to easily build and document the API. While thinking more about the frontend, I also know that I am going to want to make it look attractive and presentable. There are a lot of great CSS frameworks out there to help jump-start a platform. The *Booktracker* uses Bulma (`https://bulma.io/`) because it is both attractive and easy to work with.

For storage, I think that most of the data that I need is highly relational—that is to say that objects in my datastore will relate to each other through ordinary foreign-key relationships. Since I am most comfortable using PostgreSQL for this type of data, that is what I will use. However, I also foresee the need to use more transient data for caching, so I will also include a Redis backend.

I decided that I do not want to handle managing passwords on my own. First of all, it would be much simpler for the user if they do not need to create and remember yet another login. That is certainly a more delightful experience. Second, it is also less of a liability for me if there is a data breach at some point. Of course, I plan to take reasonable steps to secure the application, but it is one more precaution I can take by not ever needing to deal with passwords to begin with. Because I need to factor in how users will interact and log in to my application via GitHub, that should be added to my stack and flow chart.

The technology stack so far looks like this:

- Bulma (CSS)
- Svelte (JS)
- Sanic Extensions (OpenAPI)
- Sanic (Python)
- GitHub (authentication)
- PostgreSQL
- Redis

It is also helpful at this point to start creating some flow charts. These will serve both as a visual representation of my application and help to develop the concept of what needs to be built. Remember, this is an iterative process, so what I create now can always be revised later.

I might create a flow chart of the *Booktracker* that looks like this:

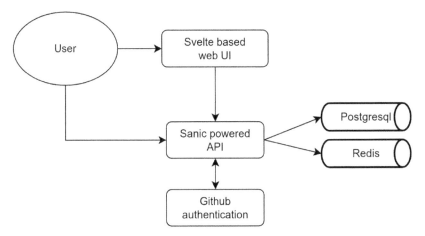

Figure 11.2 – A flowchart of the Booktracker stack

One thing to note is the directions of the arrows. These are meant to show the direction in which requests for information flows. A user requests information from the API, which, in turn, needs to request information from the database. It's worth noting that GitHub will both push and pull information from the web application.

Step 3—Architect the data structures

It is a very helpful step to start thinking about what types of objects your application will deal with. If you are dealing with extreme ironing, you might need to think about photos and locations. If you are building a restaurant review system, you will need restaurants, patrons, and menu items. Oftentimes, these types of data structures will also be some of the first endpoints that you need to build.

Designing data for the Booktracker

For the *Booktracker*, the most obvious three entities that I can think of are users, books, and authors. When I actually look at my physical bookshelf, however, I see that I also have a lot of books that belong to a similar series of books, and therefore, they are arranged to sit next to each other. I will need to think of a way to tie books together.

The schema that I have ultimately decided that will work for my application looks like this:

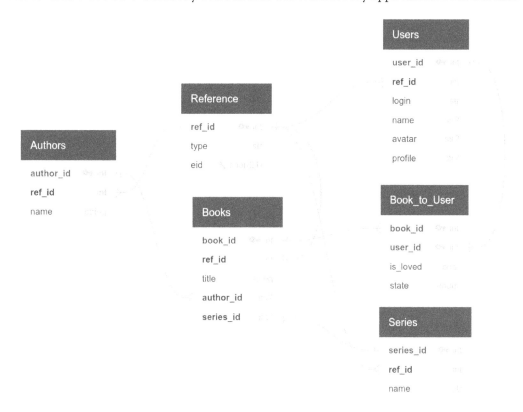

Figure 11.3 – A schema for the Booktracker database

There is perhaps little that is surprising going on in this simple schema between users, authors, books, and series. But, there were a few interesting decisions that I needed to make. For example, what if both you and I are using the *Booktracker* and we both own the same book? Let's call that book *Python Web Development with Sanic,* or *PyWDS* for short. How many instances of *PyWDS* should be in the database—one or two? From a user's perspective, what is important is that they can control their relationship to the book. As the operator of the application, it might be helpful to have a single instance of *PyWDS* so that I know how many users across the platform have it on their shelves. Therefore, I need to make a relationship from user to book through a third many-to-many table that contains information about whether the user loves the book and what state of reading the book is in.

Another interesting feature of my schema for the *Booktracker* to note is that I am going to make an effort to not leak any state about the application to users via object IDs. A common practice in relational databases is to increment rows with a sequential primary key. This allows very easy ordering and referencing.

Let's suppose we have an endpoint that looks like this:

```
@app.get("/book/<book_id:int>)
async def book_details(request, book_id):
    ...
```

Now, imagine that you have just added two books to the application. You can access details about that book as /book/1234 and /book/1236. Just by looking at book_id, you now know that there must be a book with an ID of 1235 and presumably for 1233, 1232, and so on. By allowing access to users to see and interact with sequential primary keys, information about the state of your application has been leaked to the public. This is generally considered a bad practice. At best, it is inappropriate for users to have that information. At worst, it could raise a business or security concern.

To overcome this issue, the frontend will exclusively deal with what I am calling an eid or **external ID**. The idea is that an eid is some set of random characters that uniquely identify an object in the system. Unlike a sequential ID—which is an incremented version of the last ID in the sequence—an eid does not identify objects in relation to any other object. To manage eids, I have created a single table in my database. Every record in one of my object tables will reference a record in this table. This way, when I want to look something up by eid, it is very simple to join from that table to the object table using the ref_id sequential. My queries will generally look like this:

```
SELECT   e.eid, b.*
FROM     books b
```

```
JOIN    eids e ON b.ref_id = e.ref_id
WHERE   e.eid = $1;
```

In the past, I have also solved this problem by creating `eids` using PostgreSQL itself to populate columns on every table with an `eid`. However, for this project, I thought that might be a little more complex than necessary. It does also make an interesting point in that I also now know that before I can go much further in building the first list of highly focused initial stand-up features, I will need a way to generate an `eid`. I should go back now to revise my list of features and add `eid` generation to the first list.

Step 4—Plan and build the user interface

Admittedly, this next step is not something we will spend much time learning about. Since this book is about building with Sanic, we will leave aside the wealth of discussion points that you might have about how to create an ideal UI and **User Experience** (**UX**). Suffice it to say that I think this is an incredibly helpful step in building a proper backend. Even if the UI is not browser-based and the main interaction is only through the API, you must spend some time planning exactly how users are going to interact with the application. The main goal of this step (as a backend developer) is to determine what interactions will need to be supported. Until you know what the frontend UI is going to do, you cannot realistically make decisions about what endpoints to build.

Designing the Booktracker

I like to think that having built web applications for a number of years, I have learned some tricks of the trade and that my ability to create a frontend application is pretty good. That does not mean that I am in any way a UX guru, and therefore, I will not attempt to claim what does and does not constitute good web design. I also would like to warn any UX mavens out there that the design of the *Booktracker* app is admittedly simplistic, since its function is meant to be very utilitarian. The web UI that I built—which is available at `https://sanicbook.com/`—is meant to be helpful as a tool in backend development learning. Therefore, I admittedly did not spend so much time designing it.

However, I do want to stress that I strongly believe that a decent frontend design—whether through mockups or wireframes—is critical for backend development. If I had built my *Booktracker* backend first and only then begun to work on the frontend, I likely would have ended up with something rather odd. These sorts of applications often feel disjointed and lead to a mess of code. They are usually very difficult to maintain, refactor, and iterate new features upon. The process of building the frontend is meant to shine a spotlight upon areas where the UI needs help from the backend. If you are not going to be the person building the web UI, then I highly suggest that you sit down with that person and get a clear idea of what their needs and expectations are. One of the most frustrating things as a web developer is when you try to mash together a frontend and backend that are not compatible.

In many cases, this is a good place for wireframing. Many people like to build out mockups with images and online wireframing platforms. For the more humble needs of the Booktracker, I found it far more practical and simple to just build out mockups in HTML and CSS. With the basic mockups in place, I was able to build out the backend to power the frontend with specific knowledge of what was needed.

One word of caution here is warranted: be careful not to fall into the trap of tightly coupling together frontend and backend functionality. Ideally, you should have endpoints that have no knowledge of the existence of the frontend. Your backend indeed should not even know that the frontend exists. It should instead present layers of information that may be needed and can be organized logically.

Step 5—Build the application infrastructure

It is not until *step 5* that we begin to write some Python code. By this point, you should have a pretty good outline of what needs to be built. The first Python code that we need to write is for all of the infrastructure—that is to say that we need to now build some of the non-business logic stuff that our application will need. For example, we have talked about the following topics throughout the book:

- The application factory
- Logging
- Exception handling
- Blueprint organization

These are all topics that you should be thinking about right now. In addition, we need to set up some basics for creating a connection pool to both of our databases and start thinking about how to tie our database schema to models. Will there be an **Object Relational Mapping** (**ORM**) involved? If yes, it is time to start setting it up. If not, what will be used instead?

At this stage, there is a lot to do. Remember, the first goal is to get an application to an operational stage. That means that the application will start without any exceptions. If there is a startup time error, fix it.

I usually use this time to start building a `HelloWorld` endpoint. It is not something that will likely end up in the final application. Nonetheless, it is something that I can trigger from `curl` to see the application working. This is where I will start testing DB connections to make sure that they perform as I expect them to:

```
bp = Blueprint("HelloWorld", url_prefix="/hello")



class HelloView(HTTPMethodView, attach=bp):
    async def get(
        request: Request,
    ):
        return json({"hello": "world"})
```

This simplistic endpoint will be the beginning of my application but will eventually be removed.

In regards to the *Booktracker* application, this stage looks very much like the *hiking* app from *Chapter 9, Best Practices to Improve Your Web Applications*. If you would like a refresher, the code for it is available in the GitHub repository: `https://github.com/PacktPublishing/Python-Web-Development-with-Sanic/tree/main/Chapter09/hikingapp`. What is important to note is that I use this step to make sure that all of my services are running. I want all of the databases to be live and for the backend application to establish connections to them. Largely, this means building out infrastructure that we have seen through this book. Logging is usually one of the first items in *step 5*, as well as structuring a factory pattern. Once that is complete, I can move on to the feature development of *step 6*.

Step 6—Prototype the minimally viable backend features

Now that we know the groundwork has been laid, it is time to start building that first layer of features. Here is where we want to focus on the very first layer of our application features. Take a look back at *Figure 11.1* to remind yourself of what this looks like. When deciding upon the features that must be a part of your MVP, you need to balance the needs for function, reliability, usability, and delightfulness. These are all important but with varying degrees of weight. Right now, the most important (although not the only) need is for features that are functional. We will start here.

From *step 6* through to *step 8*, we are about to enter into a bit of a loop. This will be an iterative process to build out our MVP features. The first time through *step 6*, we will focus on those features that are in the bottom-most category. In this first sweep through the loop, we only care about functionality—that is to say, all we want to see are results on screen with no errors. There likely will be some ugly patterns and some hard-coding of values here. Beautiful code abstraction is not important right now, since we mainly care about just getting something on screen. Remember, we are not building a product yet, just a prototype.

Prototyping the Booktracker

If you look at the *Booktracker* app, you will see that there are several endpoints. So, which was the first one that I built, and why? The core of the application is the delivery of information about books. Therefore, the most fundamental feature is provided by the GET /v1/books endpoint. It is designed to retrieve a list of books and output them as JSON. Because of its central importance to the application and its relative simplicity, it is precisely where I decided to begin building the application.

If we take a look at the code in GitHub, it looks roughly like this:

```
bp = Blueprint("Books", url_prefix="/books")


class BookListView(HTTPMethodView, attach=bp):
    @staticmethod
    @inject_user()
    async def get(
        request: Request,
        pagination: Optional[Pagination] = None,
        user: Optional[User] = None,
```

```
    ):
        executor = BookExecutor(request.app.ctx.postgres,
BookHydrator())
        kwargs = {**pagination.to_dict()} if pagination else {}
        getter: Callable[..., Awaitable[List[Book]]] =
executor.get_all_books

        if title := request.args.get("title"):
            kwargs["title"] = title
            getter = executor.get_books_by_title
        elif user:
            payload = await request.app.ctx.auth.extract_
payload(request)
            user_executor = UserExecutor(request.app.ctx.
postgres)
            user = await user_executor.get_by_
eid(eid=payload["eid"])
            kwargs["user_id"] = user.user_id
            getter = executor.get_all_books_for_user
        try:
            books = await getter(**kwargs)
        except NotFound:
            books = []
        output = [book.to_dict(include_null=False) for book in
books]
        return json({"meta": pagination, "books": output})
```

This request handler may look complicated at first glance, but it is actually very simple. It is designed to return a list of books in the application. There are a few different filters to help with different use cases. For example, you can supply a `title` query argument to filter for specific book titles. This is helpful for our frontend to be able to have an autocomplete feature for the user. The endpoint also will filter the books that are linked to a specific user. This will be used by the frontend to show a user which books they have. Once the books have been fetched, the handler will serialize them and deliver them as a JSON response.

I can assure you that the preceding code is definitely *not* the first version of this endpoint. In fact, the version that you see is probably the result of four or five iterations through my development process. As you will see, the endpoint includes the ability to easily paginate through results as to not overwhelm the database and response size. The first pass at developing this endpoint was much simpler and did not include pagination or filtering. It instead looked like this:

```python
bp = Blueprint("Books", url_prefix="/books")



class BookListView(HTTPMethodView, attach=bp):
    @staticmethod
    async def get(
        request: Request,
    ):
        executor = BookExecutor(request.app.ctx.postgres,
BookHydrator())
        books = await executor.get_all_books()
        output = [book.to_dict() for book in books]
        return json({"books": output})
```

This route handler looks somewhat similar to what we saw with the *hiking* app in *Chapter 9*, *Best Practices to Improve Your Web Applications*. It is meant to be a sort of bare minimum so that I can see data properly flowing from the database to the frontend.

As we advance through the iterations, there will be a need to add other features. For example, the next one to work on is the authentication flow. Then, once we have access to users, we can start to provide only books that are linked to users in the /books endpoint. And, again down the road, we may have the need to paginate this response so that responses are more controlled. Hopefully, you can see that just as we created a pyramid of FRUD for our application, the same could be said to exist at the micro level for each endpoint. First, we build a minimally functional version to prove the concept, and only then can we build the full FRUD version that will make it to the MVP.

It is difficult to determine exactly where to draw the line, and it is easy to get carried away too soon. Especially when I am in the early iterations of development, I like to ask myself a simple question: *Does this feature make my application better?* If the answer is *yes* and I am in one of the early iterations, then warning bells are going off in my head that I am headed in the wrong direction. At the outset of a project, I want the answer to that question to be, *No, this does not make it better; it makes it something.* In other words, I try to draw a distinction between an improvement and something without which the application does not exist.

Once there is some sort of minimal features in the application, what's next?

Step 7—Create continuous integration, deployment, and automation tools

Application development often does not exist purely within the confines of the service that you are working on. Usually, you will need to work on some tooling that helps support the life cycle of the application. Before we get too far into the development process, I want to shift gears to focus upon some of the scripts, manifests, and utilities that will be needed to operationalize and deploy the application.

Therefore, by this stage, you should have decided exactly where you want to run the application. You also need to decide how you plan to get your application to that location.

I have decided to deploy my application using a **Platform as a Service** (**PaaS**) solution, and a small VPS to run my two databases. This is mainly because this application is meant to be a **Proof of Concept** (**POC**) and will not carry much traffic or demand. This decision is therefore based upon a desire to keep the hosting costs to a minimum while still keeping it easy for me to deploy changes. To learn more about deploying a PaaS, please see the *Platform as a service* section in *Chapter 8, Running a Sanic Server.*

Step 8—Iterate, iterate, iterate

This final step is really the continuation of the process as a whole. More specifically, however, you should find yourself in a loop that starts in *step 6*. Once you have a running version of the application, you should continue to iterate on it with small incremental changes. Each change should be the addition or correction of a feature that matches your pyramid. For example, remember the two versions of the GET /v1/books endpoint I showed? There is the completed version that you can access today online and the simple version that was my first iteration. To get from one to the other was not a single iteration through the process; it instead took three passes to get to the final endpoint. Let's examine the evolution of this endpoint:

1. We will begin with a reminder of what it looked like after the first iteration:

    ```
    class BookListView(HTTPMethodView, attach=bp):
        @staticmethod
        async def get(
            request: Request,
        ):
            executor = BookExecutor(request.app.ctx.postgres,
    BookHydrator())
            books = await executor.get_all_books()
            output = [book.to_dict() for book in books]
            return json({"books": output})
    ```

2. In the first revision, there was a need to add pagination. I used a really helpful feature of Sanic extensions to simplify this. I added a setup_pagination(app) function in my create_app factory. The purpose of this was to automatically read the request's query parameters looking for limit and offset. When present, those values should be read into a model and auto-injected into a route handler. This is very convenient because it now can easily be reproduced on any endpoint that needs to handle pagination requests:

    ```
    @dataclass
    class Pagination(BaseModel):
        limit: int = field(default=15)
        offset: int = field(default=0)

        @staticmethod
        async def from_request(request: Request):
            args = {
    ```

```
        key: int(value)
        for key in ("limit", "offset")
        if (value := request.args.get(key))
    }
    return Pagination(**args)
```

```
def setup_pagination(app: Sanic):
    app.ext.add_dependency(Pagination, Pagination.from_
request)
```

What this tells Sanic to do is look for Pagination as a type annotation in the request handlers. If that exists, then it should call Pagination.from_request(request) and inject the return value of that method as the argument that was annotated as Pagination. Adding this to the handler, it now looks like this:

```
class BookListView(HTTPMethodView, attach=bp):
    @staticmethod
    async def get(
        request: Request,
        pagination: Optional[Pagination] = None,
    ):
        executor = BookExecutor(request.app.ctx.postgres,
BookHydrator())
        kwargs = {**pagination.to_dict()} if pagination
else {}
        books = await executor.get_all_books(**kwargs)
        output = [book.to_dict() for book in books]
        return json({"books": output})
```

3. In the second revision, I decided that if the user is logged in, then the endpoint should only return books related to the current user. Because my application uses Sanic JWT—more on this package later in the *Authentication flow* section—there happens to be a convenient decorator that I can use. By wrapping my handler with @inject_user(), Sanic will now automatically inject the currently authenticated user into my handler:

```
class BookListView(HTTPMethodView, attach=bp):
    @staticmethod
    @inject_user()
```

```
async def get(
    request: Request,
    pagination: Optional[Pagination] = None,
    user: Optional[User] = None,
):
    executor = BookExecutor(request.app.ctx.postgres,
BookHydrator())
    kwargs = {**pagination.to_dict()} if pagination
else {}
    getter = executor.get_all_books

    if user:
        payload = await request.app.ctx.auth.extract_
payload(request)
        user_executor = UserExecutor(request.app.ctx.
postgres)
        user = await user_executor.get_by_
eid(eid=payload["eid"])
        kwargs["user_id"] = user.user_id
        getter = executor.get_all_books_for_user

    try:
        books = await getter(**kwargs)
    except NotFound:
        books = []

    output = [book.to_dict() for book in books]
    return json({"books": output})
```

With this revision, it is now clear that I needed to add a new SQL statement to be able to fetch books based upon the user. Therefore, I needed to add a new function to my executor. That is why I assign the methods to the `getter` variable: so that the variable can be changed if needed.

4. On the final revision, I decided that this endpoint should also be able to handle incoming search queries for looking up books by titles. This means that we need a third getter. Since I already established the pattern for changing the getter, this change was simple:

```python
class BookListView(HTTPMethodView, attach=bp):
    @staticmethod
    @inject_user()
    async def get(
        request: Request,
        pagination: Optional[Pagination] = None,
        user: Optional[User] = None,
    ):
        executor = BookExecutor(request.app.ctx.postgres,
BookHydrator())
        kwargs = {**pagination.to_dict()} if pagination
else {}
        getter = executor.get_all_books

        if title := request.args.get("title"):
            kwargs["title"] = title
            getter = executor.get_books_by_title
        elif user:
            payload = await request.app.ctx.auth.extract_
payload(request)
            user_executor = UserExecutor(request.app.ctx.
postgres)
            user = await user_executor.get_by_
eid(eid=payload["eid"])
            kwargs["user_id"] = user.user_id
            getter = executor.get_all_books_for_user

        try:
            books = await getter(**kwargs)
        except NotFound:
```

```
        books = []

    output = [book.to_dict() for book in books]
    return json({"books": output})
```

The change in this revision was just to add the check for a `title` in the query arguments. There is one very important item worth mentioning here. This one endpoint is really doing lifting for three things. In many application architectures, it might actually be easier to split this up into multiple endpoints each with a much cleaner and narrow focus. It is ideal for a single handler to have a single responsibility. However, sometimes rules can be broken, and in this instance, I decided it was a better learning experience to handle all of these responsibilities together.

Sometimes, while iterating through a step, you will need to revisit a step even earlier than *step 6*. And many times, you will find that you are not yet ready to deploy that code to a production-ready server. Both of these changes to the process are good and to be expected.

The `/v1/books` process shows how you can take the concept of an MVP for an application and apply it to even a single endpoint. The first use case was displaying books. Features were iteratively added to the endpoint until there was an MVP for how I would like it to operate on the production application. Try to resist the temptation to overbuild it too early. The consequence of that is usually a bunch of overly complicated code that you do not know how to maintain in six months when you need to fix a bug or add a new feature.

With the basics of the process that I used to build the web application out of the way, let's next take a look at some of the decisions and features that were actually implemented.

Highlighting select features of the Booktracker

The *Booktracker* app attempts to approximate real-world concerns of web application development. The main goal of the project is to provide a bit of source data that can be coupled with a backend server to provide a realistic API and frontend web application. What we will do now is step through parts of the application and discuss some important challenges and how solutions were developed.

Development environment

The obvious place to start is by developing a working development environment. To determine how to structure the project directory and bring up a working application, I needed to consider what exactly it was that I needed to build. As a reminder, the *Booktracker* is an application built from the following:

- A Svelte-based frontend UI

- A Sanic-based backend Web API

- A PostgreSQL database

- A Redis datastore

Let's take a look at how I created a development environment for these services.

Organizing the application

There are four different services that I need to be concerned with. When I need to locally run a web server and a database, I almost always will reach for Docker Compose. We will do that here. However, since I intend to serve all of my static content—meaning the frontend Svelte UI—from Sanic, I only need three services. Let's quickly look at how I am organizing the project:

```
./booktracker
├── application
│   ├── booktracker
│   │   └── ...
│   ├── node_modules
│   │   └── ...
│   ├── ui
│   │   └── ...
│   ├── Dockerfile
│   ├── package.json
│   ├── requirements.txt
│   ├── rollup.config.js
│   └── yarn.lock
├── postgres
│   ├── Dockerfile
│   └── initial.sql
```

```
├── docker-compose.yml
└── README.md
```

As you can see here, in the root of the project, there are two main directories: ./application and ./postgres. This is to clearly define the two Docker images that will need to be built, and you will see that each of these folders has a Dockerfile in it. Since Sanic is powering both my frontend and backend, the application directory contains both a ./booktracker directory (the Web API) and a ./ui directory (the frontend). All of the configurations live in this directory so that the subdirectories can be devoted to the application code. Since we do not care too much about the frontend right now, please feel free to browse the source code at https://github.com/PacktPublishing/ Web-Development-with-Sanic/tree/main/Chapter11 to see more details about how I accomplished it.

The ./booktracker directory is designed almost exactly the same way that it was built way back in *Chapter 2, Organizing a Project*. Here is a general layout:

```
./booktracker
├── blueprints
│   ├── author
│   │   ├── queries
│   │   │   └── ...sql
│   │   ├── executor.py
│   │   ├── model.py
│   │   └── view.py
│   ├── book
│   │   ├── queries
│   │   │   └── ...sql
│   │   ├── executor.py
│   │   ├── hydrator.py
│   │   ├── model.py
│   │   └── view.py
│   ├── frontend
│   │   ├── reload.py
│   │   └── view.py
│   ├── user
│   │   ├── queries
│   │   │   └── ...sql
│   │   ├── executor.py
```

```
|   |   └── model.py
|   └── view.py
├── common
|   ├── auth
|   |   ├── endpoint.py
|   |   ├── handler.py
|   |   ├── model.py
|   |   └── startup.py
|   ├── dao
|   |   ├── decorator.py
|   |   ├── executor.py
|   |   └── hydrator.py
|   ├── base_model.py
|   ├── cache.py
|   ├── cookie.py
|   ├── csrf.py
|   ├── eid.py
|   ├── log.py
|   └── pagination.py
├── middleware
|   └── request_context.py
├── worker
|   ├── module.py
|   ├── postgres.py
|   ├── redis.py
|   └── request.py
└── server.py
```

There are four main divisions in the application, and I tried to keep them logically arranged, as follows:

- `./blueprints`: This directory—contrary to what the name implies—does not simply contain blueprint objects. The main goal of this is to contain all of my endpoints and views. Perhaps a better name might be `./views`, but I particularly do not like looking at `from booktracker.views.view import bp`. It is where all of my domain-specific logic belongs.

- ./common: For modules that are not domain-specific and may have more general applicability, I group them into the common module.

- ./middleware: Like ./blueprints, this directory name might be a little confusing because it is not the *only* location where you will find middleware. Instead, it contains the middleware that is of general applicability. Where a piece of middleware is limited to a blueprint, it will be grouped within that domain, or, if it is closely tied to a specific module, it lives alongside it. A good example is the middleware used for CSRF protection. That code lives in ./common/csrf.py.

- ./worker: The ./worker directory contains the modules that mainly pertain to the proper setup of a worker instance. This includes things such as setting up connection pools, but it also contains some utilities that help tie the rest of the application together.

And just how does the application get tied together? We will look at how I accomplished that using factory patterns in the *Creating a better factory pattern with setup functions* section. Next, we will review how the frontend application is being served from Sanic.

Serving Svelte from Sanic

In *Chapter 10*, *Implementing Common Use Cases with Sanic*, we reviewed how it was possible to use Sanic as a development server for Svelte (or any frontend) applications. If you recall, this was accomplished by running two Sanic applications side by side. One of them runs the main application and serves all of the Web API and frontend content, while the second is a livereload server and sends messages to the browser to refresh the page every time a file is saved.

In the earlier version, we served all of the frontend files simply using the static handler:

```
app.static("/", app.config.FRONTEND_DIR / "public")
```

For this project, I wanted something a little more robust. Particularly, I wanted the ability to serve the index.html content from a bare directory. I began by setting up a frontend module inside of the ./blueprints directory. The view module for the frontend looks like this:

```
from logging import getLogger
from pathlib import Path

from sanic import Blueprint, Request
from sanic.response import file
```

```python
from .reload import setup_livereload

logger = getLogger("booktracker")
bp = Blueprint("Frontend")
setup_livereload(bp)


@bp.get("/<path:path>")
async def index(request: Request, path: str):
    base: Path = request.app.config.UI_DIR / "public"
    requested_path = base / path
    logger.debug(f"Checking for {requested_path}")
    html = (
        requested_path
        if path and requested_path.exists() and not requested_
path.is_dir()
        else base / "index.html"
    )
    return await file(html)
```

So, what is going on? First, I define my blueprint as usual. I then call a function called `setup_livereload`. This function does most of the same work that we saw in the *Powering a progressive web application* section in *Chapter 10, Implementing Common Use Cases with Sanic*. You can also check the source code of the *Booktracker* app to remind yourself how that works.

To achieve the goal of serving `index.html` from a directory, I set up an endpoint on the blueprint that will act as a catch-all. Using `/<path:path>` will allow this handler to catch any calls to the application that do not have another specified route. This means I will be relying upon the frontend to catch anything that does not exist in the backend. This is really important, especially for a PWA. I want all of my routing to be controlled by the frontend application, and therefore we need to send all non-matched requests to it.

Creating a better factory pattern with setup functions

We have explored the factory pattern for creating Sanic applications earlier in this book a few times. If you recall, the idea is that we create a function that generates the Sanic application instance and then do all of the importing afterward, so that objects such as routes, middleware, signals, and listeners can attach properly. For example, the *hiking* app that we saw in *Chapter 10, Implementing Common Use Cases with Sanic*, looks like this:

```
from sanic import Sanic
from hiking.common.log import setup_logging


def create_app():
    app = Sanic(__name__)
    setup_logging(app)

    from hiking.middleware import request_context  # noqa
    from hiking.blueprints.view import bp  # noqa
    from hiking.worker import postgres  # noqa
    from hiking.worker import redis  # noqa

    app.blueprint(bp)

    return app
```

This, of course, does work, but it is kind of ugly to have all those imports inside of the function. Note how each line has a `# noqa` comment at the end. That is because without it, Python linters will throw up errors because your imports are not at the top of the file. Unfortunately, if you place those imports where they should be (at the top of the file), then the objects cannot attach, since the Sanic application instance does not exist yet.

One alternative I have seen to overcome this is to *not* use Sanic's decorators when making an object. For example, the handler would be a function without a decorator:

```
async def some_handler(request: Request):
    ...
```

Then, it is attached inside of the factory:

```
from somewhere import some_handler


def create_app():
    app = Sanic(__name__)
    app.add_route(some_handler, "/something")
```

Again, this works, but it has two problems in my opinion:

- It removes some information that is pertinent to the object (whether it is a route, middleware, and so on) and places it in a separate location. You can no longer easily identify what a function is without looking it up inside of the factory as well.

- This pattern tends to lead to very lengthy factories that are nothing more than a bunch of add_route and register_middleware calls, and so on. These long lists can become very difficult to maintain and even harder to track down a bug when there is a problem.

I am going to show you my preferred pattern that solves all of these problems. Before doing that, however, I want to remind you how all of the blueprints are being linked up.

Organizing blueprints

There are three main blueprints in the *Booktracker* application: Authors, Books, and Frontend. There could be 50 more, and I still think this pattern works best. Each of the blueprints is very similar in how they are organized. As shown earlier when reviewing the folder organization, there is a file called ./view.py that contains the blueprint instance. Let's take a look at the Authors blueprint as an example.

Inside of ./blueprints/author/view.py, there is the following:

```
bp = Blueprint("Authors", url_prefix="/authors")
```

Attached to that, there is a view. I tend to prefer **Class-Based Views** (**CBVs**), but regular functional handlers are perfectly acceptable. In our case, it looks something like this:

```
class AuthorListView(HTTPMethodView, attach=bp):
    @staticmethod
    async def get(request: Request, pagination: Pagination):
        ...
```

Now, each of these blueprints is ultimately loaded into `./blueprints/view.py`. That file looks like this in its entirety:

```
from sanic import Blueprint


from .author.view import bp as author_bp
from .book.view import bp as book_bp
from .frontend.view import bp as frontend_bp


api = Blueprint.group(author_bp, book_bp, version=1, version_
prefix="/api/v")
bp = Blueprint.group(frontend_bp, api)
```

As you can see, the idea is that there is a single location where all of the blueprints are loaded. All of the API endpoints are bundled into a group that gets a version number and a prefix. This will make it so that all of my API endpoints will begin with `/api/v1`. Then, both the API group and my frontend blueprint are loaded into another blueprint group called `bp`. I tend to use `bp` consistently to mean the blueprint that I intend to import somewhere else. For example, take a look at how I imported all of the blueprints and renamed them. The single blueprint group called `bp` in this module will become important soon.

Attaching objects in the factory

We now turn our attention toward our application factory. It has a couple of key features to take notice of:

- All regular imports take place at the top of the file, where they should be.

- No actual Sanic objects (routes, middleware, listeners, signals, and so on) are actually imported at all.

- Instead, there are `setup_*` functions that do the work of creating and linking some of the objects.

- Alternatively, objects that exist in the global scope will be imported programmatically.

- The factory pattern is highly testable.

Here is what it looks like:

```python
from pathlib import Path
from typing import Optional, Sequence, Tuple

from sanic import Sanic

from booktracker.common.auth.startup import setup_auth
from booktracker.common.csrf import setup_csrf
from booktracker.common.log import setup_logging
from booktracker.common.pagination import setup_pagination
from booktracker.worker.module import setup_modules
from booktracker.worker.request import BooktrackerRequest

DEFAULT: Tuple[str, ...] = (
    "booktracker.blueprints.view",
    "booktracker.middleware.request_context",
    "booktracker.worker.postgres",
    "booktracker.worker.redis",
)


def create_app(module_names: Optional[Sequence[str]] = None) ->
Sanic:
    if module_names is None:
        module_names = DEFAULT

    app = Sanic("BooktrackerApp", request_
class=BooktrackerRequest)
    app.config.UI_DIR = Path(__file__).parent.parent / "ui"
    app.config.CSRF_REF_PADDING = 12
    app.config.CSRF_REF_LENGTH = 18

    if not app.config.get("CORS_ORIGINS"):
        app.config.CORS_ORIGINS = "http://localhost:7777"

    setup_logging(app)
```

```
        setup_pagination(app)
        setup_auth(app)
        setup_modules(app, *module_names)
        setup_csrf(app)


        return app
```

The list of module names is important. When running the application, the server will just fall back and use the predefined DEFAULT modules. However, allowing for the list of modules to be imported dynamically by passing them to create_app, I have made it much simpler to perform very targeted unit tests.

Let's take a quick look at one of these modules, booktracker.middleware. request_context, to see what is happening:

```
from contextvars import ContextVar

from sanic import Request, Sanic

app = Sanic.get_app("BooktrackerApp")


@app.after_server_start
async def setup_request_context(app, _):
    app.ctx.request = ContextVar("request")


@app.on_request
async def attach_request(request: Request):
    request.app.ctx.request.set(request)
```

As you can see, we are using Sanic.get_app("BooktrackerApp") to fetch our application instance. This will work fine because this module will not be loaded until after the application instance is created. If you mess up your import ordering, then you will end up with an exception that looks like this:

```
Traceback (most recent call last):
  File "/path/to/module.py", line 5, in <module>
    app = Sanic.get_app("BooktrackerApp")
```

```
File "/path/to/sanic/app.py", line 1676, in get_app
    raise SanicException(f'Sanic app name "{name}" not found.')
sanic.exceptions.SanicException: Sanic app name
"BooktrackerApp" not found.
```

The way that this actually gets imported is by loading this module—and all of the other module strings, including the blueprint view we saw earlier—dynamically with the setup_modules function. This function is very simple:

```
from importlib import import_module


from sanic import Sanic



def setup_modules(app: Sanic, *module_names: str) -> None:
    for module_name in module_names:
        module = import_module(module_name)
        if bp := getattr(module, "bp", None):
            app.blueprint(bp)
```

Its job is to simply import the module from a string. As we already know, Sanic will take care of linking up the objects, since we have an application instance present to use the built-in decorators. The one place we need to take an additional step is with blueprints, since they need to be manually attached to the application instance by calling app. blueprint(bp). This is why I said earlier that it is important that I chose bp as the name for all blueprints that I intended to attach directly to the application instance.

I really like this pattern for defining modules and importing them dynamically because it leads to orderly application organization. I can easily open up a module and see exactly what is available, and adding new modules is relatively simple. There is, however, another pattern that I sometimes find useful.

Somewhat similar to setup_modules, there is a setup_csrf function. Take a look at the factory pattern we are working through now to remind yourself where this function is being called. I pass the app instance to the function, which means that I can do something like this:

```
def setup_csrf(app: Sanic) -> None:
    @app.on_request
    async def check_request(request: Request):
        request.ctx.from_browser = (
```

```
                "origin" in request.headers or "browser_check" in
    request.cookies
            )


        @app.on_response
        async def mark_browser(_, response: HTTPResponse):
            set_cookie(
                response=response, key="browser_check", value="1",
    httponly=True
            )
```

Now, instead of using `Sanic.get_app` and defining everything in the global scope of the module, we define it all inside of the local scope of the function. In the *Booktracker* application, I used both of these options. Some modules are imported dynamically by string, and some are set up via functions such as `setup_csrf`. I would encourage you to pause reading right now and to head to the factory source at `https://github.com/ahopkins/wds-finalapp/blob/main/application/booktracker/common/csrf.py`. Afterward, I suggest that you look at each of the `setup_*` functions and their source code to see what they are doing. In my opinion, both of these are good options, and I urge you to experiment with them both to see which you prefer in your own application development.

Just as an example, remember my `./blueprints/view.py`? It can just as easily look like this:

```
from sanic import Blueprint, Sanic


from .author.view import bp as author_bp
from .book.view import bp as book_bp
from .frontend.view import bp as frontend_bp


api = Blueprint.group(author_bp, book_bp, version=1, version_
prefix="/api/v")
bp = Blueprint.group(frontend_bp, api)


# this next part is all new (you will not find it in the repo
source)
def setup_views(app: Sanic):
    app.blueprint(bp)
```

In this case, the module would not need to be imported via string using `setup_modules`.

The data access layer

When discussing the database schema, I mentioned that the *Booktracker* application attempts to avoid data leakage by not ever showing an incremental object identifier to the client. This means that the backend API must have some other form of communication with a client to reference a specific object in the database. There are certainly many forms that this could take.

A common use case—especially for blogs and news websites—is to concatenate an article title to a **slug**. This bit of text is a helpful way to have a human-friendly string that uniquely identifies a specific article. This usually takes the form of an all-lowercase string, where all non-characters have been converted to a hyphen—for example, *this-is-a-slug*. While this method is great for objects that have titles (especially long ones), it is not suitable for general applicability.

A widely accepted practice is to use a **Universally Unique Identifier** (**UUID**). You have probably seen (and likely used) them before. They are simple to create and extremely unlikely to cause a name collision. You can safely assume that if you run the following code, every single UUID generated will be unique:

```
from uuid import uuid4


print(uuid4())
```

You may be wondering, why UUID version 4? What is even the difference between the UUID versions? A good reference material is this website: `https://www.uuidtools.com/what-is-uuid`. If you are interested, please feel free to read up on the other versions that allow you to add namespaces to them. For our use case, we will use UUID version 4 because it represents a completely random (so far as any computer is *random*) set of characters. But a UUID is not just a set of random characters; it actually represents a number that has been formatted to meet a very specific specification. If it is a number, then we know that there are some useful ways that we can represent it.

Given the UUID `06adf00c-0f43-47d7-941a-ce76346f3fb3`, we can express that in a few ways:

- As a hexadecimal value: `06adf00c0f4347d7941ace76346f3fb3`
- As an integer value: `8878504065002459431209101741502971827`

- As a binary value: 01101010110111110000000011000000111101000011010 00011111010111100101000001101011001110011101100011010001 1011110011111110110011

We can use this fact to solve what I feel is the most annoying part of UUIDs: they are not user-friendly. I have three main problems with them:

- They are very long.

- They are hard to copy and paste.

- They are not human-friendly and are difficult to remember.

As a developer who is debugging a platform and using UUIDs to reference objects or requests, it is harder to copy and paste a value with multiple hyphens than a single value. Usually, you cannot just double-click a UUID value to select the whole string. A single value without hyphens usually can be double-clicked to select the whole thing. To make my life easier, and to save space when sending UUIDs, I decided to use the numeric property of them to shorten the value by expanding the characters they can use.

In `booktracker.common.eid`, you will find this function:

```python
import uuid
from string import ascii_letters, digits


REQUEST_ID_ALPHABET = ascii_letters + digits
REQUEST_ID_ALPHABET_LENGTH = len(REQUEST_ID_ALPHABET)



def generate(width: int = 0, fillchar: str = "x") -> str:
    """
    Generate a UUID and make it smaller
    """
    output = ""
    uid = uuid.uuid4()
    num = uid.int
    while num:
        num, pos = divmod(num, REQUEST_ID_ALPHABET_LENGTH)
        output += REQUEST_ID_ALPHABET[pos]
    eid = output[::-1]
    if width:
```

```
        eid = eid.rjust(width, fillchar)
    return eid
```

What this does is generate a UUID and then convert it into a shorter string, using all of the ASCII letters and digits. This will turn `06adf00c-0f43-47d7-941a-ce76346f3fb3` into `mLBFLEq8vnXHPHzbvN7ap`. This, in turn, brings a 36-character string down to 21 characters. It saves space and, in my opinion, is easier to use. The only problem with this algorithm is that the length of the output might vary in length. For example, it can generate strings of 19 or 20 characters. I would prefer that all `eids` have a constant length in my database. Therefore, I will left-pad them to the desired length if needed. My database stores `eids` as 24 characters. The UUID that we have been working with will be stored in my database as `xxxmLBFLEq8vnXHPHzbvN7ap`.

Now that you understand how the application passes these `eids` around, you should be able to piece together the flow from the endpoint, through the executor, to the database. Let's take a look at how this happens by getting a book. Here is the endpoint for getting details about a book from the API:

```python
class BookDetailsView(HTTPMethodView, attach=bp, uri="/<eid>"):
    @staticmethod
    @inject_user()
    async def get(request: Request, eid: str, user:
Optional[User]):
        executor = BookExecutor(request.app.ctx.postgres,
BookHydrator())
        getter: Callable[..., Awaitable[Book]] = executor.get_
book_by_eid
        kwargs: Dict[str, Any] = {"eid": eid}
        if user:
            getter = executor.get_book_by_eid_for_user
            kwargs["user_id"] = user.user_id
        book = await getter(**kwargs)
        return json({"book": book.to_dict(include_null=False)})
```

The first thing that we notice is that this is a CBV that is attached to the blueprint instance stored as the bp variable. We also see that it is capturing the eid from the URL and then injecting that into the handler arguments. What about the decorators? A CBV can have its methods defined either as a regular instance method or a static method. Both of the following solutions are okay:

```
class SomeView(HTTPMethodView, attach=bp):
    @staticmethod
    async def get(request: Request):

        ...


    async def post(self, request: Request):

        ...
```

I tend to prefer using a staticmethod if I am not going to need the self argument, since the CBV is more a tool to encapsulate similar endpoint handlers as opposed to more traditional object-oriented programming style encapsulation. The @inject_user decorator comes from another project of mine called Sanic JWT that provides some helpful utilities around authentication. It is not important for the current discussion, other than to say that if a request comes in from an authenticated user, it will automatically query the database and inject the User object.

The next main thing to note is that we are setting up an executor. If you recall from the *Hiking* app, the executor is responsible for connecting call methods with raw SQL. It then executes them and hydrates the data into a model instance.

Tip

We can improve upon this endpoint further with Sanic extensions. One of the features there allows us to define commonly used items (such as BookExecutor) upfront in our application and auto-inject them into the route handler when necessary. That looks something like this:

```
def getter(request):

    return BookExecutor(request.app.ctx.postgres,
BookHydrator())


app.ext.add_dependency(BookExecutor, getter)
```

It will be appropriate to set these up in a setup_* function from the factory pattern previously discussed.

For the most part, you will find that the data access layer in the *Booktracker* app is nearly identical in implementation to the *Hiking* app. It is probably not so critical to our discussion here, but I would suggest taking a look if you are not interested in a full ORM.

Authentication flow

When thinking about user authentication for this application, I knew that I wanted it to be simple. After all, there is no reason to make someone go through the process of registering, selecting yet another username and password, validating an email, and so on. That flow is valid and necessary in some locations but clearly too complicated for this use case. Implementing a social media sign-on was clearly the best option for me. For the intended audience of my application (such as you, reading this book), a GitHub account seems like the perfect prerequisite.

Knowing that GitHub would be used to verify users, I needed to think about how I wanted to actually handle authentication in the application. In *Chapter 7, Dealing with Security Concerns*, there is a lengthy discussion in the *Protecting your Sanic app with authentication* section about which authentication systems are good in different situations. For my use case, I decided a simple JWT would be best. Specifically, I decided to implement the split-JWT cookie discussed in that chapter's *Solving for JWTs in browser-based applications* section. This will be very easy to set up using Sanic JWT. This tool will create the authentication endpoints and provide me with the tooling (such as decorators) that I need to implement my authentication scheme.

Because JWTs are best when they expire quickly, I knew that I would need to implement refresh tokens. The frontend application was going to need to be responsible for keeping the token fresh. With a basic understanding of what my requirements were, I was ready to start building the authentication flow. Let's look at the first part, the setup_auth method:

```
from sanic_jwt import Initialize


def setup_auth(app: Sanic):
    Initialize(
        app,
        url_prefix="/api/v1/auth",
        authenticate=authenticate,
        retrieve_user=retrieve_user,
        extend_payload=payload_extender,
        store_refresh_token=store_refresh_token,
        retrieve_refresh_token=retrieve_refresh_token,
```

```
class_views=[("/github", GitHubOAuthLogin)],
    ...
)
```

Here is only a portion of the configuration. You can see the full thing in the repository. The point that I want to discuss here is that I needed to create a bunch of handlers that will be responsible for hooking into Sanic JWT. In addition, I needed to add a new endpoint: `/api/v1/auth/github`.

The first step that a user takes is to click a link to `/api/v1/auth/github`. This new endpoint sets some CSRF cookies and then forwards the user to the GitHub **Single Sign-On** (**SSO**) page. The user is presented with the option to sign into my application and provide access to read the user profile. Once they click the button, they are redirected back to the application, where they are presented with a screen like this:

Figure 11.4 – A screenshot of the Booktracker authentication page

This page provides two options: an authorization code to run directly against the API or a button that the user can use to follow via the UI. By clicking the **Continue** button, the user is doing the same action as running the `curl` command: sending the code that GitHub generated to the `/api/v1/auth` endpoint. That endpoint executes the authenticate handler. You can see the full source of it here: `https://github.com/ahopkins/wds-finalapp/blob/main/application/booktracker/common/auth/handler.py`.

Let's step through that handler to see what is going on:

1. First, the handler does some checking to make sure it has the correct context and
    the GitHub code:

    ```python
    async def authenticate(request: Request) -> User:
        invalid = Unauthorized("Missing or invalid
    authorization code")
        auth_header = request.headers.get("authorization",
    "")

        if not auth_header or not auth_header.lower().
    startswith("code"):
            raise invalid

        _, code = auth_header.split(" ")
    ```

2. Once it has that code, it can take the next step in the GitHub authentication flow by
    sending that code along with the client_id and client_secret that GitHub
    provided. Upon a successful exchange, GitHub will issue an access_token that
    can be used to send authenticated requests to the GitHub API:

    ```python
    # Exchange the authorization code for an access token
    async with httpx.AsyncClient() as session:
        response = await session.post(
            "https://github.com/login/oauth/access_
    token",
            json={
                "client_id": request.app.config.GITHUB_
    OAUTH_CLIENT_ID,
                "client_secret": request.app.config.
    GITHUB_OAUTH_SECRET,
                "code": code,
            },
            headers={"accept": "application/json"},
        )

        if b"error" in response.content or response.status_
    code != 200:
    ```

```
logger.error(response.content)
raise invalid
```

3. After this request is complete, the application needs to send a request to fetch the user details from GitHub:

```
async with httpx.AsyncClient() as session:
    response = await session.get(
        "https://api.github.com/user",
        headers={
            "Authorization": f"token {response.json()
['access_token']}"
        },
    )


    if b"error" in response.content or response.status_
code != 200:
        logger.error(response.content)
        raise invalid


    data = response.json()
```

4. With the user data in hand, the authenticate handler can take its final step—fetch a user from the database or create a new user:

```
executor = UserExecutor(request.app.ctx.postgres)
try:
    user = await executor.get_by_
login(login=data["login"])
    logger.info(f"Found existing user: {user=}")
except NotFound:
    user = await executor.create_user(
        login=data["login"],
        name=data["name"],
```

```
            avatar=data["avatar_url"],
            profile=data["html_url"],
        )
        logger.info(f"Created new user: {user=}")

    return user
```

When this is complete, we now have the ability to issue tokens from Sanic JWT. These tokens will provide access to the application, both directly to the API using an `Authorization` header and through secure cookies. If you have not yet, I highly encourage you to follow the example on the live web application. You should then obtain a working access token and use it to explore the API. I find it incredibly helpful to learn something when I can both interact with it and take it apart to see its component parts. This is exactly the reason why I felt it necessary to maintain a live-running version of this web application. I want you to review the API documentation, play with the application in the browser, and interact with the API directly. This should be done while you simultaneously review the relevant source code. In this way, I hope that you will learn some techniques that will be successful for you throughout your journey.

Summary

The finished web application is available for usage at `https://sanicbook.com`. Perhaps most importantly, you can take a look at the OpenAPI documentation at `https://sanicbook.com/docs`. You should find some really helpful information there about how you can interact with the API. I hope that you feel very comfortable making direct HTTP requests using `curl` or another tool and will take some time to explore the API. I highly suggest that you open the GitHub repository, the OpenAPI documentation, and a terminal to start playing with the API.

The source code will be memorialized at the time of publishing this book in the GitHub repository. This is for your benefit so that you can always see exactly what the code was like when this book was written. But as we know, web applications are constantly evolving. There are bugs to fix, features to implement, upgrades to perform, and other reasons to change code. Therefore, the actual deployment of the application will be from my personal GitHub repository: `https://github.com/ahopkins/sanicbook`. If you want to see the version of the code that is deployed, you should look there.

Putting it all together

Web development is a journey. That is true on many levels. On the surface, we can say that developing a web application requires you to journey through a process of iterating on features and bug fixes. As we just saw, web development is a process that aims to balance the FRUD demands of functionality, reliability, and usability, and provide a delightful experience.

On a deeper level, web development is a journey through society. It is an ever-changing field that has progressed over the last few decades to keep pace with the world around us. In many ways, the pace at which the world advances is directly tied to the technologies we are building for the web.

But it is on the personal level that I find web development's journey to be most impactful. I began building websites in high school, at a completely amateur level. And even though I embarked upon a career as a lawyer, it was ultimately web development that I journeyed back to. I can look back at the past 25 years and not only marvel at the advancements we have made but also the changes that I have made, as a husband, a father, a son, a brother, and a human being.

And this brings me to my final point that I hope to impart upon you. Web development is a journey. That means there is no end. There is always more to build, more to improve, and more to learn. So please, do not stop learning with this book. I urge you to pick up another book, find some articles or online courses, and attend meetups or conferences, all so that you can continue to learn. And while you are in that process of learning, take some time to give back to the community. One of the most worthwhile efforts I ever made was to have the courage to contribute to open source software. That too has been a wild and amazing journey. Along the way, I have met good people from all over the globe. And, once again, the most important thing is that I have learned a lot. To become a great developer, you must become a great learner.

The Sanic project maintains a Discord server and community forums. If you are not already a member, come join us. I hope you will introduce yourself to me and the community, and share your experiences with Sanic and web development. Whether you have a question, a complaint, need some help, or just want to say *hi*, please share it. As the Sanic website makes clear: *"The project is maintained and run by the community for the community."*

Index

S

Packt.com

Subscribe to our online digital library for full access to over 7,000 books and videos, as well as industry leading tools to help you plan your personal development and advance your career. For more information, please visit our website.

Why subscribe?

- Spend less time learning and more time coding with practical eBooks and Videos from over 4,000 industry professionals

- Improve your learning with Skill Plans built especially for you

- Get a free eBook or video every month

- Fully searchable for easy access to vital information

- Copy and paste, print, and bookmark content

Did you know that Packt offers eBook versions of every book published, with PDF and ePub files available? You can upgrade to the eBook version at packt.com and as a print book customer, you are entitled to a discount on the eBook copy. Get in touch with us at customercare@packtpub.com for more details.

At www.packt.com, you can also read a collection of free technical articles, sign up for a range of free newsletters, and receive exclusive discounts and offers on Packt books and eBooks.

Other Books You May Enjoy

If you enjoyed this book, you may be interested in these other books by Packt:

Django 3 By Example - Third Edition

Antonio Melé

ISBN: 9781838981952

- Build real-world web applications

- Learn Django essentials, including models, views, ORM, templates, URLs, forms, and authentication

- Implement advanced features such as custom model fields, custom template tags, cache, middleware, localization, and more

- Create complex functionalities, such as AJAX interactions, social authentication, a full-text search engine, a payment system, a CMS, a RESTful API, and more

- Integrate other technologies, including Redis, Celery, RabbitMQ, PostgreSQL, and Channels, into your projects

- Deploy Django projects in production using NGINX, uWSGI, and Daphne

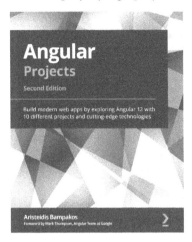

Angular Projects - Second Edition

Aristeidis Bampakos

ISBN: 9781800205260

- Set up Angular applications using Angular CLI and Nx Console

- Create a personal blog with Jamstack and SPA techniques

- Build desktop applications with Angular and Electron

- Enhance user experience (UX) in offline mode with PWA techniques

- Make web pages SEO-friendly with server-side rendering

- Create a monorepo application using Nx tools and NgRx for state management

- Focus on mobile application development using Ionic

- Develop custom schematics by extending Angular CLI

Packt is searching for authors like you

If you're interested in becoming an author for Packt, please visit `authors.packtpub.com` and apply today. We have worked with thousands of developers and tech professionals, just like you, to help them share their insight with the global tech community. You can make a general application, apply for a specific hot topic that we are recruiting an author for, or submit your own idea.

Share Your Thoughts

Now you've finished *Python Web Development with Sanic*, we'd love to hear your thoughts! Scan the QR code below to go straight to the Amazon review page for this book and share your feedback or leave a review on the site that you purchased it from.

`https://packt.link/r/1801814414`

Your review is important to us and the tech community and will help us make sure we're delivering excellent quality content.